QUANTUM ENGINEERING

Quantum engineering – the design and fabrication of quantum coherent structures – has emerged as a field in physics with important potential applications. This book provides a self-contained presentation of the theoretical methods and experimental results in quantum engineering.

The book covers such topics as the quantum theory of electric circuits; theoretical methods of quantum optics in application to solid-state circuits; the quantum theory of noise, decoherence and measurements; Landauer formalism for quantum transport; the physics of weak superconductivity; and the physics of a 2-dimensional electron gas in semiconductor heterostructures. The theory is complemented by up-to-date experimental data to help put it into context. Aimed at graduate students in physics, the book will enable readers to start their own research and apply the theoretical methods and results to their current experimental situation.

A. M. ZAGOSKIN is a Lecturer in Physics at Loughborough University, UK. His research interests include the theory of quantum information processing in solid-state devices, mesoscopic superconductivity, mesoscopic transport, quantum statistical physics and thermodynamics.

T0335747

QUANTUM ENGINEERING

Theory and Design of Quantum Coherent Structures

A. M. ZAGOSKIN

Loughborough University

CAMBRIDGE
UNIVERSITY PRESS

CAMBRIDGE
UNIVERSITY PRESS

University Printing House, Cambridge CB2 8BS, United Kingdom

One Liberty Plaza, 20th Floor, New York, NY 10006, USA

477 Williamstown Road, Port Melbourne, VIC 3207, Australia

314-321, 3rd Floor, Plot 3, Splendor Forum, Jasola District Centre, New Delhi - 110025, India

79 Anson Road, #06-04/06, Singapore 079906

Cambridge University Press is part of the University of Cambridge.

It furthers the University's mission by disseminating knowledge in the pursuit of
education, learning and research at the highest international levels of excellence.

www.cambridge.org
Information on this title: www.cambridge.org/9780521113694

First published 2011

A catalogue record for this publication is available from the British Library

Library of Congress Cataloging in Publication data
Zagoskin, Alexandre M.
Quantum engineering : theory and design of quantum coherent structures /A. M. Zagoskin.
p. cm.
Includes bibliographical references and index.
ISBN 978-0-521-11369-4 (hardback)
1. Quantum theory.   I. Title.
QC174.12.Z34 2011
530.12–dc22   2011009227

ISBN 978-0-521-11369-4 Hardback

To my family

Contents

Preface

It is always risky to combine well-known and well-tested notions in order to describe something new, since the future usage of such combinations is unpredictable. After "quantum leaps" were appropriated by the public at large, nobody, except physicists and some chemists, seems to realize that they are exceedingly small, and that breathless descriptions of quantum leaps in policy, economy, engineering and human progress in general may actually provide an accurate, if sarcastic, picture of the reality. When the notion of the "marketplace of ideas" was embraced by academia, scientists failed to recognize that among other things this means spending 95% of your resources on marketing instead of research.[1] Nevertheless, "quantum engineering" seems a justified and necessary name for the fast-expanding field, which, in spite of their close relations and common origins, is quite distinct from both "nanotechnology" and "quantum computing" in scope, approaches and purposes. Its subject covers the theory, design, fabrication and applications of solid-state-based structures, which can maintain quantum coherence in a controlled way. In a nutshell, it is about how to build devices out of solid-state qubits, and how they can be used.

The miniaturization of electronic devices to the point where quantum effects must be taken into account produced much of the momentum behind nanotechnology, together with the need to better understand and control matter on the molecular level coming from, e.g., molecular biology and biochemistry (see, e.g., Mansoori, 2005, Chapter 1). One also often uses the term "mesoscopic physics", especially with respect to solid-state devices, meaning objects on an intermediate scale between truly microscopic (single atoms or small molecules) and truly macroscopic. Despite their comparatively large size ($\sim 10^{11}-10^{12}$ particles), mesoscopic systems maintain enough quantum coherence so that quantum effects really matter (e.g., Imry, 2002, Chapter 1). The experimental techniques and theoretical

---

[1] See, e.g., Chaize (2001); Menand (2010).

understanding developed in these fields strongly contributed to the development of quantum engineering.

Another, very strong push was delivered by quantum computing. After the original papers (Feynman, 1985, 1996; Deutsch, 1985) indicated the direction, i.e., the *essential* use of quantum properties of a system for computation, and the discoveries of Shor's (Ekert and Josza, 1996; Shor, 1997) and Grover's (Grover, 1997, 2001) algorithms brought the promise of a qualitative change in computing capabilities, the field immediately became the focus of an enormous amount of attention and funding, which produced some spectacular results at the proof-of-principle level (and in the related field of quantum communications, including quantum key distribution, commercial devices are currently available).

The physical side of the research on quantum computing is guided by DiVincenzo's criteria (DiVincenzo and Loss, 1998; DiVincenzo, 2000):

(i) A scalable physical system with well-characterized qubits.
(ii) The ability to initialize the state of the qubits to a simple fiducial state, such as $|000...\rangle$.
(iii) Long relevant decoherence times, much longer than the gate operation time.[2]
(iv) A universal set of quantum gates.
(v) A qubit-specific measurement capability.

Scalability is first for a reason: it is the hardest property to achieve in conjunction with the requirement for a long enough global decoherence time. Solid-state-based devices were natural candidates from this point of view, despite the uncomfortably high number of degrees of freedom in both the qubits and the surroundings, which threatened to make quantum decoherence times in the system uselessly short. Moreover, this large number of degrees of freedom made such devices suspect, since their operation as qubits would require the regular production of "Schrödinger's cat"-superpositions of macroscopic quantum states, an admittedly hard task for even a single experiment (Leggett, 1980). Superconducting devices and quantum dots gave some of the best promise of scalability while holding the disruption of quantum coherence to an acceptable minimum and satisfying the rest of DiVincenzo's criteria. Research results (e.g., Nakamura et al., 1999; Friedman et al., 2000; van der Wal et al., 2000; Martinis et al., 2002; Vion et al., 2002; Hayashi et al., 2003; Elzerman et al., 2004; Hanson et al., 2005) justified the expectations that quantum coherence can be preserved in these structures after all, and made the kind of experiments only hoped for by Leggett (1980) almost routine.

---

[2] The decoherence time is some characteristic time after which the system loses its specific quantum correlation; see Section 1.2.2.

Meanwhile Leggett (2002a) stressed the need to test the limits of quantum mechanics through realizing truly macroscopic "Schrödinger's cat" states. He found special reasons for optimism with the successful development in the field of super-conducting qubits (Leggett, 2002b). The operation of such devices on a large enough scale would either confirm or refute the applicability of quantum mechanics to arbitrarily large systems, with fundamental consequences for science. Therefore, putting more and more qubits together while maintaining their quantum coherence is a worthwhile task. Whether a working code-cracking or database-searching quantum computer is actually built in the process would be, of course, a minor corollary to such a momentous development.

In this book I concentrate on two directions in quantum engineering, using as building blocks superconducting qubits and qubits based on a two-dimensional electron gas (2DEG). The reason is partly subjective – this is where my expertise is – and partly objective – these two kinds of devices are well understood from the theoretical point of view and are being successfully fabricated and investigated by the experimentalists, using the rich toolset of solid-state physics. This allows me to illustrate the theoretical methods with actual experimental data, more often than not being in quantitative agreement with theoretical predictions (which is a must for any kind of "engineering"). Being a theorist, I can only discuss the theoretical part of the toolset, which helped keep the volume of this book reasonable. Quantum engineering takes its theoretical tools from many, rather distant, chapters of theoretical physics, and I restricted the content to the most useful and widely used techniques. The goal was to provide the reader with enough technical information to be able to deal with current research papers and to start out on their own work in this field.

The book can be used as the basis for one-term graduate courses, with the only prerequisite being the knowledge of basic quantum mechanics and solid-state physics. The topics can be, for example, "Superconducting qubits" (Chapters 1, 2, 4 and 5); "2DEG quantum devices" (Chapters 1, 3 and 5); "Mesoscopic quantum transport" (Chapters 1 to 4), and, of course, "Quantum engineering". The sections and paragraphs which contain additional material, not essential for understanding the rest, are marked with asterisks. The bibliography is, of course, not exhaustive, due to the fast development and already very large volume of research, but I did my best to include all the key research papers and reviews.

Knocking on wood, I was always lucky with teachers, colleagues and friends. Most of what is contained in this book I learned from them or with them; it will be my own fault if I do not convey this knowledge properly. I am greatly indebted to I. Affleck, M. Amin, S. Ashhab, O. Astafiev, D. Averin, A. Balanov, V. Barzykin, A. Blais, M. Everitt, A. Golubov, M. Grajcar, E. Il'ichev, M. Jonson, I. Kulik, F. Kusmartsev, A. Maassen van den Brink, K. Maruyama, H.-G. Meyer, F. Nori,

A. Omelyanchouk, M. Oshikawa, V. Petrashov, Yu. Pashkin, A. Rakhmanov, S. Rashkeev, S. Saveliev, R. Shekhter, I. & O. Shendrik, V. Shumeiko, A. Smirnov, P. Stamp, A.-M. Tremblay, A. Tzalenchuk and J. Young. I deeply regret that A. Rozhavsky, Z. Ivanov and A. Izmalkov are no longer with us.

I am using this opportunity to thank S. Capelin and G. Hart, who made this project happen; J. Webley, who with infinite patience copy-edited the book; and all the Cambridge University Press editorial staff for their help.

I am grateful to R. Simmonds, A. Shane and M. Znidaric for kindly providing originals of figures from their articles, and to all the authors and publishers for their generous permission to reproduce the copyrighted material in this book.

Last but not least, I am very grateful to my wife Irina and daughter Ekaterina, for their support in general, and for putting up with me over the period of writing this book in particular; and to our cat Tabs for graciously agreeing to sit for Fig. 1.1.

1

Quantum mechanics for quantum engineers

Shall I refuse my dinner because I do not fully understand the process of digestion? No, not if I am satisfied with the result.

O. Heaviside, *Electromagnetic Theory*, vol. 2, 1899

1.1 Basic notions of quantum mechanics

1.1.1 Quantum axioms

Let us start with a brief recapitulation of quantum mechanics on the "how to" level. According to the standard lore, the instantaneous state of any quantum system (that is, *everything* that can be known about it at a given moment of time) is given by its wave function (state vector)[1] – a complex-valued vector in some abstract Hilbert space; the nature of this space is determined by the system. All the *observables* (i.e., physical quantities defined for the system and determined by its state – e.g., the position or momentum of a free particle, the energy of an oscillator) are described by Hermitian operators defined in the same Hilbert space. All three elements – the Hilbert space, the state vector, and the set of observables – are necessary to describe the outcome of any experiment one could perform with the system. Since humans cannot directly observe the behaviour of quantum objects, these outcomes are also called *measurements*, being the result of using some classical apparatus in order to translate the state of a quantum system into the state of the apparatus, which can then be read out by the experimentalist. The classical (i.e., non-quantum) nature of the apparatus is essential, as we shall see in a moment.

In addition, we need to know how the state of the system changes in time, and how it determines the measured values of the observables. All of the above can be presented as four textbook "axioms of quantum theory":

1) The state of a quantum system at time t is described by a normalized[2] vector $|\psi(t)\rangle$ belonging to the Hilbert space $\mathcal{H}$, which is specific for the system in question.

---

[1] A very important generalization of the wave function, the density matrix (statistical operator), will be discussed in Section 1.2.

[2] This is not strictly necessary, but makes the explanations shorter without much loss of generality.

2) [Schrödinger equation] The state evolves with time according to $i\hbar|\dot{\psi}(t)\rangle = H|\psi(t)\rangle$, where H, the Hamiltonian, is a Hermitian operator associated with the energy of the system.

3) [Collapse of the wave function] The measured value of an observable is always one of the eigenvalues of its operator A, a_j; whatever the state of the system was *before* the measurement, immediately *after* it the state vector of the system is the corresponding normalized eigenvector of A, $|a_j\rangle$ $(A|a_j\rangle \equiv a_j|a_j\rangle)$.

4) [Born's rule] The probability of measuring a particular eigenvalue, a_j, of the observable A (collapsing the wave function $|\psi(t)\rangle$ into a given eigenvector $|a_j\rangle$) is given by the square modulus of the former's projection on the latter, $p_j(t) = |\langle a_j|\psi(t)\rangle|^2$.

The most striking features of these axioms are the special roles played by time and energy (Hamiltonian) of all the other physical quantities associated with the quantum system, and the jarring difference between the linear, reversible, unitary quantum evolution determined by the Schrödinger equation, and the nonlinear, irreversible, non-unitary measurement process. Indeed, the equation

$$i\hbar|\dot{\psi}(t)\rangle = H|\psi(t)\rangle \tag{1.1}$$

has the solution (in the sufficiently general case of a time-independent Hamiltonian)

$$|\psi(t)\rangle = U(t)|\psi(0)\rangle; \quad U(t) = e^{-\frac{i}{\hbar}Ht}, \tag{1.2}$$

and for Hermitian H the *evolution operator* is unitary, $U(t)^\dagger U(t) = U(t)U(t)^\dagger = I$ (where I is the identity operator in the Hilbert space of the system). The latter property ensures that the normalization of the state vector is preserved. On the other hand, after the measurement at moment t we find the system in a state $|a_j\rangle$, which is not related to the state $|\psi(t)\rangle$ before the measurement by any reversible transformation ($\psi(t)\rangle$ instantaneously collapses to $|a_j\rangle$). The after-collapse state $|a_j\rangle$ could be obtained from *any* state $|\psi(t)\rangle$ as long as $|a_j\rangle$ and $|\psi(t)\rangle$ are not mutually orthogonal. Taken together, axioms (3) and (4) are often called the *projection postulate*: the measurement of the observable A *projects* the state vector $|\psi(t)\rangle$ on the eigenvectors of A; the square modulus of the projection gives the probability with which the system is likely to be found in the corresponding eigenstate of A. The eigenvectors of a Hermitian operator A can be chosen to form a complete orthonormal basis of the Hilbert space; therefore the sum of squared projections of the state vector $|\psi\rangle$ on the vectors of this basis is, by Parseval's equality, the square of the norm of $|\psi\rangle$,

$$\||\psi(t)\rangle\|^2 = \langle\psi(t)|\psi(t)\rangle = \sum_j |\langle a_j|\psi(t)\rangle|^2 = \sum_j p_j. \tag{1.3}$$

Once we have chosen the state vector such that it is normalized to unity, Eq. (1.3) ensures that the probabilities of finding different outcomes of the measurement of the observable A sum to one, as they should. The unitarity of the evolution given by Eq. (1.2) ensures that the probabilities "don't leak".

1.1.2 Quantum–classical boundary: the Schrödinger's cat paradox

This picture does lead to serious questions. First of all, time is *very* different from spatial coordinates: it is not an observable, but a parameter, which governs the evolution of the state vector between the instantaneous "collapses". Fortunately, we do not have to deal with it, since our problems are strictly non relativistic. More pertinent is the question of the nature of "measurement" and "collapse" and their presumed instantaneity.

Measurement is understood as an interaction of the quantum system with a macroscopic object ("apparatus") such that the final state of the latter is determined by the value of the observable a_j, and different states of the apparatus are distinguishable and immutable (i.e., can be observed without perturbation). The terminology originates from the early days of quantum mechanics. Of course, there is no need for somebody to actually set up the apparatus; any appropriate macroscopic system will do, and now the tradition forces us to talk about, e.g., the liquid "measuring" or "observing" the quantum state of a particle travelling through it.

The classical states of the apparatus correlated to the different outcomes of the measurement are called "pointer states". (The corresponding eigenstates of the observable that is being measured we will also call pointer states, where it does not lead to confusion.) In the Copenhagen interpretation, it is the apparatus that predetermines what observables can be measured (so called "complementarity with respect to the means of observation"). In principle, for any classical variable we should be able to design an apparatus which would measure its quantum counterpart, an observable.

One problem here is that there exists no well-defined boundary between the "measured", or "observed" microscopic system, and the macroscopic "apparatus" or "observer". The Copenhagen interpretation of quantum mechanics simply posits the quantum behaviour for the one, and classical for the other, which is somewhat circular. What is worse, it does not allow a description of the system and the apparatus within a single formalism, denying us any quantitative description of the *process* of measurement. This was not too troubling when dealing with (quantum) electrons going through double slits in a (classical) screen, since it was obvious which was which, and the time of the electron's interaction with the detector can be neglected compared to all other relevant timescales. It becomes of crucial importance when the quantum systems we deal with contain huge numbers of elementary particles, almost comparable with the number of particles in our "apparatus", or are being "measured" continually. How do we describe such a situation in a consistent way?

A natural thing to do is to directly extend the quantum description (as the more fundamental one) to macroscopic systems. A difficulty of this approach lies in the fact that only very special states can be measured by a macroscopic apparatus, while at the quantum level they seem to be no different from all the rest of them. The situation is highlighted by the famous "Schrödinger's cat" paradox. Let us put together a deadly contraption: a tank of poisonous gas with an electrically controlled valve, connected to a Geiger counter. Put it in a sealed container with a live cat and a radioactive atom (e.g., ^{210}Po, which decays into the stable lead-206, ^{206}Pb, with the half-life $T \approx 138$ days), and wait (Fig. 1.1).

If we had to deal just with the radioactive atom, the description would be simple. Its wave function can be written as[3]

$$|\psi(t)\rangle = 2^{-t/2T} \left|{}^{210}\text{Po}\right\rangle + \sqrt{1 - 2^{-t/T}} \left|{}^{206}\text{Pb} + \alpha \text{ particle}\right\rangle. \qquad (1.4)$$

If at any given time we measure (observe) this atom, with the probability $1 - 2^{-t/T}$ we'll find lead-206. But putting the atom in the box with a cat makes things look bizarre, once we try to describe the cat (and the rest of the contraption) by a wave

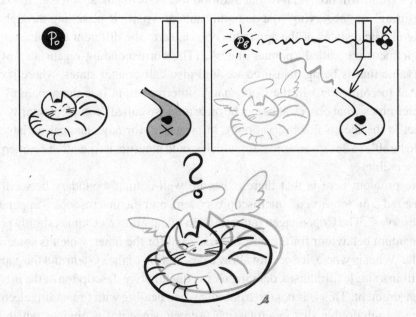

Fig. 1.1. Schrödinger's cat paradox: Should we describe macroscopic systems by wave functions, and if not, why not?

---

[3] This expression is only valid for not too short and not too long times; see § 5.5.6.

function. Now the state of the system before the observation will be

$$|\psi(t)\rangle = 2^{-t/2T} \left|^{210}\text{Po} + \text{live cat}\right\rangle + \sqrt{1 - 2^{-t/T}} \left|^{206}\text{Pb} + \alpha \text{ particle} + \text{dead cat}\right\rangle.$$
$$(1.5)$$

We don't doubt that after opening the box one will find either a live or a dead cat, with the probabilities $2^{-t/T}$ and $1 - 2^{-t/T}$ respectively. This intuitively clear outcome is in no way trivial. On the quantum side, we can always introduce an observable, which is a linear combination of other observables with real coefficients – there is nothing in the formalism to prohibit it. For example, we can build an observable, which will have as one of its eigenstates the superposition given by Eq. (1.5). Nevertheless, we evidently cannot build a classical apparatus, which would measure such a "zombie" state of the poor animal. The set of admissible classical states thus must impose some preferred set of bases on the Hilbert state of our quantum system.

Is the description (1.5) of a hybrid micro/macro system justified at all? And, what does precipitate the transition from the "live–dead" superposition to the "either–or" classical picture? The proposed answers run the full philosophical gamut from the many-worlds interpretation to the key role of the observer's conscience in the collapse. Instead of plunging into these fascinating ontological and epistemological depths, we will take a pragmatic approach and see how far it will take us.

We will not make any special difference between the "measurement" or "observation" and the interaction of our quantum system with its macroscopic surroundings.

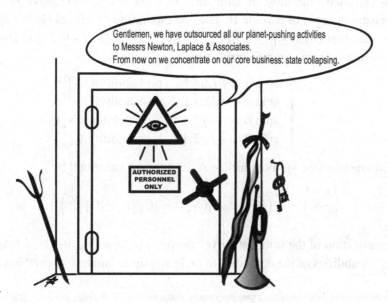

Fig. 1.2. The canonical view of the quantum–classical transition.

Indeed, explaining the "collapse" through the action of conscience would be exor-
cizing Asmodeus with Beelzebub's help, since we currently understand the latter,
if anything, less than the former.

We will assume that on the fundamental level every system is governed (and
described) by the laws of quantum mechanics, of which classical mechanics is
a limiting case – in full agreement with Bohr's *correspondence principle*. The
difference between "quantum" and "classical" systems must naturally emerge from
the formalism.[4]

1.2 Density matrix formalism

1.2.1 Justification and properties

The notion of the state vector is inadequate for our purposes, but it can be generalized
to the *density matrix*, or *statistical operator*, first introduced independently by
Landau and von Neumann in 1927. To make the motivation clear, let us start from a
quantum system, which consists of two subsystems, A and B, which we can measure
independently. Suppose, for example, that we have two particles with spin $\hbar/2$, and
have an apparatus, which can measure the z-component of spin of either particle.
Let the system be in state

$$|\Psi\rangle_{AB} = \sum_{j,k=\uparrow,\downarrow} C_{jk}|j_A k_B\rangle, \qquad (1.6)$$

which is obviously the most general form of such two-particle wave functions.
Now measure the spin of particle B. The corresponding observable is the operator
$(\hbar/2)\sigma_z$, such that $(\hbar/2)\sigma_z|\uparrow\rangle = \hbar/2|\uparrow\rangle$, and $(\hbar/2)\sigma_z|\downarrow\rangle = -\hbar/2|\downarrow\rangle$. The result
of the measurement can then be written as

$$|\Psi\rangle_{AB} \rightarrow \begin{cases} \text{spin B} = \hbar/2, |\uparrow_A\rangle, \text{probability} |C_{\uparrow\uparrow}|^2; \\ \text{spin B} = \hbar/2, |\downarrow_A\rangle, \text{probability} |C_{\downarrow\uparrow}|^2; \\ \text{spin B} = -\hbar/2, |\uparrow_A\rangle, \text{probability} |C_{\uparrow\downarrow}|^2; \\ \text{spin B} = -\hbar/2, |\downarrow_A\rangle, \text{probability} |C_{\downarrow\downarrow}|^2. \end{cases} \qquad (1.7)$$

If now we measure the spin of particle A, its average value will be

$$\left\langle \frac{\hbar}{2}\sigma_z \right\rangle = \frac{\hbar}{2}\left(|C_{\uparrow\uparrow}|^2 + |C_{\uparrow\downarrow}|^2 - |C_{\downarrow\uparrow}|^2 - |C_{\downarrow\downarrow}|^2\right). \qquad (1.8)$$

The normalization of the initial state vector, $|||\Psi\rangle_{AB}||^2 = \sum_{ij} |C_{ij}|^2 = 1$, ensures
that the probabilities of the outcomes of (1.7) add up to unity, so everything is fine,

---

[4] We are encouraged by the emergence of time-irreversible Boltzmann equations from time-reversible Newtonian
dynamics of point-like particles.

except that Eq. (1.7) is a very awkward way of dealing with even such a small system. Just imagine we had three particles!

The situation improves once the state vector is replaced with the density matrix. Let us choose a set of normalized, but not necessarily mutually orthogonal states in the Hilbert space, $\{|\Psi_j\rangle\}$, and write an operator

$$\rho = \sum_j p_j |\Psi_j\rangle\langle\Psi_j|, \tag{1.9}$$

where $p_j \geq 0$, and $\sum_j p_j = 1$. This obviously Hermitian operator can be interpreted as describing a *statistical ensemble* of quantum systems, which can be in a state $|\Psi_j\rangle$ with probability p_j. This is exactly the situation described by Eq. (1.7).

The convenience of using this operator becomes clear when calculating average values of the observables. According to rule (3) of quantum mechanics, the measured value of an observable A is always one of the eigenvalues of A, and according to rule (4), the probability of measuring a particular a_k in state $|\Psi_j\rangle$ is $|\langle a_k|\Psi_j\rangle|^2$. Therefore

$$\langle A \rangle_{\Psi_j} = \sum_k |\langle a_k|\Psi_j\rangle|^2 a_k = \sum_k \langle\Psi_j|a_k\rangle a_k \langle a_k|\Psi_j\rangle$$

$$= \sum_k \langle\Psi_j|a_k|a_k\rangle\langle a_k|\Psi_j\rangle = \langle\Psi_j|A\left(\sum_k |a_k\rangle\langle a_k|\right)|\Psi_j\rangle = \langle\Psi_j|A|\Psi_j\rangle, \tag{1.10}$$

a standard formula. Since now the system is in state $|\Psi_j\rangle$ only with some probability p_j, we must average over these too, with the result

$$\langle A \rangle = \sum_j p_j \langle A \rangle_{\Psi_j} = \sum_j p_j \langle\Psi_j|A|\Psi_j\rangle \equiv \text{tr}(\rho A). \tag{1.11}$$

The system, described by a density matrix, is said to be in a *mixed state*, instead of a *pure state* (when a single state vector suffices). One can write a density matrix for a pure state: it will include only one component, $\rho_{\text{pure}} = |\Psi\rangle\langle\Psi|$.

As a Hermitian operator, the density matrix has an orthonormal set of eigenstates, $|\rho_j\rangle$, with corresponding eigenvalues, ρ_j. It can be, therefore, written in the *spectral representation*,

$$\rho = \sum_j \rho_j |\rho_j\rangle\langle\rho_j|. \tag{1.12}$$

Since ρ_j can be interpreted as the probability of finding the system in state $|\rho_j\rangle$, all $\rho_j \geq 0$; in other words, the density matrix is *positive semidefinite*.

The density matrix can be written in any basis in the Hilbert space, and will usually have neither the explicit form (1.9) nor (1.12). We will therefore list here its invariant properties.

1) $\text{tr}\,\rho = 1$ (follows directly from the definition (1.9)).
2) $\rho^2 = \rho$ if and only if the state is pure. Indeed, if $\rho = |\Psi\rangle\langle\Psi|$, then $\rho^2 = |\Psi\rangle\langle\Psi||\Psi\rangle\langle\Psi| = |\Psi\rangle\langle\Psi| = \rho$. Conversely, if $\rho^2 = \rho$, then (from (1.12)) $\rho_j^2 = \rho_j$, and the eigenvalues of the density matrix are either zeros, or ones. Since its trace is one, then only one eigenvalue can be equal to one, i.e., there is only one term in the spectral decomposition, and the state is indeed pure.
3) $\text{tr}(\rho^2) \leq 1$, and there is equality if and only if the state is pure (therefore the quantity $\varsigma = \text{tr}(\rho^2)$ is sometimes called *purity*). Starting from (1.9) and recalling that $|\langle\Psi_j|\Psi_k\rangle| \leq 1$,

$$\text{tr}(\rho^2) = \text{tr}\left(\sum_{jk} p_j p_k |\Psi_j\rangle\langle\Psi_j|\Psi_k\rangle\langle\Psi_k|\right)$$

$$= \sum_j p_j \sum_k p_k |\langle\Psi_j|\Psi_k\rangle|^2 \leq \sum_j p_j = 1.$$

On the other hand, if $\text{tr}(\rho^2) = 1$, then $\sum_j p_j \left(\sum_k p_k |\langle\Psi_j|\Psi_k\rangle|^2\right) = 1$. Since the non negative p_j's add up to one, there is equality only if $p_j = \delta_{jq}$ for some q, that is, if the state is pure.

These properties provide us with a reliable way to check any approximate calculations of a density matrix, as well as a criterion of whether a given system is in a pure or a mixed state.

1.2.2 Averages, probabilities and coherences

Consider now, as before, a quantum-mechanical system in the pure state (1.6). Its density matrix is

$$\rho_{AB} = |\Psi\rangle_{AB}\langle\Psi|_{AB} = \sum_{j,k,l,m=\uparrow,\downarrow} C_{jk}C_{lm}^* |j_A k_B\rangle\langle l_A m_B|.$$

Let's take a partial trace of this operator over the states of particle B:

$$\rho_A \equiv \text{tr}_B \rho_{AB} = \sum_{q=\uparrow,\downarrow} \langle q_B|\Psi\rangle_{AB} \left(\langle\Psi|_{AB}\right)|q_B\rangle$$

$$= \sum_{q,j,k,l,m=\uparrow,\downarrow} C_{jk}C_{lm}^* \delta_{qk}|j_A\rangle\langle l_A|\delta_{qm} = \sum_{q,j,l=\uparrow,\downarrow} C_{jq}C_{lq}^* |j_A\rangle\langle l_A|. \quad (1.13)$$

To find the observed value of the z-component of spin of particle A, we now take the trace of the spin operator with *reduced density matrix* ρ_A:

$$\left\langle \frac{\hbar}{2}\sigma_z^A \right\rangle = \frac{\hbar}{2}\mathrm{tr}\left(\rho_A \sigma_z^A\right) = \frac{\hbar}{2}\mathrm{tr} \sum_{q,j,l=\uparrow,\downarrow} C_{jq}C_{lq}^* \sigma_z |j_A\rangle\langle l_A|$$

$$= \frac{\hbar}{2}\mathrm{tr} \sum_{q,j=\uparrow,\downarrow} |C_{jq}|^2 \langle j_A|\sigma_z|j_A\rangle. \tag{1.14}$$

This equation automatically yields the same result as Equations (1.7) and (1.8), and in a compact form. We see that as long as we are interested only in the results of measurement on one subsystem, we can use the reduced density matrix by taking the trace of the full density matrix over all the irrelevant degrees of freedom. This is an important procedure, which is crucial for describing open quantum systems, i.e., systems for which one can neither neglect their interaction with the environment nor take the latter into account explicitly.

Let us look at Eq. (1.13) in more detail. This expression contains more information than just the average value of the operator σ_z. The diagonal terms provide the probabilities of finding the system with spin up/down. What about the off-diagonal ones?

Let us write a density matrix in an orthonormal basis of eigenstates of some observable A, and we chose A such that its eigenstates are macroscopically observable pointer states – e.g., charges, positions or momenta: $\rho = \sum_{ij}\rho_{ij}|a_i\rangle\langle a_j|$. Its diagonal elements give the probabilities of finding the system in appropriate eigenstates, and if we are only performing the measurements of A this is all we care about or can extract from the experiment. The off-diagonal terms could have never been there. Nevertheless, they are crucially important.

We have seen that a diagonal density matrix describes a pure state if and only if its only element is unity (invariant property number 2). If this is the case, we will always measure the same eigenvalue a_j of the operator A and leave the system in the same state $|a_j\rangle$.[5] (This is called *quantum non demolition (QND) measurement*, and we will discuss the implications of such experiments in § 5.5.2.) If not, the density matrix must describe a mixed state, so the absence of the off-diagonal matrix elements is a tell-tale sign. These elements describe *quantum coherence*. If they are zero, mathematically there is no basis in which the density matrix contains a single diagonal element; physically, the system is in a mixed state; from the point of view of the observation, there is no observable of which the system is in an eigenstate, and its measurement will yield some eigenstate at random, with a given

---

[5] We need not bother here with the cases of degenerate eigenstates or the continuous spectrum of the operator A.

probability. This is why the evolution of the off-diagonal elements of the density matrix *("coherences")* is of special interest: their disappearance – "decoherence", due to whatever processes that cause it – means the reduction of the state of the system to a mixture, with the loss of specifically quantum correlations. The rate of this reduction can, therefore, be taken as the *decoherence rate*, and its inverse as the *decoherence time*.

1.2.3 Entanglement

We have already seen that when several distinct systems (e.g., a cat and a ^{210}Po atom) are described by a single quantum state, this can lead to quite counterintuitive conclusions. Suppose that two quantum systems (e.g., spin-1/2 particles A and B) are in a pure quantum state $|\Psi\rangle_{AB}$. Then, generally speaking, separately, neither of them can be in a definite quantum state. Consider, for example, the expectation value of σ_z for particle A, Eq. (1.14):

$$\left\langle \frac{\hbar}{2}\sigma_z^A \right\rangle = \frac{\hbar}{2}\mathrm{tr}\left(\rho_A \sigma_z^A \right).$$

The reduced density matrix $\rho_A = \mathrm{tr}_B\left[|\Psi\rangle_{AB}\langle\Psi|_{AB}\right]$ can only have the form $|\psi\rangle_A\langle\psi|_A$, corresponding to particle A being in a pure quantum state $|\psi\rangle_A$, if the quantum state of the whole system is factorized:

$$|\Psi\rangle_{AB} = |\psi\rangle_A \otimes |\psi\rangle_B. \tag{1.15}$$

Otherwise, particle A by itself is in a mixed state, even though the whole system consisting of A and B is in a pure state, and there is no interaction between A and B. The non factorized states of multipartite systems are called *entangled*, which expresses this specifically quantum correlation between its components. Essentially, it means that the properties of a quantum system in general can never be reduced to the sum of the properties of its constituents.

For a pure state $|\Psi\rangle_{ABC...}$ of a system, comprising any number of subsystems, the test for entanglement is straightforward: if all reduced density matrices correspond to pure states, the state $|\Psi\rangle_{ABC...}$ can be written as $|\psi\rangle_A \otimes |\psi\rangle_B \otimes |\psi\rangle_C ...$, and there is no entanglement. But entanglement is not limited to pure states: certain quantum correlation can exist even if the system itself is described by a density matrix. For a given density matrix one can find different numerical measures of entanglement (all of which are zero for a factorized state; see, e.g., Horodecki et al., 2009).

1.2.4 Liouville–von Neumann equation

As a generalization of the state vector, the density matrix contains all the information about the system that can possibly be obtained. It is a counterpart of the classical density function in the phase space, $f(X_a, P_a, t)$, which describes an arbitrary classical system (e.g., a non-ideal gas). The equation of motion for $f(X_a, P_a, t)$ follows from the Hamiltonian dynamics and is called the Liouville equation:

$$\frac{\partial f(X_a, P_a, t)}{\partial t} = -[\mathcal{H}, f]_P \equiv -\sum_a \left(\frac{\partial \mathcal{H}}{\partial P_a} \frac{\partial f}{\partial X_a} - \frac{\partial f}{\partial P_a} \frac{\partial \mathcal{H}}{\partial X_a} \right), \qquad (1.16)$$

where $\mathcal{H}$ is the classical Hamilton function of the system, and $[., .]_P$ is the Poisson bracket.

Similarly, the equation of motion for the density matrix (the *Liouville–von Neumann equation*) follows directly from the Schrödinger equation:

$$\frac{d\rho}{dt} = \sum_j \frac{d}{dt} \left(p_j |\Psi_j\rangle\langle\Psi_j| \right) = \sum_j p_j \frac{1}{i\hbar} \left(H|\Psi_j\rangle\langle\Psi_j| - |\Psi_j\rangle\langle\Psi_j|H \right) = \frac{1}{i\hbar}[H, \rho].$$

$$(1.17)$$

Not surprisingly, it has the same structure as the classical Eq. (1.16), with the Poisson bracket replaced by a commutator. It can be rewritten as

$$\frac{d\rho}{dt} = \mathcal{L}[\rho(t)], \qquad (1.18)$$

where the action of the *Liouvillian* on any operator A is given by $\mathcal{L}[A] \equiv \frac{1}{i\hbar}[H, A]$.

First of all, the evolution described by Eq. (1.17) is unitary. The solution can be written as (cf. (1.2))

$$\rho(t) = U(t)\rho(0)U(t)^\dagger \equiv \mathcal{U}(t, 0)[\rho(0)]; \quad U(t) = e^{-\frac{i}{\hbar}Ht}, \qquad (1.19)$$

with an obviously unitary evolution operator $U(t)$. Unitarity means, in particular, that the evolution is time-reversible, since $U^{-1} = U^\dagger$ always exists. Therefore for any $\mathcal{U}(t, 0)$ there exists the unique inverse operator: $(\mathcal{U}(t, 0))^{-1}[A(t)] = \mathcal{U}(-t, 0)[A(t)] = U(t)^\dagger A(t)U(t) = A(0)$.

There is an interesting difference between the Liouville–von Neumann equation and the Schrödinger equation from the point of view of symmetry. In the Schrödinger picture the time dependence of the energy eigenstates, $H|e_j\rangle = E_j|e_j\rangle$, is given by $\exp[-iE_j t/\hbar]$, and therefore a special role is played by the degenerate energy eigenstates (with $E_j = E_k$). For the Liouville–von Neumann equation, the corresponding phase factors, obviously, contain the energy *differences*, $\exp[-\mathrm{i}(E_j - E_k)t/\hbar]$. This yields a new (so-called Liouvillian) symmetry between the *pairs* of eigenstates with the same energy differences, which can be used, as any symmetry, to simplify the analysis (Maassen van den Brink and Zagoskin, 2002).

The purity is conserved:

$$\varsigma(t) \equiv \mathrm{tr}(\rho(t)^2) = \mathrm{tr}[U(t)\rho(0)U(t)^\dagger U(t)\rho(0)U(t)^\dagger] = \mathrm{tr}(\rho(0)^2) = \varsigma(0). \tag{1.20}$$

This means that the evolution of a pure state into a mixed state, including the process of measurement, cannot be described by (1.17) and is an essentially non-unitary process.

1.2.5 Wigner function

The relation between the classical density function and the quantum density matrix becomes more clear if we use for the latter a representation that was first proposed by Wigner (1932) for the wave function. Consider for simplicity the case of a single particle in one dimension. In the position representation its density matrix is (omitting the explicit time dependence)

$$\rho(x, x') \equiv \langle x|\rho|x'\rangle, \tag{1.21}$$

where $\hat{X}|x\rangle = x|x\rangle$. If we denote the average position and the deviation from it by $X = (x + x')/2$, $\xi = x - x'$, respectively, and take the Fourier transform over the latter, we obtain a function, which depends on *both* position and momentum:

$$W(X, P) = \frac{1}{2\pi\hbar} \int \mathrm{d}\xi \left\langle X + \frac{\xi}{2}\middle|\rho\middle|X - \frac{\xi}{2}\right\rangle e^{-\mathrm{i}P\xi/\hbar}. \tag{1.22}$$

This *Wigner function* looks similar to the classical distribution function $f(X, P)$. They indeed have common properties. For example, integrating $f(X, P)$ with respect to one variable leaves a positively defined probability distribution for the other. The same holds for $W(X, P)$:

$$\int \mathrm{d}P\, W(X, P) = \int \mathrm{d}\xi \left\langle X + \frac{\xi}{2}\middle|\rho\middle|X - \frac{\xi}{2}\right\rangle \frac{1}{2\pi\hbar} \int \mathrm{d}P\, e^{-\mathrm{i}P\xi/\hbar}$$

$$= \int \mathrm{d}\xi \left\langle X + \frac{\xi}{2}\middle|\rho\middle|X - \frac{\xi}{2}\right\rangle \delta(\xi) = \langle X|\rho|X\rangle \geq 0; \tag{1.23}$$

$$\int dX W(X, P) = \frac{1}{2\pi\hbar} \int d\xi \int dX \left\langle X + \frac{\xi}{2} \left| \rho \right| X - \frac{\xi}{2} \right\rangle e^{-iP\xi/\hbar}$$

$$= \int dx\, dx' \left\langle x | \rho | x' \right\rangle \left[\frac{1}{2\pi\hbar} e^{-iP(x-x')/\hbar} \right]$$

$$= \int dx\, dx' \left\langle x | \rho | x' \right\rangle \left\langle x' | P \right\rangle \left\langle P | x \right\rangle = \left\langle P | \rho | P \right\rangle \geq 0. \tag{1.24}$$

We took into account that the eigenfunctions of momentum in the position representation are simply plane waves, $\langle x | P \rangle = e^{iPx/\hbar}/\sqrt{2\pi\hbar}$, and that they comprise a basis of the Hilbert space (see, e.g., Messiah, 2003; Landau and Lifshitz, 2003). We also used the fact that the density matrix is positive semidefinite, Eq. (1.12).

From (1.23) and (1.24) immediately follows that the Wigner function is normalized to unity,

$$\int dX dP\, W(X, P) = 1. \tag{1.25}$$

In addition to the relations

$$\langle \hat{X}^m \rangle = \int dX dP\, W(X, P) X^m; \quad \langle \hat{P}^n \rangle = \int dX dP\, W(X, P) P^n, \tag{1.26}$$

which follow from Eqs. (1.23) and (1.24), it can be shown that, in general,

$$\int dX dP\, W(X, P) X^m P^n = \langle \{ \hat{X}^m \hat{P}^n \}_{\text{sym}} \rangle, \tag{1.27}$$

where $\{ \hat{X}^m \hat{P}^n \}_{\text{sym}}$ is a *symmetrized* product of noncommuting operators $\hat{X}$, $\hat{P}$. For example, $\{ \hat{X}\hat{P} \}_{\text{sym}} = \frac{1}{2}(\hat{X}\hat{P} + \hat{P}\hat{X})$, $\{ \hat{X}^2\hat{P}^2 \}_{\text{sym}} = \frac{1}{6}(\hat{X}^2\hat{P}^2 + \hat{P}^2\hat{X}^2 + \hat{X}\hat{P}^2\hat{X} + \hat{P}\hat{X}^2\hat{P} + \hat{X}\hat{P}\hat{X}\hat{P} + \hat{P}\hat{X}\hat{P}\hat{X})$, etc. The Wigner function, therefore, reduces the task of finding average values of a product of the observables to calculating an ordinary integral (1.27) (where X's and P's are, of course, just numbers), but imposes a specific (*Wigner*) ordering of the operators to be averaged.

Where the drastic difference between $W(X, P)$ and $f(X, P)$ does arise, it is that Eqs. (1.23) and (1.24) notwithstanding, $W(X, P)$ is not non-negative. Therefore, it is not a probability, but a *quasiprobability* distribution function, and here lies its essential quantumness. One can show, on the other hand, that the Wigner function averaged on the scale of $2\pi\hbar$ becomes non-negative and reduces to the classical distribution function $f(X, P)$ (Röpke, 1987, §2.2.2; Zubarev et al., 1996). This is another way of looking at the quantum–classical transition.[6]

---

[6] Of course, one has eventually to discuss reasons and specific mechanisms for such averaging.

The generalization to more degrees of freedom or different observables is straightforward. The similarity between the Wigner function and the classical probability density makes use of $W(X, P)$ very convenient in, e.g., quantum statistical physics and quantum kinetics (Balescu, 1975; Pitaevskii and Lifshitz, 1981).

The Wigner function for the states of a harmonic oscillator – i.e., quantized field modes – is particularly useful for our purposes and will be introduced in Section 4.4.[7]

1.2.6 Perturbation theory for density matrix. Linear response theory

Any system in the Universe is isolated only to a degree. Therefore, a reasonable theory must allow for a consistent treatment of *open systems*, i.e., systems coupled to their environment (which is weakly influenced by the system and can only be described by average macroscopic parameters, like temperature and pressure).

If the system is in equilibrium, its density matrix has the Gibbs form:

$$\rho_{CE} = e^{(F-H)/k_B T}; \quad F = -k_B T \ln \operatorname{tr} e^{-H/k_B T} \tag{1.28}$$

or

$$\rho_{GCE} = e^{(\Omega-H')/k_B T}; \quad \Omega = -k_B T \ln \operatorname{tr} e^{-H'/k_B T}. \tag{1.29}$$

The first of these equations describes the *canonical ensemble*, where the system can exchange only energy with the environment. The second corresponds to the *grand canonical ensemble*, where particles can also be exchanged. (Therefore, instead of the Hamiltonian H one uses the operator $H' = H - \mu N$, where μ is the chemical potential and N is the particle number operator; N obviously commutes with H.) These expressions are routinely derived in quantum statistical mechanics (e.g., Landau and Lifshitz, 1980, Chapter III). They are very similar to their classical analogues, and the normalization factors F and Ω are the quantum counterparts of the (Helmholtz) free energy and of the grand potential, respectively. Here we should observe that as long as the number of particles in the system is large, the two ensembles are equivalent, but in small systems, where a single particle may make a difference, one must take care to use the physically appropriate ensemble. This is precisely the case of, e.g., charge qubits.

Looking at Equations (1.28) and (1.29), we see that in equilibrium the system is in a mixed state. Nevertheless, the systems we are interested in are far from equilibrium, and the Liouville–von Neumann equations should be solved with appropriate initial conditions, which is usually impossible. The general approach to the problem

---

[7] On a more general plane, the Wigner function is a version of a non-equilibrium Green's function, which finds broad applications all across physics (see, e.g., Pitaevskii and Lifshitz, 1981, Chapter X; Zagoskin, 1998, Chapter 3).

is based on perturbation theory. Assuming that the coupling between the system and the environment is weak and the initial state of both is known, one can, e.g., iterate the equations, obtaining the answer as a series of powers of the coupling.

To demonstrate this approach and introduce (or remind the reader of) some useful ideas, we start by deriving the equations of *linear response theory*. We assume that the system is weakly perturbed by an external force and are interested in what will be the change in the average value of some observable A due to the perturbation, in the first order, in the perturbation strength.

The Hamiltonian of the problem is then

$$H(t) = H_0 + H_I(t), \tag{1.30}$$

where the unperturbed Hamiltonian H_0 is time-independent, and the perturbation $H_I(t)$ is small compared to H_0. It is convenient to use the *interaction representation*, in which every operator is put between the "brackets":

$$A \to A_{\text{int}}(t) = U_0(t)^\dagger A U_0(t), \quad U_0(t) = e^{-iH_0 t/\hbar}. \tag{1.31}$$

Performing this operation on the density matrix, we see that

$$\frac{d\rho_{\text{int}}}{dt} = -\frac{1}{i\hbar}[H_0, \rho_{\text{int}}(t)] + U_0(t)^\dagger \frac{d\rho}{dt} U_0(t)$$

$$= -\frac{1}{i\hbar}[H_0, \rho_{\text{int}}(t)] + U_0(t)^\dagger \frac{1}{i\hbar}[H_0 + H_I(t), \rho(t)]U_0(t)$$

$$\equiv \frac{1}{i\hbar}[H_{I,\text{int}}(t), \rho_{\text{int}}(t)]. \tag{1.32}$$

Integrating this from $t = -\infty$, we find

$$\rho_{\text{int}}(t) = \rho_{\text{int}}(-\infty) + \frac{1}{i\hbar}\int_{-\infty}^{t} dt'[H_{I,\text{int}}(t'), \rho_{\text{int}}(t')]. \tag{1.33}$$

The system initially being in equilibrium, $\rho_{\text{int}}(-\infty) \equiv \rho_0$ is given either by (1.29) or by (1.30). (In the case of the grand canonical ensemble we should everywhere use instead of the Hamiltonian, H, the operator $H' = H - \mu N$.) Equation (1.33) can be formally iterated:

$$\rho_{\text{int}}(t) = \rho_0 + \rho_1(t) + \rho_2(t) + \ldots;$$

$$\rho_1(t) = \frac{1}{i\hbar}\int_{-\infty}^{t} dt'[H_{I,\text{int}}(t'), \rho_0]; \quad \text{ad inf}. \tag{1.34}$$

This series usually converges only asymptotically; moreover, its higher-order terms contain unpleasant nested commutators, but the linear response of the system to the perturbation – the first term in the series – can be readily obtained.

Due to the cyclic invariance of the trace, the average $\langle A \rangle$ is the same in the interaction representation, $\langle A \rangle = \mathrm{tr}\, A\rho(t) = \mathrm{tr}\, A_{\mathrm{int}}(t)\rho_{\mathrm{int}}(t)$. The shift due to the perturbation (the linear response of the measured observable) is then

$$\Delta A(t) = \mathrm{tr}\, \rho_1(t) A_{\mathrm{int}}(t) = \frac{1}{i\hbar} \int_{-\infty}^{t} dt'\, \mathrm{tr}\, \left([H_{I,\mathrm{int}}(t'), \rho_0] A_{\mathrm{int}}(t)\right)$$

$$= \frac{1}{i\hbar} \int_{-\infty}^{t} dt'\, \langle [A_{\mathrm{int}}(t), H_{I,\mathrm{int}}(t')] \rangle_0, \qquad (1.35)$$

where $\langle \ldots \rangle_0 \equiv \mathrm{tr}(\rho_0 \ldots)$ is the equilibrium average. This expression can be used as is, but it is both convenient and insightful to transform this *Kubo formula* to the form that allows us to link the linear response of a system to the equilibrium fluctuations in it.

There are different ways of writing Kubo formulas. We will do so by assuming, without restricting the generality, that the perturbation $H_I(t)$ has the form

$$H_I(t) = -f(t) B, \qquad (1.36)$$

where B is an operator, and $f(t)$ is a conventional function called a *generalized force*. We will also introduce the *retarded Green's function* of the operators A and B,

$$\ll A(t)B(t') \gg^R \equiv \frac{1}{i\hbar} \langle [A(t), B(t')] \rangle_0 \theta(t - t'). \qquad (1.37)$$

Here $\theta(t - t')$ is Heaviside's step function, and retarded means that (1.36) is zero unless $t \geq t'$. Using this definition, we can rewrite (1.35) as

$$\Delta A(t) = - \int_{-\infty}^{\infty} dt'\, f(t') \ll A(t)B(t') \gg^R . \qquad (1.38)$$

As would be expected, from causality, only the perturbations at times earlier than t can influence $\Delta A(t)$.

The Green's function is calculated at equilibrium; therefore, it depends only on the difference $t - t'$, and the expression (1.38) is a convolution. Taking its Fourier transform, we obtain simply

$$\Delta A(\omega) = -f(\omega) \ll AB \gg_{\omega}^R . \qquad (1.39)$$

Defining the *generalized susceptibility* via

$$\chi(\omega) = \Delta A(\omega)/f(\omega), \qquad (1.40)$$

we see that it is equal to $- \ll AB \gg_{\omega}^R$. Examples of generalized susceptibilities are electrical conductivity $\sigma_{\alpha\beta}(\omega)$ (where the current density $j_\alpha(\omega) = \sigma_{\alpha\beta}(\omega) E_\beta(\omega)$), magnetic susceptibility $\chi_{\alpha\beta}(\omega)$ (with the magnetic moment $m_\alpha(\omega) = \chi_{\alpha\beta}(\omega) H_\beta(\omega)$), etc.

1.2.7 Fluctuation-dissipation theorem

Now we will, as promised, relate the susceptibility to the equilibrium fluctuations in the system. The latter are described by the *autocorrelation function* of the given observable,

$$K_A(t) = \langle A(t)A(0)\rangle_0 \equiv K_{A,s}(t) + K_{A,a}(t);$$

$$K_{A,s(a)}(t) = \frac{K_A(t) \pm K_A(-t)}{2}. \tag{1.41}$$

Since A is an operator, $A(t)$ and $A(0)$ do not necessarily commute, and the anti-symmetric part $K_{A,a}(t)$ is generally nonzero, but we are here concerned with the symmetric part of the autocorrelator. Its Fourier transform $S_{A,s}(\omega)$ (*symmetric spectral density of fluctuations*) gives the intensity of fluctuations of the observable.[8]

Let us invoke a useful *Kubo–Martin–Schwinger identity* for an equilibrium average of any two operators,

$$\langle A(\tau)B(0)\rangle_0 \equiv \mathrm{tr}\left(e^{(F_0-H_0)/k_BT}e^{iH_0\tau/\hbar}A(0)e^{iH_0\tau/\hbar}B(0)\right)$$

$$= \langle B(0)A(\tau + i\hbar/k_BT)\rangle_0, \tag{1.42}$$

which immediately follows from the cyclic invariance of the trace. Then we can write for the Fourier transforms of a commutator and an anticommutator,

$$\langle[A,A]\rangle_{0,\omega} \equiv \int_{-\infty}^{\infty} dt\, e^{i\omega t}\langle[A(t),A(0)]\rangle_0 = \left(e^{\hbar\omega/k_BT} - 1\right)\int_{-\infty}^{\infty} dt\, e^{i\omega t}\langle A(0)A(t)\rangle_0;$$

$$\langle\{A,A\}\rangle_{0,\omega} \equiv \int_{-\infty}^{\infty} dt\, e^{i\omega t}\langle\{A(t),A(0)\}\rangle_0 = \left(e^{\hbar\omega/k_BT} + 1\right)\int_{-\infty}^{\infty} dt\, e^{i\omega t}\langle A(0)A(t)\rangle_0. \tag{1.43}$$

This allows us to express the symmetric spectral density through the average commutator,

$$S_{A,s}(\omega) \equiv \langle\{A,A\}\rangle_{0,\omega}/2 = \langle[A,A]\rangle_{0,\omega}\coth(\hbar\omega/2k_BT)/2. \tag{1.44}$$

---

[8] To be more precise, in order to investigate fluctuations around the average value $\langle A\rangle_0$ one should consider the *autocovariance function*, $K'_A(t) = \langle(A(t)-\langle A\rangle_0)(A(0)-\langle A\rangle_0)\rangle_0 \equiv K_A(t) - \langle A\rangle_0^2$. In the following we assume $\langle A\rangle_0 = 0$, which makes only a trivial difference.

On the other hand, by splitting the integration interval and changing variables we get

$$
\langle [A, A] \rangle_{0,\omega} = \int_{\infty}^{0} d(-t) e^{i\omega(-t)} \langle [A(-t), A(0)] \rangle_0 + \int_{0}^{\infty} dt e^{i\omega t} \langle [A(t), A(0)] \rangle_0
$$

$$
= - \int_{0}^{\infty} dt e^{-i\omega t} \langle [A(t), A(0)] \rangle_0 + \int_{0}^{\infty} dt e^{i\omega t} \langle [A(t), A(0)] \rangle_0
$$

$$
= -2\hbar \operatorname{Im} \ll AA \gg_{\omega}^{R} \equiv 2\hbar \operatorname{Im} \chi(\omega). \tag{1.45}
$$

We have, therefore, established the relation between the intensity of *equilibrium* fluctuations in a system and its linear response to an external perturbation, known as the *fluctuation–dissipation theorem*:

$$
S_{A,s}(\omega) = \hbar \operatorname{Im} \chi(\omega) \coth(\hbar\omega/2k_B T). \tag{1.46}
$$

The relation between the equilibrium properties of the system and its nonequilibrium behaviour is not only physically insightful, but practically useful. Equilibrium Green's functions like (1.37) can be obtained perturbatively, using the well-developed and relatively simple *Matsubara formalism* (an extension of Feynman's diagrammatic approach).[9] One can also go on and calculate higher-order response functions (i.e., quadratic susceptibility, cubic susceptibility, etc.), which will again be expressed in terms of equilibrium properties of the system, but the expressions soon become unwieldy, and their practical usefulness diminishes.

1.3 Evolution of density matrix in open systems

1.3.1 Getting rid of the environment

Let us return to the Liouville–von Neumann equation (1.17) for a system and its environment described by the Hamiltonian

$$
H = H_S + H_E + H_I. \tag{1.47}
$$

Our goal is to reconcile the unitary evolution given by Eq. (1.17) with the existence of non unitary processes, like measurement, and in general, transition from a pure to a mixed state. We will see that this can be done based on several plausible assumptions.

First, we plausibly assume that H_I is a small perturbation, and that the environment is so big that the influence of the system on it is also small. As in (1.31) we will use the interaction representation with respect to the Hamiltonian $H_0 \equiv H_S + H_E$,

---

[9] See, e.g., Zagoskin (1998), Chapter 3.

so that

$$\frac{d\rho}{dt} = \frac{1}{i\hbar}[H_I(t), \rho(t)]; \quad \rho(t) = \rho(-\infty) + \frac{1}{i\hbar}\int_{-\infty}^{t} dt'[H_I(t'), \rho(t')] \quad (1.48)$$

(we dropped the subscripts "int" to unclutter the formulas). By substituting in the right-hand side of the first equation (1.48) the expression for $\rho(t)$ from the second equation, we get

$$\frac{d\rho}{dt} = \frac{1}{i\hbar}[H_I(t), \rho(-\infty)] + \frac{1}{(i\hbar)^2}\int_{-\infty}^{t} dt'[H_I(t), [H_I(t'), \rho(t')]]. \quad (1.49)$$

Now we continue making plausible assumptions. First, assume that not only initially the system and the environment were statistically independent of each other, but they will remain so:

$$\rho(-\infty) = \rho_S(-\infty) \otimes \rho_E; \quad \rho(t) = \rho_S(t) \otimes \rho_E. \quad (1.50)$$

This assumption rests on the bigness of the environment and the weakness of its interaction with the system, and is physically reasonable. (Of course, every part of the physical environment, which violates this assumption, must be included in the system.) Then we can trace out the environment and find for the reduced density matrix of the system, $\mathrm{tr}_E \rho(t) \equiv \rho_S(t)$ (compare to Eq. (1.14)),

$$\frac{d\rho_S}{dt} = \frac{1}{i\hbar}\mathrm{tr}_E([H_I(t), \rho_S(-\infty) \otimes \rho_E])$$

$$+ \frac{1}{(i\hbar)^2}\int_{-\infty}^{t} dt'\mathrm{tr}_E([H_I(t), [H_I(t'), \rho_S(t') \otimes \rho_E]]). \quad (1.51)$$

Second, realistically assume that the interaction Hamiltonian has the form (in the Schrödinger representation)

$$H_I = \sum_{ab} g_{ab} A_a B_b, \quad (1.52)$$

where the operators A act on the system, and B on the environment, and $\langle B \rangle \equiv \mathrm{tr}_E[\rho_E B] = 0$. (If $\langle B \rangle \neq 0$, the corresponding term in (1.52), $\sum_{ab} g_{ab} A_a \langle B_b \rangle$, would only depend on the operators acting on the system and should have been included in H_S.) From here it follows that the first term in (1.51) vanishes. Finally, make a really important *Markov approximation*. Namely, assume that the density matrix in the system changes more slowly than the characteristic correlation times of the environment. Then we can neglect the difference between t and t' in the argument of ρ_S and reduce the integro-differential equation (1.51) to the differential master equation

$$\frac{d\rho_S}{dt} = -\frac{1}{\hbar^2} \int_{-\infty}^{t} dt' \text{tr}_E([H_I(t),[H_I(t'),\rho_S(t)\otimes\rho_E]]), \tag{1.53}$$

which contains *only the current value* of the reduced density matrix. (Independence of the previous history is precisely what Markovian means in the mathematical theory of random processes.) We could have an additional term in (1.53), of the form $(1/i\hbar)[h(t),\rho_S(t)]$, with $h(t)$ describing some processes in the system we decided not to cancel by the transition to the interaction picture (e.g., the influence of an external field, as in Eq. (1.36)).

1.3.2 Master equation for the density matrix; Lindblad operators

Equation (1.53) does not look like a finished job yet. We should have substituted (1.52) for H_I, taken the trace over the environmental variables, and calculated the integral over t'. Nevertheless, it is more convenient to do this with a concrete Hamiltonian, to make use of the specific commutation relations of the operators A, B.

Let us consider, as an example, an environment that consists of a single bosonic mode with frequency ω_E. That is, our system is coupled to a linear oscillator, which is somehow maintained in a given stationary state ρ_E. It is usually, but not always, justified to assume that this state is an equilibrium state (1.28) with some temperature T.

The simplest form of a coupling term in the interaction representation is then (reminding us of Eq. (1.36))

$$H_I(t) = g B(t)(a^\dagger(t)+a(t)) = g B(t)\left(a^\dagger e^{i\omega_E t}+a e^{-i\omega_E t}\right). \tag{1.54}$$

We took into account that in the interaction representation the Bose operators depend on time trivially, $a(t)=a\exp[-i\omega_E t]$, $a^\dagger(t)=a^\dagger \exp[i\omega_E t]$. The time dependence of the Hermitian operator B, acting on the system, is determined by H_S, whatever it is, and B generally does not commute with the density matrix ρ_S (neither $B(t)$ and $B(t')$ should commute, unless $t'=t$). Substituting (1.54) into (1.53), we will obtain several tedious expressions, which are all written down below:

$$\frac{d\rho_S}{dt} = -\frac{g^2}{\hbar^2}\{(a)-(b)-(c)+(d)\}; \tag{1.55}$$

$$(a) = \left\langle\left(a^\dagger\right)^2\right\rangle B(t)\left[\int_{-\infty}^{t} dt' B(t')e^{i\omega_E(t'+t)}\right]\rho_S(t)$$

$$+ \langle aa^\dagger\rangle B(t)\left[\int_{-\infty}^{t} dt' B(t')e^{i\omega_E(t'-t)}\right]\rho_S(t)$$

$$+ \langle a^\dagger a \rangle B(t) \left[\int_{-\infty}^{t} dt' \, B(t') e^{-i\omega_E(t'-t)} \right] \rho_S(t)$$

$$+ \langle (a)^2 \rangle B(t) \left[\int_{-\infty}^{t} dt' \, B(t') e^{-i\omega_E(t'+t)} \right] \rho_S(t);$$

$$(b) = \left\langle \left(a^\dagger \right)^2 \right\rangle B(t) \rho_S(t) \left[\int_{-\infty}^{t} dt' \, B(t') e^{i\omega_E(t'+t)} \right]$$

$$+ \langle a^\dagger a \rangle B(t) \rho_S(t) \left[\int_{-\infty}^{t} dt' \, B(t') e^{i\omega_E(t'-t)} \right]$$

$$+ \langle aa^\dagger \rangle B(t) \rho_S(t) \left[\int_{-\infty}^{t} dt' \, B(t') e^{-i\omega_E(t'-t)} \right]$$

$$+ \langle (a)^2 \rangle B(t) \rho_S(t) \left[\int_{-\infty}^{t} dt' \, B(t') e^{-i\omega_E(t'+t)} \right];$$

$$(c) = \left\langle \left(a^\dagger \right)^2 \right\rangle \left[\int_{-\infty}^{t} dt' \, B(t') e^{i\omega_E(t'+t)} \right] \rho_S(t) B(t)$$

$$+ \langle a^\dagger a \rangle \left[\int_{-\infty}^{t} dt' \, B(t') e^{-i\omega_E(t'-t)} \right] \rho_S(t) B(t)$$

$$+ \langle aa^\dagger \rangle \left[\int_{-\infty}^{t} dt' \, B(t') e^{i\omega_E(t'-t)} \right] \rho_S(t) B(t)$$

$$+ \langle (a)^2 \rangle \left[\int_{-\infty}^{t} dt' \, B(t') e^{-i\omega_E(t'+t)} \right] \rho_S(t) B(t);$$

$$(d) = \left\langle \left(a^\dagger \right)^2 \right\rangle \rho_S(t) \left[\int_{-\infty}^{t} dt' \, B(t') e^{i\omega_E(t'+t)} \right] B(t)$$

$$+ \langle aa^\dagger \rangle \rho_S(t) \left[\int_{-\infty}^{t} dt' \, B(t') e^{-i\omega_E(t'-t)} \right] B(t)$$

$$+ \langle a^\dagger a \rangle \rho_S(t) \left[\int_{-\infty}^{t} dt' \, B(t') e^{i\omega_E(t'-t)} \right] B(t)$$

$$+ \langle (a)^2 \rangle \rho_S(t) \left[\int_{-\infty}^{t} dt' \, B(t') e^{-i\omega_E(t'+t)} \right] B(t).$$

The averages are taken over the density matrix of the environment: $\langle a^\dagger a \rangle = \text{tr}(\rho_E a^\dagger a) = \text{tr}(a \rho_E a^\dagger)$, etc. Unless the bosonic mode is in a so-called *squeezed state* (see § 4.4.4), then the off-diagonal averages are zero: $\langle (a)^2 \rangle = \langle (a^\dagger)^2 \rangle = 0$. The rest are expressed through the average values of the *number operator*,

$N = a^\dagger a; \ \langle N \rangle = n$:

$$(a) = (n+1)B(t)\bar{B}(t)\rho_S(t) + nB(t)\bar{B}^\dagger(t)\rho_S(t);$$

$$(b) = nB(t)\rho_S(t)\bar{B}(t) + (n+1)B(t)\rho_S(t)\bar{B}^\dagger(t);$$

$$(c) = n\bar{B}^\dagger(t)\rho_S(t)B(t) + (n+1)\bar{B}(t)\rho_S(t)B(t);$$

$$(d) = (n+1)\rho_S(t)\bar{B}^\dagger(t)B(t) + n\rho_S(t)\bar{B}(t)B(t),$$

where

$$\bar{B}(t) = \int_{-\infty}^{t} dt' \, B(t')e^{i\omega_E(t'-t)}. \tag{1.56}$$

A further simplification can be achieved if we assume that

$$B(t) = B_- e^{-i\omega_B t} + B_+ e^{i\omega_B t}; \quad B_+ \equiv B_-^\dagger. \tag{1.57}$$

Then the integral in (1.56) is taken explicitly,

$$\bar{B}(t) = B_- e^{-i\omega_E t}\frac{e^{i(\omega_E - \omega_B)t}}{i(\omega_E - \omega_B) + \varepsilon} + B_+ e^{-i\omega_E t}\frac{e^{i(\omega_E + \omega_B)t}}{i(\omega_E + \omega_B) + \varepsilon}.$$

Here $\varepsilon \to 0$ is the infinitesimally small term added to ensure that the integrals will converge at the lower limit (this procedure is called *regularization*). Using the well-known Weierstrass formula[10]

$$\lim_{\varepsilon \to 0} \frac{1}{x - i\varepsilon} = \mathcal{P}\frac{1}{x} + i\pi\delta(x), \tag{1.58}$$

where $\mathcal{P}$ means that when the expression is integrated around the singular point (in our case, $x = 0$), the integral must be taken in the sense of the principal value, we obtain

$$\bar{B}(t) = B_- e^{-i\omega_B t}\left(\pi\delta(\omega_E - \omega_B) - i\mathcal{P}\frac{1}{\omega_E - \omega_B}\right)$$

$$+ B_+ e^{i\omega_B t}\left(\pi\delta(\omega_E + \omega_B) - i\mathcal{P}\frac{1}{\omega_E + \omega_B}\right)$$

$$\equiv \kappa_- B_- e^{-i\omega_B t} + \kappa_+ B_+ e^{i\omega_B t}. \tag{1.59}$$

In the general case, the spectra of both the system and the environment – especially the environment! – contain many frequencies, all of which will contribute to $\kappa'_\pm(\omega) \equiv \mathrm{Re}\,\kappa_\pm(\omega)$ and $\kappa''_\pm(\omega) \equiv \mathrm{Im}\,\kappa_\pm(\omega)$. (Note that since we consider the spectra at positive frequencies only, $\kappa'_+ = 0$.)

---

[10] See, e.g., Zagoskin (1998) Eq. (2.31) or Richtmyer (1978), Section 2.9.

Substituting expression (1.59) in Eq. (1.55), we obtain some terms that explicitly depend on time via $\exp[\pm 2i\omega_B t]$. They can be dropped (the so-called *rotating wave approximation (RWA)*) on the grounds that they oscillate so fast on the scale of changes of the density matrix $\rho_S(t)$, that they will average to zero. This is usually valid (unless there are resonances!). The remaining terms are

$$\frac{d\rho_S}{dt} = \frac{1}{i\hbar}\frac{g^2}{\hbar}\left[((n+1)\kappa''_+ - n\kappa''_-)B_-B_+ + ((n+1)\kappa''_- - n\kappa''_+)B_+B, \rho_S(t)\right]$$

$$+\frac{g^2}{\hbar^2}(n+1)\left[\kappa'_-\left(2B_-\rho_S(t)B_+ - \{\rho_S(t), B_+B_-\}\right)\right.$$

$$\left. + \kappa'_+\left(2B_+\rho_S(t)B_- - \{\rho_S(t), B_-B_+\}\right)\right]$$

$$+\frac{g^2}{\hbar^2}n\left[\kappa'_-\left(2B_+\rho_S(t)B_- - \{\rho_S(t), B_-B_+\}\right)\right.$$

$$\left. + \kappa'_+\left(2B_-\rho_S(t)B_+ - \{\rho_S(t), B_+B_-\}\right)\right]. \tag{1.60}$$

The first line represents the usual unitary Hamiltonian dynamics. We could have added the two terms in the commutator with the density matrix to the system's Hamiltonian as a small correction (of order $g^2/\hbar\omega_{E,B}$).

It is the rest that we are interested in. They have the so-called *Lindblad form*

$$2L\rho_S L^\dagger - L^\dagger L\rho_S - \rho_S L^\dagger L. \tag{1.61}$$

It was indeed shown by Lindblad (1976) (see also Gardiner and Zoller, 2004) that the most general form of a linear differential equation for the density matrix, which would preserve its properties, has the form

$$\frac{d\rho}{dt} = \frac{1}{i\hbar}[H, \rho] + \sum_a (2L_a\rho L_a^\dagger - L_a^\dagger L_a\rho - \rho L_a^\dagger L_a) \tag{1.62}$$

with a Hermitian H and arbitrary operators L_a (so-called *Lindblad operators*, or simply Lindblads). It is straightforward to check (using the cyclic invariance of the trace) that such an equation indeed preserves the trace of the density matrix.

The fact that the master equation (1.60) is of the Lindblad form is not a trivial matter. We made a number of assumptions and approximations when deriving first (1.53), and then (1.60), and there was no guarantee that they will not catch up with us and lead to an equation with unphysical solutions. Fortunately this did not happen.

What makes Eq. (1.62) so important is that it models the non-unitary evolution of the density matrix of the system – the very behaviour that could not be described by our initial equation (1.17) for the whole set: system + environment. We will show this on a simple, but relevant, example of a two-level quantum system (e.g., a spin-1/2 particle fixed in space, or, if you wish, a qubit).

1.3.3 An example: a non-unitary evolution of a two-level system. Dephasing and relaxation

The density matrix of a two-level system is a two-by-two Hermitian matrix, and all calculations can easily be done explicitly. We choose the energy representation, in which the Hamiltonian of the system is

$$H_S = \begin{pmatrix} E_0 & 0 \\ 0 & E_1 \end{pmatrix},$$

and consider three examples of Lindblad operators:

$$L_1 = L_1^\dagger = \sqrt{\gamma}\sigma_+\sigma_- \equiv \sqrt{\gamma}\begin{pmatrix} 1 & 0 \\ 0 & 0 \end{pmatrix}; \qquad (1.63)$$

$$L_2 = \sqrt{\Gamma/2}\sigma_- \equiv \sqrt{\Gamma/2}\begin{pmatrix} 0 & 0 \\ 1 & 0 \end{pmatrix}; \quad L_2^\dagger = \sqrt{\Gamma/2}\sigma_+ \equiv \sqrt{\Gamma/2}\begin{pmatrix} 0 & 1 \\ 0 & 0 \end{pmatrix}; \quad (1.64)$$

$$L_3 = \sqrt{\Gamma/2}\sigma_+; \quad L_3^\dagger = \sqrt{\Gamma/2}\sigma_-. \qquad (1.65)$$

The master equation (1.62) in the first case becomes (assuming that the Hamiltonian term is removed in the interaction representation)

$$\begin{pmatrix} \dot{\rho}_{00} & \dot{\rho}_{01} \\ \dot{\rho}_{10} & \dot{\rho}_{11} \end{pmatrix} = \begin{pmatrix} 0 & -\gamma\rho_{01} \\ -\gamma\rho_{10} & 0 \end{pmatrix}, \qquad (1.66)$$

with solution

$$\rho(t) = \begin{pmatrix} \rho_{00}(0) & \rho_{01}(0)e^{-\gamma t} \\ \rho_{10}(0)e^{-\gamma t} & \rho_{11}(0) \end{pmatrix}. \qquad (1.67)$$

This is an obviously non-unitary evolution, with the density matrix being reduced to a diagonal form. The diagonal terms are not changed at all, so this choice of Lindblads realizes what is called *pure dephasing*. Equation (1.67) can be considered as a model of the measurement process, in which the upper/lower state of the system is being detected, with γ giving the rate of the state collapse. The purity, $\varsigma(t) = \mathrm{tr}\,\rho^2(t)$, also decays exponentially,

$$\varsigma(t) = \rho_{00}(0)^2 + \rho_{11}(0)^2 + |\rho_{01}(0)|^2 e^{-2\gamma t}. \qquad (1.68)$$

In the other two cases we get

$$\begin{pmatrix} \dot{\rho}_{00} & \dot{\rho}_{01} \\ \dot{\rho}_{10} & \dot{\rho}_{11} \end{pmatrix} = \begin{pmatrix} -\Gamma\rho_{00} & -(\Gamma/2)\rho_{01} \\ -(\Gamma/2)\rho_{10} & \Gamma\rho_{00} \end{pmatrix}, \qquad (1.69)$$

with solution

$$\rho(t) = \begin{pmatrix} \rho_{00}(0)e^{-\Gamma t} & \rho_{01}(0)e^{-(\Gamma/2)t} \\ \rho_{10}(0)e^{-(\Gamma/2)t} & 1 - \rho_{00}(0)e^{-\Gamma t} \end{pmatrix}, \qquad (1.70)$$

and similarly

$$\begin{pmatrix} \dot\rho_{00} & \dot\rho_{01} \\ \dot\rho_{10} & \dot\rho_{11} \end{pmatrix} = \begin{pmatrix} \Gamma\rho_{11} & -(\Gamma/2)\rho_{01} \\ -(\Gamma/2)\rho_{10} & -\Gamma\rho_{11} \end{pmatrix}, \tag{1.71}$$

$$\rho(t) = \begin{pmatrix} 1 - \rho_{11}(0)e^{-\Gamma t} & \rho_{01}(0)e^{-(\Gamma/2)t} \\ \rho_{10}(0)e^{-(\Gamma/2)t} & \rho_{11}(0)e^{-\Gamma t} \end{pmatrix}. \tag{1.72}$$

In addition to dephasing, these equations also show the *relaxation* of the diagonal elements (to the upper or lower level). Physically, this means that the Lindblads L_2, L_3 describe the competing processes of energy transmission between the system and the environment. Note that the evolution of Eqs. (1.70) and (1.72) leads eventually to a *pure* state – an eigenstate of the system's Hamiltonian. In order to make the system relax to a mixed state, the master equation must include both terms like (1.69) and (1.71), with appropriate weights.

Note also the relation between the relaxation and the dephasing rate in (1.72); this is a frequent occurrence, because Lindblads of the type (1.64) and (1.65) often appear in realistic models. The decoherence rate, which characterizes the sum total of all contributions to the loss of the off-diagonal elements of the density matrix, is, in this case, the sum of the pure dephasing rate plus half the relaxation rate.

If we include all the Lindblads (1.63, 1.64, 1.65), the solution to the master equation will have the form

$$\rho(t) = \begin{pmatrix} \overline{\rho_{00}} + (\rho_{00}(0) - \overline{\rho_{00}})e^{-\Gamma t} & \rho_{01}(0)e^{-(\Gamma/2 + \gamma)t} \\ \rho_{10}(0)e^{-(\Gamma/2 + \gamma)t} & \overline{\rho_{11}} + (\rho_{11}(0) - \overline{\rho_{11}})e^{-\Gamma t} \end{pmatrix}, \tag{1.73}$$

where the stationary values $\overline{\rho_{00}} + \overline{\rho_{11}} = 1$ are the eventual occupation probabilities of the levels. If the system relaxes towards the equilibrium at some temperature T^*, then $\overline{\rho_{11}}/\overline{\rho_{00}} = \exp[-(E_1 - E_0)/k_B T^*]$, from which one can restore the scaling factors for the Lindblad operators L_2, L_3. If, on the other hand, we know the properties of the environment, then the Lindblad terms, and therefore the stationary state of the system, can be calculated from there (see, e.g., Eq. (1.60)). If we formally ascribe to the environment a *negative* temperature, then the two-level system will undergo a *population inversion*, with a higher probability to be found on the upper, than on the lower level.[11]

The particular form of the Lindblad operators depends on the environment and on the details of its coupling to the system. Earlier in this section we found for the

---

[11] This is a common and convenient way of modelling pumping in active media, e.g., lasers.

coupling of a two-level system to a Bose mode the following Lindblad operators:

$$L_1 = \frac{g}{\hbar}\sqrt{\text{Re}(\kappa_-)(n+1)}B_-; \quad L_2 = \frac{g}{\hbar}\sqrt{\text{Re}(\kappa_+)(n+1)}B_+;$$

$$L_3 = \frac{g}{\hbar}\sqrt{\text{Re}(\kappa_-)n}B_+; \quad L_4 = \frac{g}{\hbar}\sqrt{\text{Re}(\kappa_+)n}B_-.$$

The last two disappear if the Bose mode is empty ($n = 0$). Obviously they correspond to the absorption of quanta by the system, while the first two describe the contribution of stimulated and spontaneous emission processes to the decoherence. In many cases it is impossible to provide a precise theoretical description of the environment. Then the Lindblad operators are chosen on a phenomenological basis. Very often it is sufficient to describe the situation with the simple Lindblads of Eqs. (1.63), (1.64) and (1.65) and two parameters of Eq. (1.73): the relaxation rate Γ and total dephasing (i.e., decoherence) rate $\Gamma/2 + \gamma$.

1.3.4 *Non-unitary vs. unitary evolution*

The non unitary master equation in its Lindblad form (1.62) can be written formally in the same way as the unitary Liouville–von Neumann equation (1.18), i.e.,

$$\frac{d\rho}{dt} = \mathcal{L}(t)[\rho(t)], \tag{1.74}$$

where now the Liouvillian includes the unitarity-violating Lindblad terms. Of course, there is no solution of the form (1.19), since the evolution is now irreversible: non-unitarity means that many initial states (density matrices) evolve into the same final state of the system. Let us write

$$\rho(t) = \mathcal{S}(t,0)[\rho(0)], \tag{1.75}$$

where $\mathcal{S}(t_f, t_i)[A]$ is the operator non-unitary evolution of an arbitrary operator A from the moment t_i to $T_f \geq t_i$. Obviously, $\mathcal{S}(t,t)[\cdot] = \hat{1}$, the unit operator, and

$$\frac{d}{dt}\mathcal{S}(t,t_0)[\cdot] = \mathcal{L}(t)[\mathcal{S}(t,t_0)[\cdot]]. \tag{1.76}$$

Since for any later time t' both $\rho(t') = \mathcal{S}(t',t)[\rho(t)]$ and $\rho(t') = \mathcal{S}(t',0)[\rho(0)]$ should hold, then

$$\mathcal{S}(t',0)[A] = \mathcal{S}(t',t)[\mathcal{S}(t,0)[A]], \quad t' \geq t \geq 0 \tag{1.77}$$

(the composition property): two consecutive evolutions described by the master equation form some other such evolution. For a unitary transformation of (1.19) this property obviously holds: recall that the evolution is there given by

$\mathcal{U}(t,0)[A] \equiv U(t)AU^\dagger(t)$. The difference is that the unitary transformations can be inverted: it is clear from (1.19) that identically $\mathcal{U}(t,0)[\mathcal{U}(-t,0)[A]] = A$. Therefore, from the mathematical point of view, the unitary evolutions constitute a *group*, while the non-unitary evolutions of Eq. (1.75) are only a *semigroup*.[12]

1.4 Quantum dynamics of a two-level system

1.4.1 Bloch vector and Bloch sphere

We are often – and rightly – awed by the power and beauty of universal mathematical methods. Nevertheless, all of them started as one-off tricks for solving some specific problem, and were later generalized. The approaches that defied generalization are not to be neglected, since they are often as powerful in their area and allow significant simplification. One such special trick is the representation of the density matrix of a two-level quantum system through the *Bloch vector* in a 3D space. It employs our hard-wired abilities and lifelong experience of imagining and manipulating objects in space, and has helpful analogies with classical physics. It is fortunate that all these advantages can be used to describe such important two-level systems as qubits.[13]

The density matrix in question can be written as

$$\rho = \begin{pmatrix} (1+\mathcal{R}_z)/2 & (\mathcal{R}_x - i\mathcal{R}_y)/2 \\ (\mathcal{R}_x + i\mathcal{R}_y)/2 & (1-\mathcal{R}_z)/2 \end{pmatrix} = \frac{1}{2}\left(\hat{1} + \mathcal{R}_x\sigma_x + \mathcal{R}_y\sigma_y + \mathcal{R}_z\sigma_z\right),$$

(1.78)

where $\mathcal{R}_x$, $\mathcal{R}_y$, $\mathcal{R}_z$ are real numbers, $\hat{1}$ is the unit two-by-two matrix, and σ_x, σ_y, σ_z are the Pauli matrices:

$$\sigma_x = \begin{pmatrix} 0 & 1 \\ 1 & 0 \end{pmatrix}; \ \sigma_y = \begin{pmatrix} 0 & -i \\ i & 0 \end{pmatrix}; \ \sigma_z = \begin{pmatrix} 1 & 0 \\ 0 & -1 \end{pmatrix}.$$

(1.79)

(We can check that the combinations $\sigma_\pm = (\sigma_x \pm i\sigma_y)/2$ are the same matrices that we used in Eqs. (1.64) and (1.65).)

The expression (1.78) looks like an expansion of a 3D vector with components $\mathcal{R}_x$, $\mathcal{R}_y$, $\mathcal{R}_z$ over the unit vectors $\sigma_{x,y,z}$, which it actually is: the space of all

---

[12] In common mathematical notation, the compositions of the elements of (semi)groups (i.e., evolutions $\mathcal{U}(t_2,t_1)[\cdots]$ or $\mathcal{S}(t_2,t_1)[\cdots]$) are written as, e.g., $g \circ h$, where the evolution g happens after h. The composition property, $g_2 \circ g_1 = g_3$ (i.e., two consecutive evolutions are also equivalent to an evolution), and the associativity, $(g_3 \circ g_2) \circ g_1 = g_3 \circ (g_2 \circ g_1)$, hold both for groups and semigroups. In addition, in groups for any g there exists an inverse g^{-1}: $g^{-1} \circ g = \hat{1}$.

[13] We will call an arbitrary two-level quantum system a "qubit", irrespective of whether it is an actual qubit, as long as this does not create confusion. The reason is that "qubit" is shorter than "two-level system", while the abbreviation "TLS" often refers to a special (and mostly troublesome) kind of two-state object (see Section 5.2).

Hermitian matrices with zero trace is three-dimensional, and the Pauli matrices can be chosen as its orthonormal basis. (There is, of course, also the term $\hat{1}$, but its role is only to ensure that the density matrix has the unit trace.) This 3D vector $\mathcal{R}$ is the *Bloch vector*, and

$$\rho = \frac{1}{2}\left(\hat{1} + \mathcal{R}_\alpha \sigma_\alpha\right). \tag{1.80}$$

Now recall that $\mathrm{tr}\left[\rho^2\right] \leq \mathrm{tr}\,\rho$, with equality if and only if the system is actually in a pure state. Since

$$\mathrm{tr}\left[\rho^2\right] = \mathrm{tr}\left[\frac{1}{4}\left(\begin{array}{cc} (1+\mathcal{R}_z)^2 + \mathcal{R}_x^2 + \mathcal{R}_y^2 & 2(\mathcal{R}_x - i\mathcal{R}_y) \\ 2(\mathcal{R}_x + i\mathcal{R}_y) & (1-\mathcal{R}_z)^2 + \mathcal{R}_x^2 + \mathcal{R}_y^2 \end{array}\right)\right]$$
$$= \frac{1 + \mathcal{R}_x^2 + \mathcal{R}_y^2 + \mathcal{R}_z^2}{2} = \frac{|\mathcal{R}|^2 + 1}{2} \leq 1, \tag{1.81}$$

we see that the length of the Bloch vector cannot exceed unity. Therefore it is convenient to represent the density matrix using the unit *Bloch sphere*, Fig. 1.3. Any point on its surface corresponds to some pure state, and any point inside to some mixed state. The pure states on the opposite sides of the sphere's diameters are mutually orthogonal. Indeed, for two pure states with opposite Bloch vectors

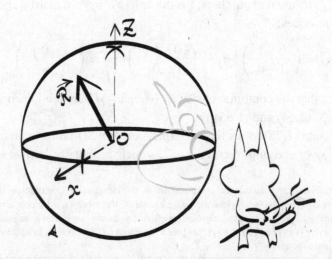

Fig. 1.3. Bloch sphere. Points on its surface correspond to the pure states of a qubit (quantum two-level system), points inside to mixed states.

the density matrices are

$$\rho_1 = |\Psi_1\rangle\langle\Psi_1| = \frac{\hat{1} + \mathcal{R}_\alpha \sigma_\alpha}{2}; \quad \rho_2 = |\Psi_2\rangle\langle\Psi_2| = \frac{\hat{1} - \mathcal{R}_\alpha \sigma_\alpha}{2}; \quad (1.82)$$

$$\text{tr}\,(\rho_1\rho_2) = |\langle\Psi_1|\Psi_2\rangle|^2 = \text{tr}\frac{1}{4}\left(\hat{1} - \mathcal{R}_\alpha\mathcal{R}_\beta\sigma_\alpha\sigma_\beta\right) = \frac{1}{2}\left(1 - |\mathcal{R}|^2\right) = 0.$$

For example, if we choose the eigenstates of the unperturbed Hamiltonian, $|0\rangle$ and $|1\rangle$, as the basis of the Hilbert space, then the north and south poles of the Bloch sphere would correspond to the pure states $|0\rangle$ and $|1\rangle$. At equilibrium at temperature T, the density matrix has the Gibbs form $a_0|0\rangle\langle0| + a_1|1\rangle\langle1|$, with the Boltzmann ratio $a_1/a_0 = \exp(-(E_1 - E_0)/k_B T)$. Therefore, all the equilibrium states lie on the positive vertical axis of the Bloch sphere, with the ground state ($\rho = |0\rangle\langle0|$) at its north pole, and the state at infinite temperature (when both qubit states are equipopulated) in the centre of the sphere. The "equilibrium states with negative temperatures", for which $a_1 > a_0$, fill the negative vertical axis; they terminate at the pure excited state $|1\rangle\langle1|$ at the south pole. The points away from the vertical axes represent mixed states with nonzero coherences.

The evolution of the qubit state is visually represented by the hodograph of the Bloch vector. In the presence of decoherence (dephasing and relaxation), it will approach the positive vertical axis and terminate at some point there, determined by the temperature of the bath with which the qubit eventually equilibrates.

Using the Bloch sphere we can give a clear geometrical meaning to the density matrix representation as a weighted sum of projectors on pure states, Eq. (1.9). Take a section of the Bloch sphere by a plane, containing the Bloch vector $\mathcal{R} \equiv \vec{OR}$ (Fig.1.4). (There is a continuum of such sections, of course.) Choose an *arbitrary* point A on the boundary and draw the chord AB through the point R. Then the Bloch vector $\vec{OR}$ can be written as[14]

$$\vec{OR} = \frac{|RB|}{|AB|}\vec{OA} + \frac{|RA|}{|AB|}\vec{OB} \equiv a\mathcal{A} + b\mathcal{B} \quad (1.83)$$

(obviously $a, b \geq 0$, $a + b = 1$), and the density matrix can be *unravelled* (e.g., Percival, 2008, 3.2) as

$$\rho = \frac{1}{2}\left[(a+b)\hat{1} + (a\mathcal{A}_\alpha + b\mathcal{B}_\alpha)\sigma_\alpha\right] = a|\Psi_A\rangle\langle\Psi_A| + b|\Psi_B\rangle\langle\Psi_B|. \quad (1.84)$$

There is a continuum of such unravellings of the density matrix into a sum of projectors: we are free to choose both a point on the great circle, and the great

---

[14] This is easy to see if we write $\vec{OR} = a\vec{OA} + b\vec{OB}$ and notice that R lies on the line AB if and only if $\vec{AR}\|\vec{AB}$ (or $\vec{RB}\|\vec{AB}$), and that these conditions can be written as $\vec{AR} \times \vec{AB} = 0$ ($\vec{RB} \times \vec{AB} = 0$). Here $\times$ denotes the vector product – another very useful operation, which only works in the special case of 3D space!

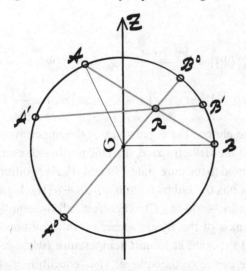

Fig. 1.4. Section of the Bloch sphere, which contains the Bloch vector $\mathcal{R} = \vec{OR}$. Different chords AB, $A'B'$,... represent different unravellings of the density matrix, $\rho = W_A|\Psi_A\rangle\langle\Psi_A| + W_B|\Psi_B\rangle\langle\Psi_B|$. The chord passing through the centre of the sphere, A^0B^0, corresponds to the spectral representation of the density matrix. Only in this representation are the constituent states orthogonal, $\langle\Psi_{A^0}|\Psi_{B^0}\rangle = 0$.

circle itself. There exists, though, a special unravelling, represented by the chord A^0B^0, which passes through the centre of the Bloch sphere. According to Eq. (1.83), the states $|\Psi_{A^0}\rangle$ and $|\Psi_{B^0}\rangle$ are mutually orthogonal, being on the opposite ends of the diameter, and such an unravelling is obviously unique. This is the spectral representation (1.12) of the density matrix, and the state vectors $|\Psi_{A^0}\rangle$ and $|\Psi_{B^0}\rangle$ are the eigenvectors of ρ.

1.4.2 Bloch equations and quantum beats

Now consider a qubit in an external field, described in the so-called diabatic or "physical" basis by the Hamiltonian

$$H(t) = -\frac{1}{2}\left(\Delta\sigma_x + \epsilon(t)\sigma_z\right). \tag{1.85}$$

The basis is called "physical", since we assume that the two eigenstates of the operator σ_z, $|0\rangle$ and $|1\rangle$, correspond to the pointer states, which can eventually be read out. For example, if the qubit is a spin, then these states are "up" and "down"; if it is a superconducting loop with current, they are "clockwise" and "anticlockwise"; if it is a quantum dot, then they are "charge zero" and "charge one" etc. For clarity, we will call these states "left", $|L\rangle$, and "right", $|R\rangle$, whatever their actual physical meaning.

The σ_x-term describes quantum-mechanical tunnelling with amplitude Δ between $|L\rangle$ and $|R\rangle$, and the term $\epsilon(t) = (\epsilon_0 + \epsilon_1(t))\sigma_z$ is the time-dependent bias. (Without losing generality, we can assume that the average $\langle\epsilon_1(t)\rangle = 0$.)

To begin, let us write the Liouville–von Neumann equation,

$$\frac{d\rho}{dt} = \frac{1}{i\hbar}[H, \rho],$$

in terms of the Bloch vector as

$$\frac{d}{dt}\mathcal{R} = \mathbf{M}(t)\mathcal{R}, \tag{1.86}$$

where the antisymmetric matrix $\mathbf{M}$

$$\mathbf{M} = \begin{pmatrix} 0 & \Delta & 0 \\ -\Delta & 0 & \epsilon(t) \\ 0 & -\epsilon(t) & 0 \end{pmatrix} \tag{1.87}$$

describes the unitary evolution. This equation can be rewritten as

$$\frac{d\mathcal{R}}{dt} = \frac{2}{\hbar}\mathcal{H}(t) \times \mathcal{R}, \tag{1.88}$$

where the vector $\mathcal{H}(t) = (\Delta/2, 0, \epsilon(t)/2)^{\mathrm{T}}$. (In other words, $\mathcal{H}$ is the Hamiltonian (1.85) considered as another vector in the space of traceless two-by-two Hermitian matrices.) We can also check, for the sake of completeness, that Eq. (1.88) holds if the Hamiltonian (1.85) also contains the y-term, $-\Delta_y\sigma_y/2$ (that is, $\mathcal{H}(t) = (\Delta/2, \Delta_y/2, \epsilon(t)/2)^{\mathrm{T}}$).

If our qubit was actually a spin-1/2, and we had N such independent elementary magnets per unit volume, then the α-th component of the macroscopic magnetization of the system would be

$$M_\alpha = \frac{Ng\hbar}{2}\langle\sigma_\alpha\rangle, \quad \alpha = x, y, z, \tag{1.89}$$

where g is the gyromagnetic ratio. In the external magnetic field $\vec{H}(t) = \vec{H}_0 + \vec{H}_1(t)$ (assuming that the stationary field component is in the z-direction, and $\langle\vec{H}_1(t)\rangle = 0$) the magnetization vector would rotate around the direction of the field, while simultaneously relaxing to the stationary state with $M_x = M_y = 0$ (in the absence of a time-dependent field) and $M_z = \bar{M}_z$ (determined by the temperature and the energy splitting between the "up" and "down" states in the field), which is described by the *Bloch equations*

$$\frac{dM_x}{dt} = g\left[\vec{M} \times \vec{H}(t)\right]_x - \frac{M_x}{T_2};$$

$$\frac{dM_y}{dt} = g\left[\vec{M} \times \vec{H}(t)\right]_y - \frac{M_y}{T_2}; \qquad (1.90)$$

$$\frac{dM_z}{dt} = g\left[\vec{M} \times \vec{H}(t)\right]_z - \frac{M_z - \bar{M}_z}{T_1}$$

of nuclear magnetic resonance (NMR) theory (see, e.g., Blum, 2010, Section 8.4). Here $T_{1,2}$ are the longitudinal and transverse relaxation times. These equations describe the Larmor precession of the average magnetic moment in the external magnetic field and its relaxation. On the other hand, mathematically they are – *mutatis mutandis* – the same as Eq. (1.88) for the components of the Bloch vector. Moreover, the simplest Lindblad operators (1.69) and (1.71) lead (in a basis of the energy eigenstates) to the relaxation terms exactly like in (1.90), with the relaxation rate Γ and the dephasing rate $\Gamma/2 + \gamma$ corresponding to the rates $1/T_1$ and $1/T_2$.

This is why Eq. (1.90) using the Bloch vector is also called the Bloch equations:

$$\frac{d\mathcal{R}_x}{dt} = \frac{2}{\hbar}[\mathcal{H}(t) \times \mathcal{R}]_x - \frac{\mathcal{R}_x}{T_2};$$

$$\frac{d\mathcal{R}_y}{dt} = \frac{2}{\hbar}[\mathcal{H}(t) \times \mathcal{R}]_y - \frac{\mathcal{R}_y}{T_2}; \qquad (1.91)$$

$$\frac{d\mathcal{R}_z}{dt} = \frac{2}{\hbar}[\mathcal{H}(t) \times \mathcal{R}]_z - \frac{\mathcal{R}_z - \bar{\mathcal{R}}_z}{T_1}.$$

Eqs. (1.88) and (1.91) describe a rotation of the vector $\mathcal{R}$ with the instantaneous angular velocity $2\mathcal{H}(t)/\hbar$. The absolute value of this velocity is $\Omega(t) = 2|\mathcal{H}(t)|/\hbar = \sqrt{\Delta^2 + \epsilon^2(t)}/\hbar$, that is, the transition frequency between the ground state and the excited state of the qubit, produced by the tunnelling Δ and bias ϵ between its two physical states.

For a time-independent bias, the rotation (1.88) represents *quantum beats* between the eigenstates of the system. If it was initially in the "left"/"right" state, it will rotate around $\mathcal{H}$ between the north/south pole and a point close to the opposite pole of the Bloch sphere (Fig. 1.5); in the limit of zero bias, the rotation is around the x-axis, and the hodograph is the great circle through both poles.

Quantum beats are a natural way of switching the qubit between its "left" and "right" states. If we start from any other point of the Bloch sphere, they will carry the system around the corresponding trajectory, as shown in Fig.1.5. To be precise,

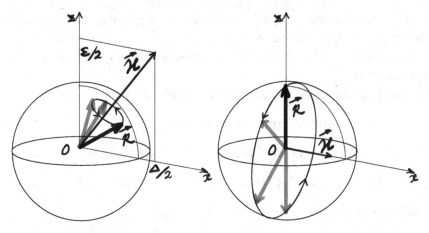

Fig. 1.5. Quantum beats: the Bloch vector $\mathcal{R}$ rotates around the "field vector" $\mathcal{H}$ with the angular velocity $\Omega = \sqrt{\Delta^2 + \epsilon^2}/\hbar$ (left). For zero bias the rotation axis coincides with Ox, and the trajectory of a pure state is a great circle (right).

the *probability* of finding the system in either state will be a periodic function of time.[15]

Nevertheless, if the system was initially in an equilibrium state, it will remain there, since then its Bloch vector is parallel to $\mathcal{H}$. Transitions between such states can be produced by applying a harmonic resonant field to the system. This effect is called *Rabi oscillations*.

1.4.3 Rabi oscillations

Let us act on the qubit with a harmonic perturbation, which in the diabatic ("physical") basis has the form $H_1(t) = \hbar\eta\cos\omega t\,\sigma_z$. In the absence of the static bias ($\epsilon_0 = 0$) the physical states $|L\rangle$, $|R\rangle$ have the same energy, so the qubit is at its *degeneracy point*. Then the unperturbed Hamiltonian is

$$H_0 = -\frac{1}{2}\Delta\sigma_x \equiv -\frac{\hbar\Omega}{2}\sigma_x, \tag{1.92}$$

and in the eigenbasis of H_0 (note that now the z- and x-directions change places)

$$H(t) = H_0 + H_1(t) = -\frac{\hbar\Omega}{2}\sigma_z + \hbar\eta\cos\omega t\,\sigma_x. \tag{1.93}$$

---

[15] A simpler way to come to the same conclusion for a pure state is to write the wave function as $|\Psi(t)\rangle = a_0|0\rangle\exp[-iE_0 t/\hbar] + a_1|1\rangle\exp[-iE_1 t/\hbar] = \exp[-iE_0 t/\hbar]\,(a_0|0\rangle + a_1|1\rangle\exp[-i\Omega t])$ and the density matrix as $\rho(t) = |\Psi(t)\rangle\langle\Psi(t)|$.

If we do not assume that the qubit is at the degeneracy point, the perturbation in (1.93) will also have a component proportional to σ_z. This does not actually make the solution any harder, but the formulas get messier (see, e.g., Scully and Zubairy, 1997, Chapter 5).

Without the perturbation, the unitary evolution of the qubit is given by

$$\rho(t) = \exp[-iH_0 t/\hbar]\rho(0)\exp[iH_0 t/\hbar]. \tag{1.94}$$

Expanding the exponents in series and using the properties of the Pauli matrices,

$$[\sigma_x, \sigma_y] = 2\sigma_z, \; [\sigma_y, \sigma_z] = 2\sigma_x, \; [\sigma_z, \sigma_x] = 2\sigma_y, \; (\sigma_x)^2 = (\sigma_y)^2 = (\sigma_z)^2 = \hat{1}, \tag{1.95}$$

it is easy to see that

$$\exp[\pm iH_0 t/\hbar] = \hat{1}\cos(\Omega t/2) \pm i\sigma_z \sin(\Omega t/2). \tag{1.96}$$

Here $\hat{1}$ is a unit two-by-two matrix. The corresponding rotation of the Bloch vector $\mathcal{R}$ with the angular velocity Ω around the axis Oz is given by

$$\mathcal{R}(t) = \mathbf{U}(t)^T \mathcal{R}(0), \tag{1.97}$$

where

$$\mathbf{U}(t)^T = \begin{pmatrix} \cos\Omega t & -\sin\Omega t & 0 \\ \sin\Omega t & \cos\Omega t & 0 \\ 0 & 0 & 1 \end{pmatrix}. \tag{1.98}$$

We chose to write it this way using the operation of matrix transposition; we also use the fact that the matrix $\mathbf{U}$ is real unitary (that is, orthogonal, as it should be, since it describes rotations in a 3D space):

$$\mathbf{U}\mathbf{U}^\dagger = \mathbf{U}\mathbf{U}^T = 1. \tag{1.99}$$

Eq. (1.97) describes quantum beats, which in our 3D picture are nothing else but a Larmor precession of the fictitious "magnetic moment" $\mathcal{R}$ in the fictitious "magnetic field" $\mathcal{H}$. To cancel the effects of this evolution, which we after all have already discussed, it is convenient to switch to the rotating frame of reference, which is equivalent to using the standard quantum-mechanical interaction representation. To do so, we represent the Bloch vector as

$$\mathcal{R}(t) = \mathbf{U}(t)^T \mathcal{R}_I(t), \tag{1.100}$$

with the time dependence of $\mathcal{R}_I(t)$ caused by everything else apart from the unperturbed Hamiltonian H_0. Using Eq. (1.99), we obtain

$$\frac{d\mathcal{R}_I(t)}{dt} = \frac{d[\mathbf{U}(t)\mathcal{R}(t)]}{dt} = \mathbf{M}_I(t)\mathcal{R}_I(t), \tag{1.101}$$

where

$$\mathbf{M}_I(t) = \mathbf{U}(t)\mathbf{M}(t)\mathbf{U}(t)^{\mathrm{T}} - \mathbf{U}(t)\frac{\mathrm{d}\mathbf{U}(t)^{\mathrm{T}}}{\mathrm{d}t} \tag{1.102}$$

describes the evolution of the Bloch vector with respect to the rotating frame. It is only due to the harmonic perturbation. (We will deal with the non-unitary terms in a moment.) The matrix $\mathbf{M}$ (we have dropped the subscript I to unclutter the formulas) is now

$$\mathbf{M}(t) = \begin{pmatrix} 0 & 0 & \eta(\sin(2\Omega+\delta)t - \sin\delta t) \\ 0 & 0 & -\eta(\cos(2\Omega+\delta)t - \cos\delta t) \\ -\eta(\sin(2\Omega+\delta)t - \sin\delta t) & \eta(\cos(2\Omega+\delta)t - \cos\delta t) & 0 \end{pmatrix}. \tag{1.103}$$

Here we have introduced the *detuning*

$$\delta = \omega - \Omega \tag{1.104}$$

between the driving frequency ω and the interlevel distance.

If we neglect the "fast" terms (with frequency $2\Omega+\delta \sim 2\omega$), which will average to almost zero, the matrix (1.103) becomes

$$\mathbf{M}(t) \approx \begin{pmatrix} 0 & 0 & -\eta\sin\delta t \\ 0 & 0 & \eta\cos\delta t \\ \eta\sin\delta t & -\eta\cos\delta t & 0 \end{pmatrix}. \tag{1.105}$$

It describes the rotation of the Bloch vector with the angular velocity $\mathcal{Y}(t)$

$$\frac{\mathrm{d}\mathcal{R}}{\mathrm{d}t} = \mathbf{M}(t)\mathcal{R}(t) \approx \mathcal{Y}(t) \times \mathcal{R}(t); \quad \mathcal{Y}(t) = -\eta \begin{pmatrix} \cos\delta t \\ \sin\delta t \\ 0 \end{pmatrix}, \tag{1.106}$$

which itself slowly rotates around the axis Oz.

Dropping the fast terms is called the *rotating wave approximation* (RWA). The name can be better understood with the magnetic analogy and considering a magnetic moment in a constant magnetic field parallel to the axis Oz and a small harmonic field along the axis Ox. The harmonic field can always be represented as a sum of two fields with opposite circular polarizations: "rotating waves", one co-rotating and the other counter-rotating with the Larmor precession of the magnetic moment. When the perturbation frequency is close to the Larmor frequency, we can neglect the counter-rotating wave, since in the frame of reference of the rotating magnetic moment it changes too fast to produce a significant effect.

Clearly, RWA is justified only as far as the evolution of the Bloch vector in the rotating frame is slow, that is, near the resonance, $\delta \ll \omega$, but this is exactly where we need it. More formally, we can average the equations over the period $\tau_\omega = 2\pi/\omega$ of fast drive oscillations: $\langle \cdots \rangle_{\tau_\omega} \equiv \tau_\omega^{-1} \int_t^{t+\tau_\omega} (\cdots) dt$. The slow terms vary little during this time and can be approximated by constants taken at any point inside the interval $[t, t + \tau_\omega]$ (for example, at the moment t). Then integration eliminates the fast terms.

Now let us return to Eq. (1.106). The rotation of the vector of angular velocity $\mathcal{Y}(t)$ can be compensated by the counter-rotation

$$
\mathbf{U}'(t) = \begin{pmatrix} \cos \delta t & -\sin \delta t & 0 \\ \sin \delta t & \cos \delta t & 0 \\ 0 & 0 & 1 \end{pmatrix}. \tag{1.107}
$$

In this new frame of reference the rotation matrix becomes (cf. Eq. (1.102)),

$$
\mathbf{M}(t) \rightarrow \mathbf{U}'(t)\mathbf{M}(t)\mathbf{U}'(t)^{\mathrm{T}} - \mathbf{U}'(t)\frac{d\mathbf{U}'(t)^{\mathrm{T}}}{dt}
$$

$$
= \begin{pmatrix} 0 & -\delta & \eta \sin 2\omega t \\ \delta & 0 & -2\eta \cos^2 \omega t \\ -\eta \sin 2\omega t & 2\eta \cos^2 \omega t & 0 \end{pmatrix}. \tag{1.108}
$$

The Bloch vector now satisfies the equation

$$
\frac{d\mathcal{R}}{dt} = \mathbf{M}(t)\mathcal{R}(t) \approx \mathcal{Y}' \times \mathcal{R}(t); \quad \mathcal{Y}' = \begin{pmatrix} \eta \\ 0 \\ \delta \end{pmatrix}. \tag{1.109}
$$

Here we again used RWA and dropped the fast terms. We see already that in the new frame of reference the Bloch vector rotates with the angular velocity

$$
\omega_R = \sqrt{\delta^2 + \eta^2}; \tag{1.110}
$$

ω_R is the *Rabi frequency*. If the harmonic field is exactly in resonance with the qubit, this frequency is directly proportional to the amplitude of the applied field, $\hbar\eta$.

To tidy things up, we align the x-axis of the rotating frame with the vector $\mathcal{Y}'$:

$$\mathbf{U}'' = \begin{pmatrix} \dfrac{\eta}{\omega_R} & 0 & \dfrac{\delta}{\omega_R} \\ 0 & 1 & 0 \\ -\dfrac{\delta}{\omega_R} & 0 & \dfrac{\eta}{\omega_R} \end{pmatrix}; \quad \mathbf{U}''\mathcal{Y}' = \begin{pmatrix} \omega_R \\ 0 \\ 0 \end{pmatrix}. \tag{1.111}$$

The Bloch equations become

$$\frac{d\mathcal{R}}{dt} \approx \omega_R \hat{\mathbf{x}} \times \mathcal{R}(t). \tag{1.112}$$

This is what we were looking for: we found the frame of reference in which the evolution of the Bloch vector is a simple precession with a known angular velocity, and a solution with any initial conditions is obvious, for example,

$$\mathcal{R}(t) = \begin{pmatrix} R_{\parallel} \\ R_{\perp}\cos\omega_R t \\ R_{\perp}\sin\omega_R t \end{pmatrix}. \tag{1.113}$$

All that remains is to transform it back to the laboratory frame of reference (i.e., the energy basis). Since the Bloch vector in the rotating frame was obtained by the consecutive application of three rotations,

$$\mathcal{R}(t) \to \mathbf{U}''\mathbf{U}'(t)\mathbf{U}(t)\mathcal{R}(t), \tag{1.114}$$

we now need to invert the matrix

$$\mathbf{W}(t) \equiv \mathbf{U}''\mathbf{U}'(t)\mathbf{U}(t), \tag{1.115}$$

which transforms the Bloch vector from the laboratory to the rotating frame of reference, that is,

$$\mathbf{W}(t) = \begin{pmatrix} (\eta/\omega_R)\cos\omega t & -(\eta/\omega_R)\sin\omega t & \delta/\omega_R \\ \sin\omega t & \cos\omega t & 0 \\ -(\delta/\omega_R)\cos\omega t & (\delta/\omega_R)\sin\omega t & \eta/\omega_R \end{pmatrix}. \tag{1.116}$$

Note that we did *not* make any approximations in our frame-to-frame transformations $\mathbf{U}$, $\mathbf{U}'$, $\mathbf{U}''$. Nevertheless in the final matrix $\mathbf{W}(t)$ all the complicated time dependences of separate rotations miraculously reduce to the single frequency ω. But it would not be easy to come up with this simple transformation, if we did not use our intuition for 3D rotations.

The matrix $\mathbf{W}(t)$ provides a direct transformation for the matrix $\mathbf{M}(t)$ between the laboratory and rotating frames. After making the "formal" RWA in the final result, we obtain

$$\left\langle \mathbf{W}(t)\mathbf{M}(t)\mathbf{W}^{\mathrm{T}}(t) - \mathbf{W}(t)\frac{d\mathbf{W}(t)^{\mathrm{T}}}{dt} \right\rangle_{\tau_\omega} = \begin{pmatrix} 0 & 0 & 0 \\ 0 & 0 & -\omega_R \\ 0 & \omega_R & 0 \end{pmatrix}, \quad (1.117)$$

so that the Bloch equations in the rotating frame are indeed Eq. (1.112). The transformation is unitary (even orthogonal), as it should be:

$$[\mathbf{W}(t)]^{-1} = \mathbf{W}(t)^{\dagger} = \mathbf{W}(t)^{\mathrm{T}}. \quad (1.118)$$

Applying $\mathbf{W}(t)^{\mathrm{T}}$ to the solution Eq. (1.113), we find

$$\mathcal{R}(t) = \begin{pmatrix} [(\eta/\omega_R)R_{\parallel} - (\delta/\omega_R)R_{\perp}\sin\omega_R t]\cos\omega t + R_{\perp}\cos\omega_R t\sin\omega t \\ -[(\eta/\omega_R)R_{\parallel} - (\delta/\omega_R)R_{\perp}\sin\omega_R t]\sin\omega t + R_{\perp}\cos\omega_R t\cos\omega t \\ (\delta/\omega_R)R_{\parallel} + (\eta/\omega_R)R_{\perp}\sin\omega_R t \end{pmatrix}.$$

$$(1.119)$$

Note the behaviour of the z-component of the Bloch vector in the laboratory frame, which describes the populations of the energy levels. While the other components are oscillating with the fast driving-field frequency (plus slow modulations at ω_R), the z-component oscillates at the Rabi frequency only.

The fact that the Rabi frequency, $\omega_R = \sqrt{\delta^2 + \eta^2}$, can be controlled by varying the field amplitude and detuning, and that it is small compared to the drive frequency, can be (and is) used for the coherent control of the quantum state of a qubit. Suppose, for example, that a qubit is initially in its ground state. After applying an external field at the resonance frequency $\omega = \Delta$ and with the amplitude $\hbar\eta$ and waiting for the interval $\tau = \pi/\eta$, we will find the qubit in its excited state with probability 1; generally this probability will behave as $P_{\mathrm{up}}(t) = \sin\eta t$. On the other hand, the telltale periodic oscillations of $P_{\mathrm{up}}(t)$ in the external field, with the frequency behaving like $\sqrt{\delta^2 + \eta^2}$, indicate that the system in question is indeed a quantum-mechanical two-level system. In the following chapters we will see a number of examples of how this works in real hardware.

1.4.4 *Rabi oscillations in the presence of dissipation*

Now let us consider a more general case, when the density matrix evolution is non unitary due to the effects of dissipation. Using the simple Lindblad operators (1.69) and (1.71), we find in the energy basis (which is, in our case of zero static bias, the same as the laboratory frame of reference for the Bloch vector) the Bloch equations

$$\frac{\mathrm{d}\mathcal{R}_x}{\mathrm{d}t} = [\text{unitary evolution}] - \Gamma_2 \mathcal{R}_x;$$

$$\frac{\mathrm{d}\mathcal{R}_y}{\mathrm{d}t} = [\text{unitary evolution}] - \Gamma_2 \mathcal{R}_y; \qquad (1.120)$$

$$\frac{\mathrm{d}\mathcal{R}_z}{\mathrm{d}t} = [\text{unitary evolution}] - \Gamma_1 (\mathcal{R}_z - Z_T).$$

Here $Z_T = \tanh(\Delta/2k_B T)$ is the equilibrium value of the z-component of the Bloch vector, and $\Gamma_{1,2}$ are the rates of the longitudinal and transverse relaxation (i.e., relaxation proper and dephasing), respectively.

It is convenient to write the non-unitary part of the evolution operator as

$$\mathbf{L}[\mathcal{R}] = -\Gamma_2 \mathcal{R} - (\Gamma_1 - \Gamma_2) \begin{pmatrix} 0 & 0 & 0 \\ 0 & 0 & 0 \\ 0 & 0 & 1 \end{pmatrix} \mathcal{R} + \Gamma_1 Z_T \begin{pmatrix} 0 \\ 0 \\ 1 \end{pmatrix}. \qquad (1.121)$$

Then in the rotating frame we will have

$$\mathbf{L}[\mathcal{R}] \to \mathbf{W}(t)\mathbf{L}[\mathcal{R}] = -\Gamma_2 \mathcal{R} - (\Gamma_1 - \Gamma_2)\mathbf{W}(t) \begin{pmatrix} 0 & 0 & 0 \\ 0 & 0 & 0 \\ 0 & 0 & 1 \end{pmatrix} \mathbf{W}(t)^{\mathsf{T}} \mathcal{R}$$

$$+ \Gamma_1 Z_T \mathbf{W}(t) \begin{pmatrix} 0 \\ 0 \\ 1 \end{pmatrix};$$

$$\mathcal{R} \to \mathbf{W}(t)\mathcal{R}. \qquad (1.122)$$

Explicitly, in the rotating frame

$$\mathbf{L}[\mathcal{R}] = -\Gamma_2 \mathcal{R} - (\Gamma_1 - \Gamma_2) \begin{pmatrix} (\delta/\omega_R)^2 & 0 & \delta\eta/\omega_R^2 \\ 0 & 0 & 0 \\ \delta\eta/\omega_R^2 & 0 & (\eta/\omega_R)^2 \end{pmatrix} \mathcal{R} + \Gamma_1 Z_T \begin{pmatrix} \delta/\omega_R \\ 0 \\ \eta/\omega_R \end{pmatrix}.$$

$$\qquad (1.123)$$

Therefore, the Bloch equations in the rotating frame and in the RWA are:

$$\frac{d\mathcal{R}_x}{dt} = -\Gamma_2 \mathcal{R}_x - (\Gamma_1 - \Gamma_2)\left[\left(\frac{\delta}{\omega_R}\right)^2 \mathcal{R}_x + \frac{\eta\delta}{\omega_R^2}\mathcal{R}_z\right] + \Gamma_1 Z_T\left(\frac{\delta}{\omega_R}\right);$$

$$\frac{d\mathcal{R}_y}{dt} = -\Gamma_2 \mathcal{R}_y - \omega_R \mathcal{R}_z; \qquad (1.124)$$

$$\frac{d\mathcal{R}_z}{dt} = \omega_R \mathcal{R}_y - \Gamma_2 \mathcal{R}_z - (\Gamma_1 - \Gamma_2)\left[\left(\frac{\eta}{\omega_R}\right)^2 \mathcal{R}_z + \frac{\eta\delta}{\omega_R^2}\mathcal{R}_x\right] + \Gamma_1 Z_T\left(\frac{\eta}{\omega_R}\right).$$

These equations are valid in the rotating wave approximation (i.e., when $|\delta|, \omega_R \ll \omega$) and under the assumption of weak non-unitarity ($\Gamma_1, \Gamma_2 \ll \omega_R, \omega$).

Let us consider for simplicity the special case $\Gamma_2 = \Gamma_1 = \Gamma$. Now in (1.124) the equations for $\mathcal{R}_x$ and for $\mathcal{R}_y, \mathcal{R}_z$ decouple:

$$\frac{d\mathcal{R}_x}{dt} = -\Gamma \mathcal{R}_x(t) + \Gamma Z_T \frac{\delta}{\omega_R};$$

$$\frac{d\mathcal{R}_y}{dt} = -\Gamma \mathcal{R}_y(t) - \omega_R \mathcal{R}_z(t); \qquad (1.125)$$

$$\frac{d\mathcal{R}_z}{dt} = -\Gamma \mathcal{R}_z(t) + \omega_R \mathcal{R}_y(t) + \Gamma Z_T \frac{\eta}{\omega_R}.$$

The first equation describes the pure decay of $\mathcal{R}_x$, while the other two correspond to the rotation in the y-z-plane with decaying amplitude, so that a solution can be sought in the form:

$$\mathcal{R}_x(t) = a + be^{-\Gamma t};$$

$$\mathcal{R}_y(t) = c + re^{-\Gamma t}\cos(\omega_R t + \alpha); \qquad (1.126)$$

$$\mathcal{R}_z(t) = d + re^{-\Gamma t}\sin(\omega_R t + \alpha).$$

It is straightforward to check that the expressions (1.126) satisfy Eq. (1.125). The constants a, c and d are determined from the stationary limit of the Bloch equations (i.e., taking $d\mathcal{R}_{x,y,z}/dt = 0$ and $t \to \infty$):

$$a = Z_T\delta/\omega_R; \quad c = Z_T\eta\frac{\Gamma}{\Gamma^2 + \omega_R^2}; \quad d = Z_T\frac{\eta}{\omega_R}\frac{\Gamma^2}{\Gamma^2 + \omega_R^2}; \qquad (1.127)$$

the rest must be chosen to satisfy the initial conditions.

Suppose we start from equilibrium. Then in the laboratory frame of reference the Bloch vector has only a z-component, $\mathcal{R}_z(0) = Z_T = \tanh(\Delta/2k_B T)$. After applying the transformation matrix $\mathbf{W}(0)$, we find in the rotating frame

$$\mathcal{R}(0) = Z_T \begin{pmatrix} \delta/\omega_R \\ 0 \\ \eta/\omega_R \end{pmatrix}. \tag{1.128}$$

Therefore, we must choose in Eq. (1.126)

$$b = 0; \quad r = Z_T \frac{\eta}{\sqrt{\omega_R^2 + \Gamma^2}}; \quad \alpha = \arctan \frac{\omega_R}{\Gamma}. \tag{1.129}$$

Transforming the solution back to the laboratory frame, we see that the Bloch vector participates in two distinct motions: forced oscillations of its x-,y-components with frequency ω (plus exponentially decaying transients at frequencies $\omega \pm \omega_R \approx \omega$), and exponentially decaying Rabi oscillations of its z-component:

$$\mathcal{R}(t) = \begin{pmatrix} \text{fast terms} \\ \text{fast terms} \\ Z_T \frac{\delta^2 + \Gamma^2}{\omega_R^2 + \Gamma^2} + r \exp(-\Gamma t) \sin(\omega_R t + \alpha) \end{pmatrix}. \tag{1.130}$$

The quantity

$$R(t) = r \exp(-\Gamma t) \tag{1.131}$$

is the instantaneous amplitude of decaying Rabi oscillations.

1.5 Slow evolution of a quantum system

1.5.1 Adiabatic theorem

After having investigated the evolution of a qubit under the influence of a fast resonant perturbation, we will now consider a different important limiting case. Let us return to the qubit Hamiltonian (1.85):

$$H(t) = -\frac{1}{2}(\Delta \sigma_x + \epsilon(t)\sigma_z) \tag{1.85}$$

and consider an arbitrary $\epsilon(t)$.

Suppose that initially, at $t = 0$, the qubit was in a pure state. Direct diagonalization of (1.85) yields the *instantaneous eigenstates* $|g(t)\rangle$, $|e(t)\rangle$:

$$H(t)|g(t)\rangle = -\frac{\hbar\Omega(t)}{2}|g(t)\rangle; \quad H(t)|e(t)\rangle = \frac{\hbar\Omega(t)}{2}|e(t)\rangle;$$

$$\hbar\Omega(t) = \sqrt{\Delta^2 + \epsilon(t)^2}; \tag{1.132}$$

$$|g(t)\rangle = \begin{pmatrix} \sqrt{\frac{1}{2}(1 + \epsilon(t)/\Omega(t))} \\ \sqrt{\frac{1}{2}(1 - \epsilon(t)/\Omega(t))} \end{pmatrix}; \quad |e(t)\rangle = \begin{pmatrix} \sqrt{\frac{1}{2}(1 - \epsilon(t)/\Omega(t))} \\ -\sqrt{\frac{1}{2}(1 + \epsilon(t)/\Omega(t))} \end{pmatrix}.$$

The states $|g(t)\rangle$, $|e(t)\rangle$ at any moment form an orthonormal basis – the so-called *adiabatic basis*[16] of the qubit's Hilbert space, i.e., the instantaneous eigenbasis of the time-dependent Hamiltonian. In the limit of infinitely strong bias, obviously,

$$\lim_{\epsilon \to \infty} |g\rangle = \begin{pmatrix} 1 \\ 0 \end{pmatrix} \equiv |0\rangle; \quad \lim_{\epsilon \to \infty} |e\rangle = \begin{pmatrix} 0 \\ 1 \end{pmatrix} \equiv |1\rangle, \tag{1.133}$$

and the adiabatic basis coincides with the diabatic ("physical") one. If we change the polarity ($\epsilon \to -\infty$), then the limits of $|g(t)\rangle$, $|e(t)\rangle$ change places (see Fig. 1.6).

The applicability of the simple Hamiltonian (1.85) is actually much wider than just qubit dynamics. In a quantum system with discrete energy levels and a Hamiltonian dependent on some external parameter λ, only two levels can get close together at any given value of λ (unless the system has a special symmetry). Moreover, in

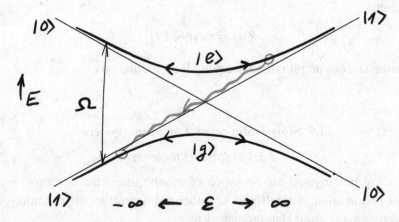

Fig. 1.6. Adiabatic evolution of a two-level system and the Landau–Zener–Stückelberg effect. Adiabatic transitions between the diabatic lines ("physical states") $|0\rangle$ and $|1\rangle$ and Landau–Zener tunnelling (wavy arrow) between the excited state $|e\rangle$ and the ground state $|g\rangle$ in the adiabatic basis are shown.

---

[16] Why it is called adiabatic will become clear directly.

the generic case the levels do not cross (see, e.g. Stöckmann, 1999, Chapter 3). The Hamiltonian (1.85), with λ replacing time, thus describes the one-parametric *level anticrossing* in a system with a discrete energy spectrum.[17]

The solution to the Schrödinger equation,

$$i\hbar \frac{d}{dt}|\psi(t)\rangle = H(t)|\psi(t)\rangle, \qquad (1.134)$$

can still be written as in (1.2),

$$|\psi(t)\rangle = U(t)|\psi(0)\rangle,$$

only this time the unitary operator $U(t)$ is given by the standard *time-ordered* exponent (e.g. Zagoskin, 1998, § 1.3):

$$U(t) = \mathrm{T}e^{-\frac{i}{\hbar}\int_0^t H(t')dt'}. \qquad (1.135)$$

The time-ordering operator T arranges all the operators standing to the right of it in chronological order; for example, $\mathrm{T}A(t_1)B(t_2) = A(t_1)B(t_2)$ if $t_1 \geq t_2$, and $\mathrm{T}A(t_1)B(t_2) = B(t_2)A(t_1)$ otherwise. (The operator exponent is to be understood, as usual, as its Taylor expansion.) The nontriviality of the time ordering is due to the fact that $H(t_1)$ and $H(t_2)$ do not necessarily commute. Otherwise, the action of $U(t)$ would reduce to producing phase factors multiplying the instantaneous basis states: $U(t)|g(0)\rangle \to \exp[-\frac{i}{2}\int_0^t E_0(t')\,dt']|g(t)\rangle$, $U(t)|e(0)\rangle \to \exp[-\frac{i}{2}\int_0^t E_1(t')\,dt']|e(t)\rangle$, where $E_{0,1}(t) = \mp\hbar\Omega(t)/2$. These formulas are, of course, wrong, but they are almost right in an important special case when $H(t)$ changes *adiabatically*. (In this context "adiabatic" means, as in classical mechanics, "slow enough for our purposes".)

A pure state of the qubit can always be written as

$$|\psi(t)\rangle = C_0(t)|g(t)\rangle + C_1(t)|e(t)\rangle. \qquad (1.136)$$

Taking the matrix elements of the Schrödinger equation:

$$i\hbar\left[\dot{C}_0(t)|g(t)\rangle + C_0(t)|\dot{g}(t)\rangle + \dot{C}_1(t)|e(t)\rangle + C_1(t)|\dot{e}(t)\rangle\right]$$
$$= E_0(t)C_0(t)|g(t)\rangle + E_1(t)C_1(t)|e(t)\rangle, \qquad (1.137)$$

---

[17] To describe the general case, when the diabatic lines (i.e., the asymptotes of the anticrossing levels) are asymmetrically inclined, would require a more general Hamiltonian,

$$H = -\frac{1}{2}\left[\Delta\sigma_x + \epsilon(\lambda)(\alpha\sigma_z + \beta\hat{1})\right],$$

but it would lead to trivial changes only. Moreover, the Hamiltonian (1.85) accurately describes the behaviour of the systems we are actually interested in, i.e., qubits.

we find

$$\dot{C}_0(t) = -\frac{i}{\hbar}E_0(t)C_0(t) - C_1(t)\langle g(t)|\dot{e}(t)\rangle; \qquad (1.138)$$

$$\dot{C}_1(t) = -\frac{i}{\hbar}E_1(t)C_1(t) - C_0(t)\langle e(t)|\dot{g}(t)\rangle.$$

If not for the scalar products $\langle g(t)|\dot{e}(t)\rangle$ and $\langle e(t)|\dot{g}(t)\rangle$, the solutions would have, indeed, been trivial. But if the Hamiltonian changes slowly, so do its instantaneous eigenstates, therefore, their time derivatives can be neglected, and indeed

$$C_0(t) \approx C_0(0)e^{-\frac{i}{\hbar}\int_0^t E_0(t')\,dt'}; \quad C_1(t) \approx C_1(0)e^{-\frac{i}{\hbar}\int_0^t E_1(t')\,dt'}. \qquad (1.139)$$

In the case of a mixed state we obtain the elements of the density matrix, $\rho(t) = \sum_{i,j}\rho_{ij}(t)|i(t)\rangle\langle j(t)|$, in the same way:

$$\dot{\rho}_{00} = -\rho_{10}(t)\langle g(t)|\dot{e}(t)\rangle - \rho_{01}(t)\langle\dot{e}(t)|g(t)\rangle;$$

$$\dot{\rho}_{11} = -\rho_{01}(t)\langle e(t)|\dot{g}(t)\rangle - \rho_{10}(t)\langle\dot{g}(t)|e(t)\rangle;$$

$$\dot{\rho}_{01} = -\rho_{11}(t)\langle g(t)|\dot{e}(t)\rangle - \rho_{00}(t)\langle\dot{g}(t)|e(t)\rangle + \frac{i}{\hbar}(E_1(t) - E_0(t))\rho_{01}(t);$$

$$\dot{\rho}_{10} = -\rho_{00}(t)\langle e(t)|\dot{g}(t)\rangle - \rho_{11}(t)\langle\dot{e}(t)|g(t)\rangle - \frac{i}{\hbar}(E_1(t) - E_0(t))\rho_{10}(t),$$

$$(1.140)$$

and the diagonal terms are approximately time-independent under the same conditions:

$$\dot{\rho}_{00} \approx 0; \quad \dot{\rho}_{11} \approx 0. \qquad (1.141)$$

It does not actually matter for our reasoning how many levels the system has, as long as they are discrete. Therefore, we have obtained the *adiabatic theorem* (see, e.g., Messiah, 2003, Chapter XVII, §§7–12):

In a system with a discrete energy spectrum under certain conditions an infinitely slow, or adiabatic, change of the Hamiltonian does not change the level populations.

The change of the *off*-diagonal elements of the density matrix under an adiabatic evolution – as well as the phase gain by a pure state – can be observed and has an important physical meaning, but this does not concern us now. What does concern us are the conditions of applicability of the "theorem".

For a qubit "slow change" should mean that $d\Omega/dt$ is small, i.e.,

$$\frac{\hbar}{\Omega^2(t)}\frac{d\Omega}{dt} \ll 1 \qquad (1.142)$$

(this is the simplest dimensionless combination of relevant variables). Indeed, if we use the explicit solutions (1.133), calculate the values of $|\langle e|\dot{g}\rangle| (= |\langle g|\dot{e}\rangle|)$, and require that they can be neglected in Eqs. (1.138) and (1.140), we come to the above condition. For a general multilevel system the adiabatic condition is usually stated as

$$\left| \hbar \frac{\langle m(t)|\dot{n}(t)\rangle}{E_n(t) - E_m(t)} \right| \equiv \left| \hbar \frac{\langle m(t)|\dot{H}(t)|n(t)\rangle}{(E_n(t) - E_m(t))^2} \right| \ll 1. \tag{1.143}$$

Here we have used

$$\langle m(t)|\dot{n}(t)\rangle = \frac{\langle m(t)|\dot{H}(t)|n(t)\rangle}{E_n(t) - E_m(t)}, \quad m \neq n, \tag{1.144}$$

which directly follows from considering the expression $\frac{d}{dt}(H(t)|n(t)\rangle)$.

It is clear from (1.143) that for the adiabatic theorem to hold the *levels must not cross*: $E_m(t) \neq E_n(t)$, if $m \neq n$. But even if this is satisfied, there are problems with (1.143). Let us, e.g., take the qubit Hamiltonian (1.85) with $\epsilon(t) = \hbar\eta\cos(\omega t)$ with $\omega \approx \Delta$ and $\hbar\eta \ll \Delta$. Then

$$\frac{\hbar}{\Omega^2(t)} \frac{d\Omega}{dt} \approx \frac{1}{2}\left(\frac{\hbar\eta}{\Delta}\right)^2 \ll 1. \tag{1.145}$$

Nevertheless, we have seen that such a small harmonic perturbation produces Rabi oscillations. Even though their frequency η is small, the periodic changes of the density matrix elements is of order unity, and the theorem as stated above breaks down. This situation is not exclusive to quantum mechanics and is not surprising: small resonant perturbations *do* produce large cumulative effects. Therefore, another condition of validity of the criterion (1.143) is the absence of such terms.

It is actually possible to formulate the condition of adiabaticity in such a way as to include the case of resonant perturbations (Amin, 2009). We will do this here considering the evolution of a density matrix rather than of a pure state. For a multilevel system, Eqs. (1.140) easily generalize to

$$\dot{\rho}_{mn} = -\frac{i}{\hbar}\rho_{mn}(E_m(t) - E_n(t)) - \sum_j \left(\rho_{jn}\langle m|\dot{j}\rangle + \rho_{mj}\langle \dot{j}|n\rangle\right), \tag{1.146}$$

where $H(t)|j(t)\rangle = E_j(t)|j(t)\rangle$. By separating the adiabatic phase gain,

$$\rho_{mn}(t) = \tilde{\rho}_{mn}(t)e^{-\frac{i}{\hbar}\int_0^t (E_m(t') - E_n(t'))\,dt'}, \tag{1.147}$$

where

$$\omega_{mn}(t) = \frac{E_m(t) - E_n(t)}{\hbar},$$

we obtain

$$\dot{\tilde{\rho}}_{mn} = -\sum_{j\neq m} \tilde{\rho}_{jn} \left(\langle m|\dot{j}\rangle e^{-i\int_0^t \omega_{jm}(t')\,dt'} \right) - \sum_{j\neq n} \tilde{\rho}_{mj} \left(\langle \dot{j}|n\rangle e^{-i\int_0^t \omega_{nj}(t')\,dt'} \right)$$

(1.148)

(obviously, $\langle m|\dot{m}\rangle = 0$, etc.).

Integrating (1.148) over the time interval $[0, \tau]$ we find the change of the diagonal density matrix element $\tilde{\rho}_{nn}$:

$$\Delta\rho_{nn}(\tau) = -2\text{Re} \int_0^\tau \left[\sum_{j\neq n} \tilde{\rho}_{jn} \langle n|\dot{j}\rangle e^{-i\int_0^t \omega_{jn}(t')\,dt'} \right] dt \equiv -2\text{Re}\sum_{j\neq n} \alpha_{jn}(\tau).$$

(1.149)

For the adiabatic theorem to hold, $|\Delta\rho_{nn}(\tau)|$ must be zero in the limit of infinitely slow evolution. Introducing the quantity

$$r_{jn}(t) = \frac{\tilde{\rho}_{jn}(t)\langle n(t)|\dot{j}(t)\rangle}{\omega_{jn}(t)} \equiv \frac{\tilde{\rho}_{jn}(t)\langle n(t)|\dot{H}(t)|j(t)\rangle}{\hbar\omega_{jn}(t)^2}$$

(1.150)

we can express the contribution from the transitions $j \leftrightarrow n$ as follows:

$$\alpha_{jn}(\tau) = i\int_0^\tau dt\, r_{jn}(t) \left[-i\omega_{jn}(t) e^{-i\int_0^t \omega_{jn}(t')\,dt'} \right]$$

$$= i\int_0^\tau dt\, r_{jn}(t) \frac{d}{dt} \left[e^{-i\int_0^t \omega_{jn}(t')\,dt'} \right].$$

(1.151)

Integrating this by parts, we find

$$\alpha_{jn}(\tau) = i\left[r_{jn}(\tau) e^{-i\int_0^\tau \omega_{jn}(t')\,dt'} - r_{jn}(0) \right] - i\int_0^\tau dt \left[\frac{dr_{jn}(t)}{dt} e^{-i\int_0^t \omega_{jn}(t')\,dt'} \right].$$

(1.152)

Looking at the definition of r_{jn}, Eq. (1.150), we see that the absolute value of the first term in (1.152) is infinitesimally small if the adiabaticity condition (1.143) is satisfied. Therefore, the violation of the adiabatic theorem can only come from the second term. Using the Fourier transform of $r_{jn}(t) = \int \frac{d\omega}{2\pi} R^*(\omega)\exp(i\omega t)$, we can write this term as

$$-i\int_0^\tau dt \left[\frac{dr_{jn}(t)}{dt} e^{-i\int_0^t \omega_{jn}(t')\,dt'} \right] = \tau \int \frac{d\omega}{2\pi} \omega R^*(\omega) \int_0^1 ds\, e^{-i\int_0^s (u(s')-v)\,ds'},$$

(1.153)

where $u = \omega_{jn}\tau$ and $v = \omega\tau$ are dimensionless frequencies.

If the Hamiltonian does not contain resonant terms for the transition $j \leftrightarrow n$, then this integral is small due to fast oscillations of the integrand. Indeed, integrating by parts:

$$\tau \int \frac{d\omega}{2\pi} \omega R^*(\omega) \int_0^1 ds \, e^{-i\int_0^s (u(s')-v)\,ds'}$$

$$= \tau \int \frac{d\omega}{2\pi} \omega R^*(\omega) \int_0^1 ds \left[\frac{d}{ds} \left(\frac{e^{-i\int_0^s (u(s')-v)\,ds'}}{-i(u(s)-v)} \right) - \frac{e^{-i\int_0^s (u(s')-v)\,ds'}}{i(u(s)-v)^2} \frac{du}{ds} \right]$$

$$= \int \frac{d\omega}{2\pi} \omega R^*(\omega) \left[\frac{e^{-i\int_0^\tau (\omega_{jn}(t')-\omega)\,dt'}}{-i(\omega_{jn}(\tau)-\omega)} - \frac{1}{-i(\omega_{jn}(0)-\omega)} \right] + \cdots$$

we see that the dominating term in the expansion is infinitesimally small if the condition (1.143) holds, independently of the length of the evolution interval τ.

On the other hand, if $R^*(\omega)$ is nonzero at frequencies close to $\omega_{jn}(t)$, the main contribution to the integral will be due to these resonant frequencies, and

$$-i \int_0^\tau dt \left[\frac{dr_{jn}(t)}{dt} e^{-i\int_0^t \omega_{jn}(t')\,dt'} \right] \approx \tau \int \frac{d\omega}{2\pi} \omega R^*(\omega) \approx -i\tau \left. \frac{d}{dt} r_{jn}(t) \right|_{t=t^*, \, 0\leq t^* \leq \tau}.$$

The error is proportional to the duration of the evolution, and however small is r_{jn} or its derivative, for large enough τ the adiabatic theorem will break down.[18]

Therefore, for resonance there is an upper limit on the duration of the adiabatic evolution. It is easy to see that both cases are incorporated if instead of (1.143) we use (Amin, 2009)

$$\max_{0\leq s\leq 1} \left| \hbar \frac{\langle m(s)|dH/ds|n(s)\rangle}{(E_n(s)-E_m(s))^2} \right| \ll \tau, \quad s = t/\tau. \tag{1.154}$$

1.5.2 Landau–Zener–Stückelberg effect

The adiabatic theorem shows a different way of controlling the quantum state. Instead of relying on a free evolution of an initial superposition of eigenstates of a system (as in the case of quantum beats), or applying a harmonic resonant perturbation to induce the transition between them (as in the case of Rabi oscillations), a slow and smooth variation of the Hamiltonian changes the eigenstates themselves without changing their occupation numbers. As we have seen, this ideal situation can be achieved only in the unrealistic limit of infinitesimally slow change. For any finite speed of the variation of $H(t)$, even in the absence of resonant terms,

---

[18] Another, extrinsic mechanism, which limits the duration of an adiabatic evolution of a system, is through its interaction with the environment (Sarandy and Lidar, 2005a).

the adiabatic theorem will be violated. The corresponding process is called the Landau–Zener–Stückelberg effect, or – more often – the Landau–Zener transition, or Landau–Zener tunnelling. In addition to being an important physical process, showing up in such different areas as molecular collisions, quantum control and, of course, quantum computing, this is a nice example of an exactly solvable problem with a nonanalytical dependence on the parameters.

Let us return to the qubit Hamiltonian:

$$H(t) = -\frac{1}{2}(\Delta \sigma_x + \epsilon(t)\sigma_z), \tag{1.85}$$

which, up to trivial changes, can be considered as representing two levels of any multilevel system near an anticrossing, if $\epsilon(t)$ changes monotonically between a large negative and a large positive value.

Despite what was done in the previous subsection, we expand the wave function over the diabatic basis,

$$|\psi(t)\rangle = C_0(t)e^{-\frac{i}{\hbar}\int_{-\infty}^{t} E_0(t')\,dt'}|0\rangle + C_1(t)e^{-\frac{i}{\hbar}\int_{-\infty}^{t} E_1(t')\,dt'}|1\rangle$$

(we have explicitly written the phase factors). Given that $E_1(t) - E_0(t) = \epsilon(t)$, we find from the Schrödinger equation:

$$\dot{C}_0 = i\frac{\Delta}{2\hbar}C_1 e^{-\frac{i}{\hbar}\int_{-\infty}^{t}\epsilon(t')\,dt'}, \quad \dot{C}_1 = i\frac{\Delta}{2\hbar}C_0 e^{\frac{i}{\hbar}\int_{-\infty}^{t}\epsilon(t')\,dt'}. \tag{1.155}$$

Differentiating these equations by time, making obvious substitutions and taking into account that $E_1(t) + E_0(t) = 0$, we decouple the system:

$$\ddot{C}_0 - \frac{i}{\hbar}\epsilon(t)\dot{C}_0 + \left(\frac{\Delta}{2\hbar}\right)^2 C_0 = 0; \tag{1.156}$$

$$\ddot{C}_1 + \frac{i}{\hbar}\epsilon(t)\dot{C}_1 + \left(\frac{\Delta}{2\hbar}\right)^2 C_1 = 0. \tag{1.157}$$

Notably, for a linear sweep,

$$\epsilon(t) = vt,$$

these equations allow an explicit solution. We will see that the effect is dominated by the time when the bias $|\epsilon(t)| \leq \Delta$, where such a linear approximation is very plausible.[19] Instead of following Zener (1932) in reducing (1.156) to a known in the mathematical physics Weber equation, solvable using special functions, we use the elegant result of Wittig (2005), where solving the equations is avoided altogether.

---

[19] The case of nonlinear sweep is considered in detail, e.g., by Garanin and Schilling (2002).

To begin, let us find the answer in the limit of weak tunnelling, when Δ is small. Consider a system, which is at $t = -\infty$ in the excited state (that is, $C_1(-\infty) = 1$). In the absence of the tunnelling term in the Hamiltonian, at $t = \infty$ the system would remain in this state, which – of course – would be the ground state (see Fig. 1.6), i.e., $C_1(\infty) = 1$ as well. Therefore, we expect that in the presence of tunnelling ($\Delta \neq 0$) this coefficient will remain close to unity, and correspondingly $C_0 \ll 1$. Substituting $C_1 = 1$ in the first equation of (1.155), we find (up to a non-essential phase factor originating from the lower integration limit by t' in the exponent)

$$\dot{C}_0 \approx i\frac{\Delta}{2\hbar}e^{-\frac{ivt^2}{2\hbar}}; \quad C_0(\infty) \approx \int_{-\infty}^{\infty} i\frac{\Delta}{2\hbar}e^{-\frac{ivt^2}{2\hbar}}\,dt = i\frac{\Delta}{2\hbar}\sqrt{\frac{2\pi\hbar}{iv}}. \tag{1.158}$$

The probability of *not* switching from the state $|1\rangle$ to the state $|0\rangle$ at the end of the evolution is, therefore (now you see why the phase factor does not matter),

$$P_{LZ} = 1 - |C_0|^2 = 1 - \frac{\pi\Delta^2}{2\hbar v}, \tag{1.159}$$

which is essentially the result obtained by Landau (1932). Note that P_{LZ} is the probability of not switching between the *diabatic* states, $|0\rangle$ and $|1\rangle$, and, therefore, it is the probability of *switching* between the *adiabatic* states (the eigenstates of the qubit Hamiltonian (1.85)), $|g\rangle$ and $|e\rangle$.

The ratio $\Delta^2/\hbar v$ can be written as $(\Delta/\hbar) \cdot (\Delta/\dot{\epsilon})$, that is, as the product of the frequency of quantum beats[20] (see Fig. 1.5) and the anticrossing passage time during which the bias changes by Δ. Note that – up to a factor of order unity – we could derive from dimensional considerations only. What we can't do this way is to find the whole functional dependence, which we will now proceed to derive.

Dividing (1.157) by $tC_1(t)$ and integrating over time from minus to plus infinity, we find

$$\int_{-\infty}^{\infty}\frac{\ddot{C}_1}{tC_1}\,dt + \frac{i}{\hbar}\int_{-\infty}^{\infty}v\frac{\dot{C}_1}{C_1}\,dt + \left(\frac{\Delta}{2\hbar}\right)^2\int_{-\infty}^{\infty}\frac{dt}{t} = 0, \tag{1.160}$$

that is,

$$\ln\left(\frac{C_1(\infty)}{C_1(-\infty)}\right) = \frac{i\hbar}{v}\left\{\left(\frac{\Delta}{2\hbar}\right)^2\int_{-\infty}^{\infty}\frac{dt}{t} + \int_{-\infty}^{\infty}\frac{\ddot{C}_1}{tC_1}\,dt\right\}. \tag{1.161}$$

---

[20] For some mysterious reason this frequency is often called the Rabi frequency. It *is not*. The Rabi frequency describes the effects of a resonant harmonic perturbation, and is nothing to do with quantum beats, which are free evolution of a superposition of eigenstates of a static Hamiltonian.

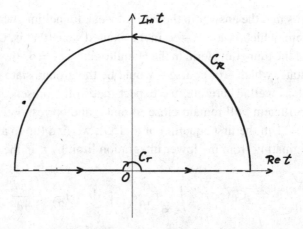

Fig. 1.7. Calculation of the integral (1.163). The closed contour $\mathcal{C}$ consists of the real positive and negative semiaxes plus an infinitesimal semicircle $\mathcal{C}_r$ around zero, and an infinitely large semicircle $\mathcal{C}_R$. The contour does not enclose the singularity at $t = 0$.

The integral of $1/t$ can be found using regularization (1.58):

$$\int_{-\infty}^{\infty} \frac{dt}{t} \to \int_{-\infty}^{\infty} \frac{dt}{t \pm i\varepsilon} \to \mathcal{P} \int_{-\infty}^{\infty} \frac{dt}{t} \mp i\pi = \mp i\pi. \tag{1.162}$$

We will fix the sign later.

The remaining integral,

$$J \equiv \int_{-\infty}^{\infty} \frac{\ddot{C}_1}{t C_1} \, dt \tag{1.163}$$

requires a little more attention. Let us consider $F(t) \equiv \ddot{C}_1(t)/(t C_1(t))$ as a function of a complex variable t (Fig. 1.7). In the general case, $F(t)$ should be an analytical function, except, possibly, at $t = 0$. Any exponential dependence on t in $C_1(t)$ – e.g., from the phase factors – which could cause trouble, will obviously cancel in the combination $\ddot{C}_1/C_1$. Therefore, the integral of $F(t)$ along the contour $\mathcal{C}$ in the complex plane is the sum of the residues in the singularities of $F(t)$, which are enclosed by this contour, in accordance with Cauchy's theorem.[21] We choose the contour as in Fig. 1.7, so that

$$J + J_R = 0, \tag{1.164}$$

and we need only to calculate the integral along the infinite semicircle $\mathcal{C}_R$ to find J. This integral is easily found. Looking at (1.157) with $\epsilon(t) = vt$ we see that, since $|C_1(t)| \le 1$, the last two terms must cancel when $t \to \pm\infty$, and the second

---

[21] See any textbook on complex analysis (e.g. Gamelin, 2001, Chapter VI).

derivative should be negligible:

$$\frac{\dot{C_1}}{C_1} \sim -\frac{i\hbar}{vt}\left(\frac{\Delta}{2\hbar}\right)^2 ; \quad C_1(t) \sim (t)^{-i\frac{\hbar}{v}\left(\frac{\Delta}{2\hbar}\right)^2}. \tag{1.165}$$

Then indeed $\ddot{C_1}/\dot{C_1} \sim 1/t$, $\ddot{C_1}/C_1 \sim 1/t^2$, and our initial assumption (that the second derivative can be neglected) holds. Therefore, for large values of $|t|$ $|F(t)| = |\ddot{C_1}/(t\dot{C_1})| \sim 1/|t|^3$. Since the integral J_R

$$J_R = \int_{C_R} F(t)\,dt \leq \int_{C_R} |F(t)|\,dt \sim \frac{\pi R}{R^3} \xrightarrow{R \to \infty} 0,$$

we see that $J = 0$ as well.

Returning to (1.161), we see that

$$\ln\left(\frac{C_1(\infty)}{C_1(-\infty)}\right) = \pm\frac{\pi\hbar}{v}\left(\frac{\Delta}{2\hbar}\right)^2. \tag{1.166}$$

Since $C_1(-\infty) = 1$ and the probability of staying in the same state is $|C_1(\infty)|^2$, we see that

$$P_{LZ} = e^{-\frac{\pi\Delta^2}{2\hbar v}}. \tag{1.167}$$

The choice of the sign was obvious: the probability cannot exceed unity. Eq. (1.167) agrees with the perturbative result (1.159) in the limit $\Delta^2 \ll \hbar v$, as it should. Note that the speed of the sweep, v, plays a crucial role: the faster the sweep, the less the probability of switching diabatic states (or, equivalently, of staying in the same adiabatic state). Moreover, the dependence of P_{LZ} on v is essentially non-analytical.

Of course, we would obtain the same result if starting from the state $|0\rangle$ instead. That particular case is of special interest, since it is sometimes easier to cool a system down to its ground state and then manoeuvre the state into something we would like to have (like, e.g., in *adiabatic quantum computing*, Section 6.2). The Landau–Zener effect would then be one of the leading sources of failure of such a scheme.[22]

Concluding remarks

Before proceeding to practical applications, let us recapitulate what has been achieved so far. First, the introduction of the density matrix provides a unified description of the quantum system before, during and after the measurement, that is, the reduction of the density matrix to an appropriate diagonal form. Second, the non-unitary evolution of the density matrix has been shown to be explainable

---

[22] This has been directly observed, e.g., in a superconducting flux qubit (Izmalkov et al., 2004a).

Fig. 1.8. Quantum–classical transition: two possibilities.

by averaging over – tracing out – the "environment", i.e., averaging over the huge number of degrees of freedom, which interact with our system but are affected by it only to a negligibly weak degree. This "restricted description" is commonly employed for the same reason in classical statistical mechanics. It is not our purpose here to establish under what strict conditions this kind of explanation works – it is good enough for our purposes. The key physical effect leading to this is the loss of correlation via its "removal" from the system – to the environment, or to infinity, as we will see when considering the effective impedance of a transmission line in Chapter 5. There appears to be no "fundamental size", exceeding which a system *must* drop all quantum behaviour in favour of classical physics, no matter what. Third, this still *does not* resolve the problem indicated after Eq. (1.5), namely, that the formalism of the density matrix still does not give us a clue as to which bases in the Hilbert state can describe macroscopic measurements, and which cannot. We can have a diagonal density matrix, but its elements may – for all we know – give the probabilities of finding the system in different Schrödinger-cat-type states.

In other words, we still must answer the question as to why we can build the apparatus for measuring eigenstates of only certain observables and not others, and how to tell the difference *from the quantum mechanics itself*. This is a fascinating and still open question, and again it is beyond the scope of this book.[23] We will be, therefore, pragmatically minded and satisfy ourselves that these special observables and states can be found from macroscopic experience. With this restriction, and evading finer ontological points of the description based on a density matrix,[24] we have now laid a sufficient foundation for dealing with the more practical side of quantum engineering.

[23] A very attractive possibility is the "quantum Darwinism" approach (Zurek, 2003), which argues that the "macroscopic" observables are the ones with the eigenstates that are the least disrupted by their interaction with different subsystems of the environment.

[24] A very good and accessible resume of the problem, together with an extensive bibliography, is given in Penrose, (2004), Chapter 29.

2

Superconducting quantum circuits

Good order is the foundation of all good things.

E. Burke, *Reflections on the Revolution in France*, 1790

2.1 Josephson effect

2.1.1 Superconductivity: A crash course

The transition from the theoretical description of hypothetical building blocks of a quantum coherent device to something which can be actually fabricated and controlled is made much easier by the existence of *superconductivity*. This phenomenon, roughly speaking, allows a *macroscopic quantum coherent* flow of electrons in a sufficiently cold piece of an appropriate material by establishing a specific long-range order among them. Due to this one can, for example, use macroscopically different states of a superconductor as quantum states of a qubit, with the obvious advantage over "microscopically quantum" systems (like actual atoms) from the point of view of control, measurement and, last but not least, fabrication of structures with the desired parameters and on the desired scale.

The phenomenon of superconductivity – the history of its discovery, experimental manifestations, theoretical explanation, open questions, relevance to other branches of physics, and technological applications – requires a thorough treatment, which can be found in any number of books. For our purposes Tinkham (2004) will provide more than sufficient background.[1]

We will not wander into the field of exotic/high-temperature superconductors for two simple reasons: the quantum coherent behaviour of the kind we need to realize qubits or other quantum coherent devices has not *yet* been properly and/or routinely

---

[1] A clear, simple and broad overview of the subject is given by Schmidt (2002). Phenomenology and the microscopic theory of superconductivity is presented in, e.g., Landau et al. (1984), Pitaevskii and Lifshitz (1980), Ketterson and Song (1999). The Josephson effect is treated in detail in Barone and Paterno (1982), Likharev (1986).

achieved in these systems; and the fabrication of such devices is not *yet* reliable enough or even feasible. Therefore, *all* of the following only refers to "standard", low-temperature superconductors, like niobium or aluminium. The temperatures necessary to preserve the quantum effects we are after (at most a few dozens of millikelvins) are well below their superconducting transition temperatures anyway.

The key property of a superconductor is the existence of the electrically charged *Cooper pair condensate* (its charge is neutralized by the charge of the crystal lattice). Cooper pairs are correlated states of two conduction electrons with opposite spins and almost opposite momenta; the resulting object has zero spin, thus being a compound boson. Cooper pairs form due to the weak residual attraction between electrons, resulting from the latter's interaction with lattice oscillations (phonons). Whether such an attraction exists depends on the material. The size of a Cooper pair (measured by its *coherence length*, ξ_0, i.e., the distance over which the correlation between them significantly drops) is much larger than the average distance between the electrons in the superconductor, so the Cooper pairs cannot be thought of as "point-like quasibosons". Nevertheless, being bosons, Cooper pairs can and do undergo Bose condensation at a low enough temperature.

Actually both pair formation and the appearance of the condensate happen simultaneously, when the normal state of the electronic system becomes unstable with respect to an arbitrarily weak electron–electron attraction. This is a very important and very beautiful example of the *second-order phase transition*, in which the symmetry of the system is changed. (For the two superconductors most successfully used in making such devices as qubits, aluminium and niobium, the transition temperatures are respectively 1.75 K and 9.25 K.)

What is important to us is that the long-range order thus established is pretty special. Consider the correlation function between two electron pairs situated at $\mathbf{r}_1$ and $\mathbf{r}_2$:

$$S_{\uparrow\downarrow}(\mathbf{r}_1, \mathbf{r}_2) = \left\langle \psi_\uparrow^\dagger(\mathbf{r}_1)\psi_\downarrow^\dagger(\mathbf{r}_1)\psi_\downarrow(\mathbf{r}_2)\psi_\uparrow(\mathbf{r}_2) \right\rangle. \tag{2.1}$$

The Fermi-operator $\psi_\uparrow^\dagger(\mathbf{r}_1) \propto \sum_{\mathbf{k}} e^{i\mathbf{k}\mathbf{r}_1} a_{\uparrow\mathbf{k}}^\dagger$ creates an electron with spin up at $\mathbf{r}_1$, etc. The average, of course, means the trace of the averaged expression with the density matrix of the whole system, and at zero temperature reduces to the average value in the ground state of the superconductor. As the distance $|\mathbf{r}_1 - \mathbf{r}_2|$ grows to infinity, one expects such a quantity will factorize into local averages,

$$S_{\uparrow\downarrow}(\mathbf{r}_1, \mathbf{r}_2) \to \left\langle \psi_\uparrow^\dagger(\mathbf{r}_1)\psi_\downarrow^\dagger(\mathbf{r}_1) \right\rangle \left\langle \psi_\downarrow(\mathbf{r}_2)\psi_\uparrow(\mathbf{r}_2) \right\rangle. \tag{2.2}$$

For this expression to be nonzero, the *anomalous* averages, $\left\langle \psi_\uparrow^\dagger(\mathbf{r}_1)\psi_\downarrow^\dagger(\mathbf{r}_1) \right\rangle$ and $\left\langle \psi_\downarrow(\mathbf{r}_2)\psi_\uparrow(\mathbf{r}_2) \right\rangle$ should also be nonzero. In a normal state this is impossible: the operator $\psi_\uparrow^\dagger(\mathbf{r}_1)\psi_\downarrow^\dagger(\mathbf{r}_1)$ (or, if you wish, $a_{\uparrow\mathbf{k}}^\dagger a_{\downarrow\mathbf{k}'}^\dagger$) changes the number of electrons by two and has, therefore, a zero average value in any state with a definite number of particles. Nevertheless, the superconducting state "in the bulk" (that is, in an infinitely large superconductor – otherwise how could we have $|\mathbf{r}_1 - \mathbf{r}_2| \to \infty$?) is *not* such a state. Due to the fact that it contains the Bose condensate of a *macroscopic* number of Cooper pairs, an extra pair of electrons entering or leaving the system does not really matter. The anomalous average multiplied by a scalar g,

$$g \langle \psi_\downarrow(\mathbf{r})\psi_\uparrow(\mathbf{r}) \rangle \equiv \Delta(\mathbf{r}) = |\Delta(\mathbf{r})|e^{i\phi(\mathbf{r})} = g \langle \psi_\uparrow^\dagger(\mathbf{r})\psi_\downarrow^\dagger(\mathbf{r}) \rangle^* \qquad (2.3)$$

is a complex number with the dimensionality of energy and is called the *order parameter* of the superconductor. It describes the quantum coherent behaviour of the electrons in the system. The coupling constant g is a measure of the binding energy of electrons in a Cooper pair; thus $2|\Delta|$ is the energy necessary to break a Cooper pair and create two quasiparticles, i.e., conventional conducting electrons. The ground state of a superconductor does not contain any occupied quasiparticle states, and is, therefore, separated from the excited states by a finite energy gap (usually given per electron, and not per Cooper pair, that is, $|\Delta|$).

Using the basis of Fock states, $|n\rangle$, a.k.a. "number states" (i.e., states with a definite number n of electrons), we can write

$$\Delta(\mathbf{r}) = g \operatorname{tr}\left[\rho\psi_\downarrow(\mathbf{r})\psi_\uparrow(\mathbf{r})\right] = g \sum_n \langle n+2|\rho|n\rangle \langle n|\psi_\downarrow(\mathbf{r})\psi_\uparrow(\mathbf{r})|n+2\rangle, \quad (2.4)$$

and for this to be nonzero the equilibrium density matrix ρ must contain off-diagonal matrix elements in this basis. Superconductivity is thus an example of an *off-diagonal long-range order* (ODLRO), a concept introduced by Yang (1962).

The ground state of a superconductor, obtained by Bardeen, Cooper and Schrieffer' (BCS) under some simplifying assumptions about the electron–electron interaction ("pairing Hamiltonian": see, e.g., Tinkham, 2004, Chapter 3) is given by

$$|\mathrm{BCS}\rangle = \prod_\mathbf{k} \left(u_\mathbf{k} + v_\mathbf{k} a_{\uparrow\mathbf{k}}^\dagger a_{\downarrow,-\mathbf{k}}^\dagger\right)|0\rangle, \qquad (2.5)$$

where $|0\rangle$ is the vacuum, and the complex numbers $u_\mathbf{k}$, $v_\mathbf{k}$ (*coherence factors*) are normalized:

$$|u_\mathbf{k}|^2 + |v_\mathbf{k}|^2 = 1 \qquad (2.6)$$

and satisfy certain (Bogoliubov–de Gennes) equations, which we do not need to discuss here.[2] The product in Eq. (2.5) is taken over all one-electron states (labelled by their momenta, $\hbar\mathbf{k}$), and it is clear that the BCS ground state is indeed a coherent

---

[2] These details can be found in, e.g., Tinkham (2004), Schmidt (2002), Zagoskin (1998), Ketterson and Song (1999).

superposition of states all with even numbers of electrons, from zero to infinity. As any wave function, the one given by Eq. (2.5) is defined up to an overall phase factor. What matters is the relative phase between different terms in the superposition, and importantly, for any **k** the phase shift between the coherence factors is the same, so that one can write the BCS wave function as

$$|\text{BCS}\rangle = \prod_{\mathbf{k}} \left(|u_{\mathbf{k}}| + |v_{\mathbf{k}}| e^{i\phi} a_{\uparrow\mathbf{k}}^{\dagger} a_{\downarrow,-\mathbf{k}}^{\dagger} \right) |0\rangle.$$

The order parameter $\Delta(\mathbf{r})$, up to the factor g, can be considered as a *macroscopic wave function* of a Cooper pair, normalized to the density of Cooper pairs (i.e., *half* the density of electrons in the condensate, n_s):

$$\Delta(\mathbf{r}) \propto \Psi_s(\mathbf{r}) = \sqrt{n_s(\mathbf{r})/2}\, e^{i\phi(\mathbf{r})}. \tag{2.7}$$

Note that the phase ϕ is here the same as in the BCS wave function (if we cared to write its generalization for the case of a nonuniform and moving – current-carrying – condensate).

Borrowing the expression for the electric current density from one-particle quantum mechanics (Landau and Lifshitz, 2003, § 19), we obtain the superconducting current (*supercurrent*) density,

$$\mathbf{j}_s(\mathbf{r}) = \frac{2e\hbar}{2im_e} \Psi_s^*(\mathbf{r}) \nabla \Psi_s(\mathbf{r}) = n_s(\mathbf{r}) e v_s(\mathbf{r}), \tag{2.8}$$

where the *superfluid velocity*

$$\mathbf{v}_s(\mathbf{r}) \equiv \frac{\hbar}{2m_e} \nabla \phi(\mathbf{r}). \tag{2.9}$$

(We took into account that the Cooper pair has twice the charge and twice the mass of an electron.) The current flow is determined by the spatial dependence of the superconducting phase. This superconducting current involves a macroscopic number of electrons: it is a quantum coherent motion of the condensate as a whole with the velocity $\mathbf{v}_c$, since all the Cooper pairs have the same quantum phase.

The Cooper pair momentum operator is, in agreement with (2.8),

$$\mathbf{p}_s = -i\hbar\nabla. \tag{2.10}$$

The superfluid velocity can be defined as a quantum average of this operator, or more precisely,

$$2m_e\mathbf{v}_s(\mathbf{r}) = \frac{2}{n_s}\langle\Psi_s(\mathbf{r})|\left[-i\hbar\nabla - \frac{2e}{c}\mathbf{A}(\mathbf{r})\right]|\Psi_s(\mathbf{r})\rangle = \hbar\nabla\phi(\mathbf{r}) - \frac{2e}{c}\mathbf{A}(\mathbf{r}). \tag{2.11}$$

The last term takes into account the extra phase that a Cooper pair, as any electrically charged quantum particle, will acquire in an electromagnetic field, defined by its vector potential **A** (Landau and Lifshitz, 2003, § 111).

The electromagnetic field is expelled from a bulk superconductor beyond the material- and temperature-dependent *penetration length* (the *Meissner–Ochsenfeld effect*, see Tinkham, 2004, § 1.1); the supercurrent and the superfluid velocity, accordingly, must also disappear in the bulk. Nevertheless, the vector potential can be nonzero where the field is absent, and thus affect the behaviour of the condensate. Consider, for example, a hole in a massive superconductor, or a closed loop made of a superconducting wire, which is much thicker than the penetration depth (for most usual superconductors at $T = 0$ the penetration depth $\lambda \sim 500$ Å). Consider the magnetic flux Φ passing through the hole and integrate Eq. (2.11) along a closed contour around the hole (Fig. 2.1):

$$\oint 2m_e \mathbf{v}_s(\mathbf{r}) \cdot d\mathbf{r} = \oint \hbar \nabla \phi(\mathbf{r}) \cdot d\mathbf{r} - \oint \frac{2e}{c} \mathbf{A}(\mathbf{r}) \cdot d\mathbf{r}. \qquad (2.12)$$

We can shift the contour into the bulk of the superconductor, where $\mathbf{v}_s = 0$. Then the left-hand side is zero, and the change in the phase along the closed loop is

$$\Delta\phi = \frac{2e}{\hbar c} \int \nabla \times \mathbf{B}(\mathbf{r}) \cdot d\mathbf{s} = 2\pi \frac{\Phi}{\Phi_0}. \qquad (2.13)$$

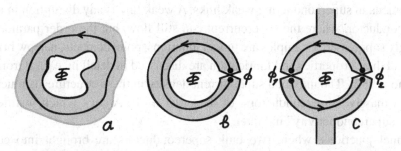

Fig. 2.1. Magnetic flux quantization in a superconductor. (a) If the magnetic flux Φ passes through a hole in a bulk superconductor, the requirement of single-valuedness of the superconducting order parameter imposes the quantization condition $\Phi = n\Phi_0$. (b) The same requirement in the case of a superconducting loop interrupted by a weak link (rf SQUID configuration) leads to Eq. (2.28). (c) In a dc SQUID, the flux quantization condition (2.29) allows us to tune the critical current through the device using the external magnetic field, Eq. (2.32). Note that in all cases the thickness of the superconducting leads is assumed to exceed the penetration depth, which allows us to draw the integration contour through the region where the superfluid velocity is zero.

Here the *superconducting flux quantum* $\Phi_0 = hc/2e$ (which takes into account the Cooper pair charge $2e$).[3] Since the order parameter must be single-valued, the phase can only change by a multiple of 2π for each round trip along the loop. Therefore, the magnetic flux through a hole in a massive superconductor must be quantized in the units of Φ_0.

The supercurrent cannot lose energy and momentum unless the Cooper pairs start breaking up, which requires a finite energy ($|\Delta|$ per electron). Therefore, the supercurrent is non-dissipative, and the states with finite supercurrent are *metastable* states, with exceedingly long relaxation times. Of course, one cannot make the supercurrent too large, otherwise the superconducting state will break down altogether (see, e.g., Tinkham, 2004, § 4.4; Zagoskin, 1998, § 4.4.5). The *critical current density* depends on the material and on the geometry of the sample, and is typically of the order of 10^5 to 10^6 A/cm^2, that is, much higher than what occurs in the situations we will be interested in. In other words, at small current densities, and at temperatures well below the gap ($k_B T \ll |\Delta|$, that is, at dozens of millikelvins), one does not need to worry about decoherence from the thermal excitations in the superconductor: there will be none.[4] Superconductors, in a word, might have been created specifically for the purpose of making quantum-coherent devices, like qubits.

2.1.2 Weak superconductivity

One of the most striking manifestations of the quantum character of superconductivity is so-called *weak superconductivity*, alias Josephson effects and related phenomena in superconducting weak links. A weak link is any disruption in a bulk superconductor, where the supercurrent can still flow, but the order parameter is strongly suppressed. Examples are tunnel junctions, point contacts, narrow bridges, etc., and their properties and fabrication are discussed in detail in, e.g., Barone and Paterno (1982). The maximal supercurrent density in these structures is much less than in massive superconductors, of the order of 10 A/cm^2, which justifies the "weak superconductivity" moniker.

Tunnel junctions, where two bulk superconductors are brought into contact through a thin tunnelling barrier, are the most relevant weak links for our purposes (see Fig. 2.2). Due to quantum tunnelling electrons can penetrate through the

---

[3] You have noticed the factor of c, of course. We will be using the CGS units in all the formulas. A minor disadvantage – the need to convert everything into SI units at the end – is far outweighed by the transparency of the expressions and simplicity of checking the dimensionalities (to say nothing of never thinking of such nonentities as the permittivity and permeability of a vacuum). For example, capacitance and inductance have the dimensionality of length, and it is immediately obvious that the magnetic flux has the same dimensionality as the electric charge (since $e^2/\hbar c \approx 1/137$ is just a number, the fine structure constant).

[4] In the following, unless the opposite is stated, we always assume $k_B T \ll |\Delta|$.

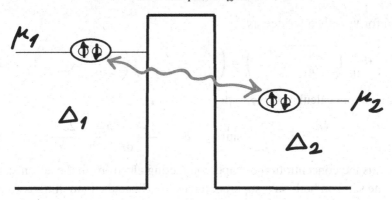

Fig. 2.2. Josephson effects in a tunnelling junction.

barrier. As a result, the wave functions of the condensates overlap and hybridize, leading to the two Josephson effects. Owing to the small transparency of the barrier we can neglect the perturbation of the bulk superconductors due to the contact. For example, even though the electric current may flow through the barrier, its density is so much smaller than the critical current density in the bulk, that the current in the bulk can be neglected. Therefore, each of the bulk superconductors can be character-ized by its own *spatially uniform* order parameter $\Delta_j = \sqrt{n_{sj}/2}\exp(i\phi_j)$; $j = 1, 2$. They can also have different chemical potentials, μ_j.

Following the approach due originally to Feynman (see, e.g., Schmidt, 2002, § 20), let us describe the system by a two-component "wave function":

$$\hat{\Psi} = \begin{pmatrix} \sqrt{n_{s1}/2}\exp(i\phi_1) \\ \sqrt{n_{s2}/2}\exp(i\phi_2) \end{pmatrix}. \tag{2.14}$$

Its evolution is governed by the Hamiltonian

$$H = \begin{pmatrix} eV & K \\ K & -eV \end{pmatrix}. \tag{2.15}$$

Here the off-diagonal terms are responsible for transferring a Cooper pair from one bank of the junction to the other.[5] The diagonal terms account for the potential bias between the banks, $eV = \mu_1 - \mu_2$; in Eq. (2.15) we take into account that the energy difference per Cooper pair is twice this much. This is a version of the so-called *tunnelling Hamiltonian* approximation (e.g. Barone and Paterno, 1982; Zagoskin, 1998), which can be justified by a microscopic theory. The Schrödinger

---

[5] It is a crucial feature of the Josephson effect that the electrons in a Cooper pair tunnel coherently and not independently. Therefore, the superconducting current across the barrier turns out to be proportional to the barrier transparency $D \propto K$ and not $D^2 \ll D \ll 1$, which would make the effect vanishingly small.

equation for $\hat{\Psi}$ is then written as

$$i\hbar\frac{d}{dt}\begin{pmatrix} \sqrt{n_{s1}/2}e^{i\phi_1} \\ \sqrt{n_{s2}/2}e^{i\phi_2} \end{pmatrix} = \begin{pmatrix} eV & K \\ K & -eV \end{pmatrix}\begin{pmatrix} \sqrt{n_{s1}/2}e^{i\phi_1} \\ \sqrt{n_{s2}/2}e^{i\phi_2} \end{pmatrix}. \qquad (2.16)$$

After a simple calculation we find:

$$\frac{dn_{s1}}{dt} = -\frac{dn_{s2}}{dt} = \frac{2Kn_{s0}}{\hbar}\sin(\phi_2 - \phi_1); \quad \frac{d(\phi_1 - \phi_2)}{dt} = \frac{2eV}{\hbar}, \qquad (2.17)$$

where n_{s0} is the concentration of superconducting electrons in the absence of tunnelling (the same in both bulk superconductors). (Since the tunnelling is small, we can neglect the change in it in the right-hand side of the first equation.)

If we recall that $e\,dn_{s1}/dt$ is nothing else but the rate of electric charge flow from the first to the second superconductor across the barrier, the first equation of (2.17) can be rewritten as

$$I_J = I_c\sin(\phi_1 - \phi_2). \qquad (2.18)$$

It describes the *dc Josephson effect*, a non dissipative, equilibrium, coherent flow of electric current – *Josephson current* – through the barrier, which is determined only by the phase difference between the superconductors and the properties of the junction. All of the latter are swept into a single coefficient, the *critical current I_c* of a Josephson junction. The sinusoidal current dependence on the phase difference is not fundamental: it appears as the lowest-order approximation in the barrier transparency. A Josephson current in weak links other than a tunnel junction can have a different functional form, but the essence of the effect does not change. Unless stated otherwise, we will always assume the dependence (2.18).

It is convenient to rewrite the second equation of (2.17) as

$$V = \frac{\hbar}{2e}\cdot\frac{d}{dt}(\phi_2 - \phi_1). \qquad (2.19)$$

This relation describes the *ac Josephson effect*: if somehow voltage is applied between the two superconducting banks, then the superconducting phase difference between them will grow. Conversely, if the phase difference between the banks depends on time, there will be an electric voltage between them.

Both these effects are of seminal importance. Note for the future, that in his theoretical prediction of these effects Josephson recognized that the superconducting order is eventually determined by the nonzero anomalous average, $\langle\psi_\downarrow\psi_\uparrow\rangle$, and not by $\Delta = g\langle\psi_\downarrow\psi_\uparrow\rangle$. After all, you can have $\Delta = 0$ in some part of the system just because the coupling constant g is zero there – such as inside the tunnelling barrier. Nevertheless, the superconducting ordering survives there and is responsible for the Josephson effect.

The stationary Josephson current (2.18) is an equilibrium phenomenon. It can, therefore, be obtained directly by differentiating the appropriate thermodynamic potential of the system:

$$I_J = c \frac{\partial F}{\partial \Phi},$$ (2.20)

where $\Phi = \Phi_0 \phi / 2\pi$ has the dimensionality of the magnetic flux, and $\phi = \phi_1 - \phi_2$. Therefore, we should include in energy of the system the *Josephson energy*,[6]

$$U(\phi) = \frac{1}{c} \int d\Phi I_J = -\frac{I_c \Phi_0}{2\pi c} \cos \phi \equiv -E_J \cos \phi.$$ (2.21)

On the other hand, the ac Josephson effect (2.19) is clearly a non-equilibrium phenomenon: the electric voltage is somehow applied to the Josephson junction. This means that in addition to the supercurrent there is a normal electric current through the junction. This current is carried by the excitations (quasiparticles). How can this happen, if quasiparticles with energies below $|\Delta|$ do not exist in the superconductor? Returning to Eq. (2.18), we see that the magnitude of the superconducting current through the junction is limited. If we try to pass through it any current $I > I_c$, then the excess will have to be carried by the quasiparticles, in a normal, dissipative way. To investigate this situation, we will use a convenient and usually pretty precise *resistively shunted junction (RSJ)* model of a Josephson junction, where parallel to the Josephson junction proper there is a finite resistance R. For all $I < I_c$ it is shorted by the supercurrent. Then for $I > I_c$

$$I = I_c \sin \phi(t) + \frac{V(t)}{R} = I_c \sin \phi(t) + \frac{\hbar}{2eR} \frac{d\phi}{dt},$$ (2.22)

with the implicit solution in the form

$$t(\phi) = \frac{\hbar}{2eRI_c} \int^{\phi} \frac{d\phi}{I/I_c - \sin \phi}$$

$$= \frac{\hbar}{2eRI_c} \frac{2}{\sqrt{(I/I_c)^2 - 1}} \arctan \left[\frac{(I/I_c)\tan(\phi/2) - 1}{\sqrt{(I/I_c)^2 - 1}} \right].$$ (2.23)

The phase difference, and, therefore, the voltage, will be a periodic function of time, with period

$$T_{RSJ} = \frac{\hbar}{2eRI_c} \int_0^{2\pi} \frac{d\phi}{I/I_c - \sin \phi} = \frac{2\pi}{\sqrt{(I/I_c)^2 - 1}} \cdot \frac{\hbar}{2eRI_c} \equiv \frac{2\pi}{\omega_{RSJ}}.$$ (2.24)

---

[6] The cosine dependence is valid for the tunnelling Josephson junction; for a long SNS (Superconductor - Normal metal - Superconductor) junction at zero temperature, with $I_J(\phi)$ given by Eq. (2.165), the Josephson energy $U(\phi)$ will be a periodically extended section of a parabola, etc.

The average Josephson voltage will be accordingly

$$\bar{V}_J = \frac{\hbar \omega_{RSJ}}{2e} = R\sqrt{I^2 - I_c^2}. \tag{2.25}$$

The explicit solution for the voltage can be obtained from (2.23) in the form

$$V(t) = \frac{\hbar}{2e} \frac{d\phi}{dt} = \frac{R(I^2 - I_c^2)}{I + I_c \cos \omega_{RSJ} t}, \tag{2.26}$$

and such oscillations have indeed been observed (see Barone and Paterno, 1982; Schmidt, 2002; Tinkham, 2004). From Eq. (2.19) also follows another interesting and useful property. According to the general definition of inductance, $V = L\dot{I}/c^2$, the Josephson junction can be considered as a *nonlinear* inductance L_J:

$$L_J(\phi) = \frac{\hbar c^2}{2e I_c \cos \phi}. \tag{2.27}$$

The fact that this inductance can be tuned (and even change sign) by fixing a stationary phase difference across the junction is very useful for various applications.

2.1.3 rf SQUID

As an example, consider an rf SQUID (Fig. 2.1b), i.e., a superconducting loop of inductance L interrupted by a single Josephson junction.[7] This is the simplest of SQUIDs (superconducting quantum interference devices). The single-valuedness of the order parameter will require, instead of Eq. (2.13), that

$$\phi + 2\pi \frac{\Phi_{tot}}{\Phi_0} = \phi + 2\pi \frac{\widetilde{\Phi} - L I_J(\phi)/c}{\Phi_0} = 2\pi n. \tag{2.28}$$

Here ϕ is the phase drop across the Josephson junction, $\widetilde{\Phi}$ is the external magnetic flux through the loop, and L is the (magnetic) self-inductance of the loop. The response of the current in the loop to small changes in $\widetilde{\Phi}$, and, therefore, the effective inductance of the SQUID, depends on the value of $\widetilde{\Phi}$.

Note that, unlike a thick, solid superconducting ring (or a hole in a bulk superconductor), the rf SQUID allows an arbitrary flux through it: the superconducting phase

---

[7] Generally, it is necessary to distinguish between the magnetic, kinetic and full inductance of a superconductor. The full inductance accounts for the kinetic energy of the electric current I via $E_{kin} = L_{tot} I^2/2c^2 = (L_m + L_k) I^2/2c^2$, which includes the energy of the magnetic field created by the current and the kinetic energy of electrons. In normal conductors the second term (*kinetic inductance*) is not negligible only in the above-THz region, while in superconductors it can be significant already at low frequencies (see, e.g., Schmidt, 2002, §10). Unless stated otherwise, we will understand inductance to mean the magnetic (geometric) inductance.

can drop across the Josephson junction and makes the left-hand side of Eq. (2.28) a multiple of 2π. But this is precisely where the advantage of this device lies. A small change in the external flux (a fraction of $\Phi_0 \approx 2.068 \times 10^{-15}$ Wb in not-so-convenient SI units) will significantly shift the Josephson phase, and with it the effective inductance of the loop (2.27). By coupling the rf SQUID to a resonance circuit, one can precisely measure this shift by the change in the resonant frequency, which makes the rf SQUID a sensitive detector of weak magnetic fields (Tinkham, 2004, § 6.5.3). We will see that this method works very well for measuring the quantum states of qubits.

2.1.4 dc SQUID

A slightly more complex device is a dc SQUID, a loop with two Josephson junctions in parallel inserted in a (superconducting) wire (Fig. 2.1c). Now the quantization condition is

$$\phi_1 + \phi_2 + 2\pi \frac{\Phi_{\text{tot}}}{\Phi_0} = \phi_1 + \phi_2 + 2\pi \frac{\widetilde{\Phi} - L I_{\text{loop}}(\phi)/c}{\Phi_0} = 2\pi n, \qquad (2.29)$$

where $I_{\text{loop}} < \min(I_{c1}, I_{c2})$ is the current around the loop. Taking this into account, the superconducting current through the loop is

$$I = I_{c1} \sin \phi_1 - I_{c2} \sin \phi_2$$

$$= (I_{c1} + I_{c2}) \cos \left[\pi \frac{\Phi_{\text{tot}}}{\Phi_0} \right] \sin \frac{\phi_1 - \phi_2}{2} - (I_{c1} - I_{c2}) \sin \left[\pi \frac{\Phi_{\text{tot}}}{\Phi_0} \right] \cos \frac{\phi_1 - \phi_2}{2}.$$

$$(2.30)$$

If the SQUID is symmetric, $I_{c1} = I_{c2}$, and the self-inductance flux can be neglected compared to the external flux $\widetilde{\Phi}$, this equation simplifies to

$$I(\delta\phi, \widetilde{\Phi}) = I_{c,\text{eff}}(\widetilde{\Phi}) \sin \delta\phi, \qquad (2.31)$$

where the effective critical current

$$I_{c,\text{eff}}(\widetilde{\Phi}) = 2I_c \cos \left[\pi \frac{\widetilde{\Phi}}{\Phi_0} \right] \qquad (2.32)$$

can be tuned by the external flux to zero. In measurement applications, this property is used to detect very weak magnetic fields (see, e.g., Schmidt, 2002, § 23). To us, the primary importance of this device is as a *tunable Josephson junction*.

2.1.5 Current-biased Josephson junction

Let us look again at a current-biased Josephson junction. For $I < I_c$ all the current $I = I_c \sin\phi$ is the equilibrium superconducting current, and the total energy of the system should be at a minimum. Therefore, the total energy of the system is (cf. (2.21))

$$U(\phi; I) = -E_J \cos\phi - I\Phi \equiv -\frac{I_c \Phi_0}{2\pi c} \cos\phi - \frac{I\Phi_0}{2\pi c}\phi. \qquad (2.33)$$

If we consider the phase difference ϕ (or the equivalent flux variable Φ) as a position, this is the potential energy inasmuch as it depends only on ϕ.

Suppose now that the phase difference across the junction does change in time, but sufficiently slowly, so that

$$|eV| = \frac{\hbar}{2}|\dot{\phi}| < |\Delta| \qquad (2.34)$$

(it is good to remember that $I_c \propto |\Delta|$ (Tinkham, 2004, § 6.2.2; Barone and Paterno, 1982, § 2.5).[8] Then the current through the junction, even if no longer in equilibrium, remains quantum coherent, because the voltage produced is not sufficient to break Cooper pairs (in other words, there is not enough energy to create quasiparticles above the superconducting gap). But the finite voltage will contribute to the energy of the system the term

$$K(\dot{\phi}) = \frac{CV^2}{2} = \frac{C\dot{\phi}^2}{2}\left(\frac{\hbar}{2e}\right)^2, \qquad (2.35)$$

which looks like the kinetic energy of a classical particle. Together, (2.33) and (2.35) give the energy of a particle in a *washboard potential* (Fig.2.3),

$$E(\phi, \dot{\phi}; I) = \frac{C\dot{\phi}^2}{2}\left(\frac{\hbar}{2e}\right)^2 + U(\phi; I), \qquad (2.36)$$

with the "mass" proportional to the electrical capacitance C of the junction. (Any weak link will have some capacitance; in the case of a tunnelling junction it is straightforward, since the junction is essentially a capacitor, with superconducting electrodes and a layer of dielectric in between.)

---

[8] For a symmetric tunnel junction with the normal (i.e., when not in the superconducting state) electric resistance R_N the *Ambegaokar–Baratoff formula* (Tinkham, 2004, § 6.2.2) yields

$$I_c R_N = (\pi|\Delta(T)/2e|)\tanh(|\Delta(T)|/2k_B T).$$

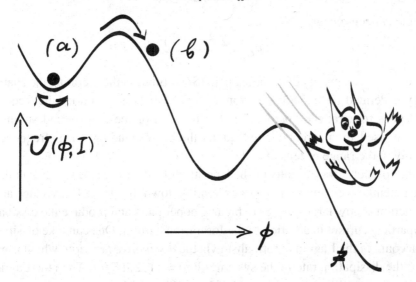

Fig. 2.3. Current-biased Josephson junction: classical picture. Depending on the capacitance and the initial conditions the phase difference $\phi(t)$ can perform small oscillations around a local minimum of the Josephson potential $U_J(\phi, I) = -E_J \cos \phi - [(I\Phi_0)/(2\pi c)]\phi$ (a) or grow indefinitely (b) (resistive state). If the bias current I exceeds the critical current, the minima disappear, and only case (b) is possible.

It is easy to write as well the Lagrange function for the system,

$$\mathcal{L}(\phi, \dot{\phi}; I) = K - U = \frac{C\dot{\phi}^2}{2}\left(\frac{\hbar}{2e}\right)^2 + \frac{I_c\Phi_0}{2\pi c}\cos\phi + \frac{I\Phi_0}{2\pi c}\phi, \qquad (2.37)$$

from which the equations of motion can be directly obtained (see, e.g., Landau and Lifshitz, 1976, § 2):

$$\frac{d}{dt}\frac{\partial}{\partial\dot{\phi}}\mathcal{L} - \frac{\partial}{\partial\phi}\mathcal{L} = \left(\frac{\hbar}{2e}\right)^2 C\ddot{\phi} + \frac{I_c\Phi_0}{2\pi c}\sin\phi - \frac{I\Phi_0}{2\pi c} = 0. \qquad (2.38)$$

It is clear from Eqs (2.36), (2.38) that for small biases, $I \ll I_c$, and sufficiently small initial values of $\dot{\phi}$, the phase will perform small oscillations around equilibrium,[9] with the *Josephson plasma frequency*

$$\omega \approx \omega_0 = \sqrt{\frac{E_J}{C(\hbar/2e)^2}} \equiv \sqrt{\frac{2E_J E_C}{\hbar^2}}. \qquad (2.39)$$

---

[9] Which will *eventually* die out due to the relaxation processes other than Cooper pair breaking: e.g., radiative losses.

Here the *charging energy*

$$E_C = \frac{2e^2}{C} \equiv \frac{(2e)^2}{2C} \tag{2.40}$$

is the electrostatic energy of a Cooper pair. (Sometimes in the literature the charging energy is defined for a single electron, $e^2/2C$.) For large junctions the frequency (2.39) goes to zero: the "particle" is too heavy. For the micron and submicron junctions used in superconducting qubits the corresponding linear frequency is typically in the range of tens of GHz.

If the "particle" was initially pushed too strongly, it will be able to run over the lip of the potential well and will keep accelerating down the slope. The voltage across the junction is now large enough to break Cooper pairs and produce the dissipative quasiparticle current in addition to the Josephson current. One can take dissipation into account in the Lagrange formalism via the *dissipative function*, which is equal to half the dissipation rate in the system: $\mathcal{Q}(\dot{\phi}) = (1/2)\mathrm{d}E/\mathrm{d}t$. Then the Lagrange equations are modified (Landau and Lifshitz, 1976, § 25):

$$\frac{\mathrm{d}}{\mathrm{d}t}\frac{\partial}{\partial\dot{\phi}}\mathcal{L} - \frac{\partial}{\partial\phi}\mathcal{L} = -\frac{\partial}{\partial\dot{\phi}}\mathcal{Q}. \tag{2.41}$$

In the RSJ model $\mathcal{Q} = (1/2)(V^2/R) = (1/2R)(\hbar\dot{\phi}/2e)^2$, which leads to the equation of motion

$$\left(\frac{\hbar}{2e}\right)^2 C\ddot{\phi} + \frac{I_c\Phi_0}{2\pi c}\sin\phi - \frac{I\Phi_0}{2\pi c} = -\frac{\hbar}{2eR}\frac{\Phi_0}{2\pi c}\dot{\phi}. \tag{2.42}$$

The new term is analogous to viscous friction and will eventually stabilize the average "velocity" $\dot{\phi}$ and, therefore, the voltage. For a large junction, $E_C \ll E_J$, the second derivative of the phase, must be negligibly small (otherwise the first term in (2.42) will dominate and cannot be compensated by the rest), and we return to Eq. (2.22). This equation describes the *resistive state* of a Josephson junction.

Note that even though the Josephson phase difference is determined modulo 2π, in the presence of current bias the states of the system with $\phi = \phi_1$ and $\phi = \phi_1 + 2\pi$ are different and can be, in principle, distinguished. For example, energy was dissipated (which can be measured) that was supplied by the current source (the change of state of which also can be observed).

The usual way of producing a resistive state is to increase the bias current. Then the local minima in the washboard become shallower, and disappear as it achieves the critical value, forcing the phase to run down the slope. At finite temperatures, the thermal fluctuations can catapult the "particle" across the lip of the well for $I = I_{r1} < I_c$. The current-voltage characteristics of the junction will then show an instant increase of voltage from zero to a finite average value (cf. (2.25)). Retrapping the "particle" would usually require a drop in the bias to $I_2 < I_1$, to account for the extra kinetic energy of the "particle", proportional to $\bar{V}_J^2$, resulting in a hysteretic loop (Tinkham, 2004; Barone and Paterno, 1982). Since the resistive state is essentially dissipative, it is to be avoided as long as we want to maintain quantum coherence in the system.

2.2 Quantum effects in Josephson junctions. Phase and flux qubits

2.2.1 Number and phase as quantum observables

So far we have considered the parameters of a superconductor, Δ (or n_s) and ϕ, as classical variables. This was justified given the presumably infinitely large size of the system, so that the BCS ground state (2.5) with its coherent superposition of states with all possible numbers of Cooper pairs looked quite plausible. Nevertheless, once we start dealing with an essentially finite piece of a superconductor (like an rf SQUID), and especially when we start reducing its size, such an assumption no longer looks natural.

Using the method due to Anderson (see Tinkham, 2004, § 3.3), one can project out of the state |BCS⟩ a state with a fixed number N of Cooper pairs, but indefinite superconducting phase. To do so, let us integrate (2.5) over the phase:

$$|N\rangle = \int_0^\phi d\phi\, e^{-iN\phi}|BCS\rangle = \int_0^\phi d\phi\, e^{-iN\phi} \prod_{\mathbf{k}} \left(|u_\mathbf{k}| + |v_\mathbf{k}|e^{i\phi}a_{\uparrow\mathbf{k}}^\dagger a_{\downarrow,-\mathbf{k}}^\dagger \right)|0\rangle.$$

$$(2.43)$$

Expansion of the product under the integral yields terms of the form

$$e^{iM\phi}a_{\uparrow\mathbf{k}_1}^\dagger a_{\downarrow,-\mathbf{k}_1}^\dagger \ldots a_{\uparrow\mathbf{k}_M}^\dagger a_{\downarrow,-\mathbf{k}_M}^\dagger |0\rangle,$$

which contain $M = 0, 1, \ldots$ Cooper pairs. Only the terms with $M = N$ will survive the integration over the phase, and the state $|N\rangle$ of (2.43) indeed contains a fixed number $(2N)$ of electrons. The price to pay is that the superconducting phase is now totally indefinite. Since the superconducting current is determined by the phase *gradient*, this does not really matter for a solid superconductor: a uniform shift in the phase will not affect the observable effects.

The situation is more delicate for a Josephson junction: there is no reason to expect that two weakly connected condensates will fluctuate in step, and any difference between their phases would immediately produce an observable Josephson current. But, of course, the number of electrons is no longer fixed: they can tunnel back and forth and save the situation by introducing an uncertainty in N. Nevertheless, there are important cases when, in the presence of Josephson tunnelling, the number of particles still cannot fluctuate freely, so that neither Eq. (2.5) nor (2.43) applies. In order to deal with the situation it is necessary to recognize that the number of Cooper pairs and the superconducting phase should be treated as quantum-mechanical, rather than classical, observables.

We will do this using an example of a harmonic oscillator. Its Hamiltonian can be written as

$$H = \hbar\omega_0 \left(b^\dagger b + \frac{1}{2} \right) \equiv \hbar\omega_0 \left(\hat{N} + \frac{1}{2} \right). \tag{2.44}$$

Here $b^\dagger$, b are the creation/annihilation Bose operators, and $\hat{N} = b^\dagger b$ is the particle number operator. When the oscillator is in its eigenstate (a Fock state), $\hat{N}|n\rangle = n|n\rangle$, the energy of the oscillator is exactly known as well, $E = (n+1/2)\hbar\omega_0$. However, the phase of the oscillation is indefinite: the average value of both position, $\hat{x} \propto (b + b^\dagger)$ and momentum, $\hat{p} \propto (b - b^\dagger)/i$, operators in such a state are zero, since $\langle n|b|n\rangle = \langle n|b^\dagger|n\rangle = 0$, which simply means that both take positive and negative values with equal probability. Is it possible to introduce a quantum mechanical observable, that is, a Hermitian phase operator, $\hat{\phi}$, which would be conjugate to $\mathcal{N}$, and have eigenstates with definite phase?

Since the number operator is – up to a constant shift and a factor of $\hbar\omega_0$ – proportional to the Hamiltonian (2.44), and we know from basic quantum mechanics that there is no conjugate operator to a Hamiltonian (time is not an operator!), the answer should be negative. Nevertheless, let us look at the situation in some detail. Suppose such a $\hat{\phi}$ exists. Then it would satisfy the commutation relation

$$[\hat{N}, \hat{\phi}] = -i \tag{2.45}$$

(the commutator is $-i$ and not $-i\hbar$, because $\hat{N}$ is already the operator of the number of energy quanta $\hbar\omega_0$ in the oscillator). Then the Heisenberg equation of motion for this operator is

$$i\hbar\dot{\hat{\phi}} = [\hat{\phi}, H] = i\hbar\omega_0, \tag{2.46}$$

and $\hat{\phi} = \omega_0 t$, which is, indeed, the phase of the oscillator.

There are two problems here. First, it looks as if time – a real number – somehow becomes an operator, $t = \hat{\phi}/\omega_0$. Second, from (2.45) the uncertainty relation would follow in a standard way

$$\Delta N \Delta \phi \geq \frac{1}{2}. \tag{2.47}$$

While for $\Delta N = \infty$ this indeed yields an exactly determined phase, similar to what we had in the BCS ground state (2.5); in order to obtain a state with a fixed number of bosons, according to (2.47) one needs $\Delta\phi = \infty$. This is, firstly, impossible, since the phase is only defined on an interval of length 2π; and, secondly, unnecessary, since we did obtain from the BCS state a state with a fixed number of Cooper pairs, Eq. (2.43), by integrating over the phase on the finite interval $[0, 2\pi]$.

Carruthers and Nieto (1968) gave a detailed analysis of the situation. Using the hypothetical operator $\hat{\phi}$ we could write the Bose operators in (2.44) as

$$b = e^{-i\hat{\phi}}\sqrt{\hat{N}}; \quad b^\dagger = \sqrt{\hat{N}}e^{i\hat{\phi}}. \qquad (2.48)$$

Obviously, $b^\dagger b = \hat{N}$, as it should, while $bb^\dagger = e^{-i\hat{\phi}}\hat{N}e^{i\hat{\phi}}$. Expanding the exponents in series and commuting $\hat{N}$ with each term using the relation (2.45), we find that

$$\left[e^{-i\hat{\phi}}, \hat{N}\right] = e^{-i\hat{\phi}}, \qquad (2.49)$$

and therefore $bb^\dagger = \hat{N} + 1$. Everything seems right, until we look at the operator $U = e^{-i\hat{\phi}}$. Since $\hat{\phi}$ is presumably Hermitian, then U must be unitary, $UU^\dagger = U^\dagger U = \hat{1}$. It then follows from Eq. (2.49) that $U^\dagger U\hat{N} - U^\dagger\hat{N}U = U^\dagger U = \hat{1}$, i.e.,

$$U^\dagger \hat{N} U = \hat{N} - \hat{1}.$$

Therefore, while the eigenvalues of $\hat{N}$ are $\{0, 1, 2, 3 \ldots\}$, those of $U^\dagger\hat{N}U$ are $\{-1, 0, 1, 2, 3 \ldots\}$. This is, of course, impossible: a unitary transformation does not change the set of eigenvalues of an observable, to say nothing of the appearance of a negative occupation number. The problem would not arise if the spectrum of eigenvalues of $\hat{N}$ extended to negative infinity, since then a shift by one would not change anything. On the bright side, when the average number of quanta $\langle N \rangle \gg 1$ we could pretend that the number of quanta can go both to plus and minus infinity, and hopefully use the "phase operator", satisfying Eq. (2.45), as a convenient approximation.[10] This is indeed the case. For large N, in the Fock state $|N\rangle$ the probability distribution of phase is almost uniform in the interval $[0, 2\pi]$. Also, as long as the average number of quanta is large and $\Delta\phi \ll 2\pi$, the uncertainty relation (2.47) approximately holds.

From (2.45), the action of the phase and number operators on the eigenfunction of the phase should be

$$\hat{\phi}|\phi\rangle = \phi|\phi\rangle; \quad \hat{N}|\phi\rangle = \frac{1}{i}\frac{\partial}{\partial\phi}|\phi\rangle, \qquad (2.50)$$

while

$$\hat{\phi}|N\rangle = -\frac{1}{i}\frac{\partial}{\partial N}|N\rangle \quad \hat{N}|N\rangle = N|N\rangle, \qquad (2.51)$$

and

$$|\phi\rangle = \sum_N e^{iN\phi}|N\rangle; \quad |N\rangle = \int_0^{2\pi} \frac{d\phi}{2\pi}e^{-iN\phi}|\phi\rangle. \qquad (2.52)$$

---

[10] Rigorously, one can introduce, in place of the non existing phase operator $\hat{\phi}$, two Hermitian operators, $\widehat{\cos\phi}$ and $\widehat{\sin\phi}$ (Carruthers and Nieto, 1968).

The latter relation is essentially the same as (2.43).

There remains a question of how to describe a state intermediate between $|N\rangle$ and $|\phi\rangle$. We defer this to §4.4.4 and proceed to the description of quantum behaviour of phase in a Josephson junction.

2.2.2 Phase qubit: Current-biased Josephson junction in quantum limit

Now recall that the number of Cooper pairs in the condensate is very large and therefore Eq. (2.45) is almost exact. Therefore, we can treat our system described by the Lagrangian (2.37) as any other one-dimensional quantum system we would like to quantize (Landau and Lifshitz, 2003; Messiah, 2003). First we obtain the canonical momentum,

$$\Theta = \frac{\partial}{\partial \dot{\phi}} \mathcal{L} = \left(\frac{\hbar}{2e}\right)^2 C \dot{\phi}. \tag{2.53}$$

Since $C(\hbar/2e)\dot{\phi} = CV = Q$ (the extra electric charge on the junction), we see that, quite correctly, $\Theta = \hbar N$ (the number of extra Cooper pairs corresponding to this charge times $\hbar$, to maintain the correct dimensionality – action – of the product $\Theta\phi$). The Hamilton function then is

$$\mathcal{H}(\Theta, \phi) = \Theta\dot{\phi} - \mathcal{L} = \frac{1}{2C}\left(\frac{2e\Theta}{\hbar}\right)^2 - E_J \cos\phi - \frac{I\Phi_0}{2\pi c}\phi$$

$$\equiv E_C \left(\frac{\Theta}{\hbar}\right)^2 - E_J \cos\phi - \frac{I\Phi_0}{2\pi c}\phi. \tag{2.54}$$

Now it remains to replace Θ with the operator $(\hbar/\mathrm{i})\partial_\phi = \hbar\hat{N}$ to obtain the Hamiltonian (in the "position" representation)[11]

$$H = -E_C \frac{\partial^2}{\partial\phi^2} - E_J \cos\phi - \frac{I\Phi_0}{2\pi c}\phi \equiv -E_C \frac{\partial^2}{\partial\phi^2} - E_J \left(\cos\phi + \frac{I}{I_c}\phi\right). \tag{2.55}$$

From Eq. (2.55) directly follows the Josephson plasma frequency (2.39) for small oscillations around the equilibrium phase $\phi_{\min} = \arcsin(I/I_c)$. The Heisenberg

---

[11] We could have arrived at Eq. (2.55) directly by writing the potential energy of the system, (2.33), plus the charging term $Q^2/2C = (2eN)^2/2C$, and replacing in the latter the number of Cooper pairs N with the operator $\hat{N}$. We did it the way we did because the Lagrangian-based approach is better for analysing more complex circuits.

equations of motion for the phase and number operators give

$$i\hbar\dot{\phi} = [\phi, H] = -E_C\left[\phi, \frac{\partial^2}{\partial\phi^2}\right] = 2E_C\frac{\partial}{\partial\phi} = 2E_C \times i\hat{N}; \tag{2.56}$$

$$i\hbar\dot{\hat{N}} = [\hat{N}, H] = -E_J\left[\frac{1}{i}\frac{\partial}{\partial\phi}, \cos\phi + \frac{I}{I_c}\phi\right] = \frac{E_J}{i}\left(\sin\phi - \frac{I}{I_c}\right), \tag{2.57}$$

and, therefore (introducing $\eta = \phi - \phi_{min}$),

$$\ddot{\eta} \approx -\frac{2E_C E_J \cos\phi_{min}}{\hbar^2}\eta = -\frac{2E_C E_J}{\hbar^2}\left[1 - (I/I_c)^2\right]^{1/2}\eta = -\left[1 - (I/I_c)^2\right]^{1/2}\omega_0^2\eta. \tag{2.58}$$

We see that a finite bias current softens the potential, reducing the plasma frequency to zero as $I \to I_c$.

In the *phase regime* ($E_J \gg E_C$) the Josephson term dominates the Hamiltonian (2.55). Then in the absence of bias ($I = 0$) each potential well ($\phi = 2\pi q$, $q = 0, \pm1, \ldots$) contains many almost equidistant quantized levels, and the tunnelling between the minima with different q's near the bottom is negligible. In order to realize quantum behaviour, the potential must be strongly tilted, with the bias current typically in the range $0.95I_c$ to $0.98I_c$. Then there remain only a few quantized levels in each potential well. Strictly speaking, they are only metastable: there is a finite probability for each of them to tunnel out, into the states of continuous spectrum, thereby switching the junction into the resistive state. Nevertheless, in the phase limit this tunnelling probability is still negligible for at least the lowest two states, which are used as $|0\rangle$ and $|1\rangle$ qubit states. At such biases, the potential is strongly anharmonic, which means that the transition energy $\hbar\omega_{01}$ is sufficiently different from, e.g., $\hbar\omega_{12}$ to make their evolution largely independent of whatever happens to higher levels (if they are still present in the well). This is why such biased junctions can be employed as *phase qubits* (Martinis et al., 2002, 2003).

Let us concentrate on a single well around a local minimum of the Josephson potential; see Fig. 2.4, where $\frac{d}{d\phi}\left(\cos\phi + \frac{I}{I_c}\phi\right) = 0$, and

$$\phi_{min} = \arcsin\frac{I}{I_c}, \tag{2.59}$$

the other solution corresponding to the lip of the well. Then, for $\phi = \phi_{min} + \eta$,

$$U_J(\eta) = -E_J\left(\left(1 - (I/I_c)^2\right)^{1/2}\cos\eta + \frac{I}{I_c}(\eta - \sin\eta) + \frac{I}{I_c}\arcsin\frac{I}{I_c}\right). \tag{2.60}$$

Expanding this exact expression to the first anharmonic order in η and dropping the currently irrelevant η-independent terms, and substituting this into (2.55), we

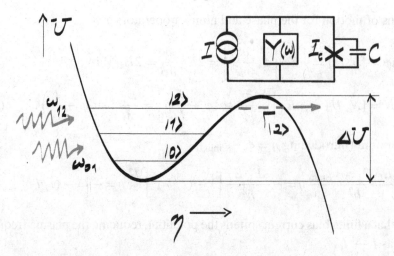

Fig. 2.4. Phase qubit. The Josephson potential near a local minimum, $\phi = \phi_{min} + \eta$, is approximated by the cubic parabola. The two lowest states are used as the qubit states. The transitions between the states are realized by applying an rf pulse with resonance frequency ω_{01} and inducing Rabi oscillations. The application of an rf pulse with $\omega_{12} \neq \omega_{01}$ induces the transition between the excited qubit state $|1\rangle$ and the readout state $|2\rangle$, which quickly tunnels into the continuum. In the inset, $Y(\omega)$ is the admittance of the control and readout circuit.

find the Hamiltonian of an oscillator with cubic anharmonicity,

$$H \approx -E_C \frac{\partial^2}{\partial \eta^2} + E_J \left(\left(1 - (I/I_c)^2 \right)^{1/2} \frac{\eta^2}{2} - \frac{I}{I_c} \frac{\eta^3}{6} \right). \tag{2.61}$$

Neglecting the anharmonicity, we find for the frequency of small oscillations, in agreement with (2.58), the "softened" plasma frequency

$$\omega_I = \omega_0 \left[1 - \left(\frac{I}{I_c} \right)^2 \right]^{1/4} \approx \omega_0 \left[2 \left(1 - \frac{I}{I_c} \right) \right]^{1/4} \tag{2.62}$$

(recalling that $I/I_c \approx 1$). It's worth noting that as $I/I_c = \sin \phi$, this frequency is exactly the frequency of an LC circuit, $\omega_I^2 = \omega_0^2 |\cos \phi| = (2eI_c|\cos \phi|/\hbar)/C = c^2/|L_J(\phi)|C$, where the Josephson inductance $L_J(\phi)$ is given by Eq. (2.27). Note also that the product $CL_J \propto C/I_c$ does not depend on the area of the junction and is therefore essentially determined by the material of the junction (the dependence on the bias current is very weak, as follows from Eq. (2.62)). The height of the potential barrier, ΔU, is in the same approximation given by

$$\Delta U = \frac{2}{3} E_J \frac{\left[1 - (I/I_c)^2 \right]^{3/2}}{(I/I_c)^2} \approx \frac{2^{5/2}}{3} E_J \left[1 - I/I_c \right]^{3/2}. \tag{2.63}$$

The phase qubits are commonly based on Nb junctions, with $\omega_{01}/2\pi \sim 5 - 10\,\text{GHz}$ with a decoherence rate of order 5 MHz at a few dozen millikelvin (Devoret et al., 2004; Devoret and Martinis, 2004; Wendin and Shumeiko, 2005).

Due to the anharmonicity of the potential the interlevel distances ($\approx \hbar\omega_I$) acquire corrections of order $\hbar\omega_I/\Delta U$ (Landau and Lifshitz, 2003, § 38):

$$\hbar\omega_{01} \approx \hbar\omega_I \left(1 - \frac{5}{36}\frac{\hbar\omega_I}{\Delta U}\right);$$

$$\hbar\omega_{12} \approx \hbar\omega_I \left(1 - \frac{5}{18}\frac{\hbar\omega_I}{\Delta U}\right). \tag{2.64}$$

The difference between these frequencies is large enough to ensure that the two lowest levels can be approximately treated as independent of the higher energy states, when excited by a resonant field with the frequency ω_{01}. In the (typical) conditions of the experiment (Martinis et al., 2002), the ratio (roughly the number of quantized levels fitting in the well) $\Delta U/\hbar\omega_I \approx 4$, and the anharmonicity $\Delta\nu \equiv (\omega_{01} - \omega_{12})/2\pi \approx 0.034\omega_{01}/2\pi \approx 0.3$ GHz. Moreover, the tunnelling rates out of these states (Martinis et al., 2002),

$$\Gamma_0 \approx 52\frac{\omega_I}{2\pi}\sqrt{\frac{\Delta U}{\hbar\omega_I}}e^{-7.2\frac{\Delta U}{\hbar\omega_I}}, \tag{2.65}$$

and $\Gamma_{n+1} \approx 1000\,\Gamma_n$, are such that while the states $|0\rangle$ and $|1\rangle$ can be considered as stable, the state $|2\rangle$ is already fast decaying into the continuum, bringing the Josephson junction into the resistive state. This, as we have noted, provides a way of measuring the state of the phase qubit.

The manipulation of the qubit is realized through the variation of the bias current,

$$I(t) = I_{dc} + \Delta I(t) \equiv I_{dc} + I_{lf}(t) + I_{rf,c}(t)\cos\omega_{01}t + I_{rf,s}(t)\sin\omega_{01}t. \tag{2.66}$$

In order for the state of the system to remain in the subspace $\{|0\rangle, |1\rangle\}$, the changes in the amplitudes $I_{lf}(t)$, $I_{rf,c}(t)$, $I_{rf,s}(t)$ must be slow compared to the anharmonicity, $\Delta\nu$. Then the relevant part of the Hamiltonian (2.55) can be written in matrix form as

$$H = \begin{pmatrix} 0 & 0 \\ 0 & \hbar\omega_{01} \end{pmatrix} + E_J\frac{\Delta I(t)}{I_c}\begin{pmatrix} \langle 0|\phi|0\rangle & \langle 0|\phi|1\rangle \\ \langle 1|\phi|0\rangle & \langle 1|\phi|1\rangle \end{pmatrix} \equiv H_0 + H_1(t). \tag{2.67}$$

After compensating for the trivial evolution with this frequency by going to the interaction representation with H_0 of Eq. (2.55) and applying the rotating wave approximation, we reduce Eq. (2.67) to (Martinis et al., 2002, 2003)

$$H = \sqrt{\frac{\hbar}{2C\omega_{01}}}\left(\frac{I_{rf,c}(t)}{2}\sigma_x + \frac{I_{rf,s}(t)}{2}\sigma_y\right) + \frac{\partial\hbar\omega_{01}}{\partial I_{dc}}\frac{I_{lf}(t)}{2}\sigma_z. \tag{2.68}$$

This is a qubit Hamiltonian, of the same kind as Eq. (1.85), and all the operations described in § 1.4 are therefore possible with the phase qubit. The evolution of the Bloch vector of the qubit is given by Eq. (1.88),

$$\frac{d}{dt}\mathcal{R}(t) = 2\mathcal{H}(t) \times \mathcal{R}(t), \tag{2.69}$$

with

$$\mathcal{H}(t) = -\left(\sqrt{\frac{\hbar}{2C\omega_{01}}} \frac{I_{\text{rf},c}(t)}{2}, \sqrt{\frac{\hbar}{2C\omega_{01}}} \frac{I_{\text{rf},c}(t)}{2}, \frac{\partial \hbar \omega_{01}}{\partial I_{\text{dc}}} \frac{I_{\text{lf}}(t)}{2} \right)^{\text{T}}. \tag{2.70}$$

For example, applying a rectangular control pulse of duration Δt, we make the Bloch vector in the rotating frame describe an arc around the direction of the vector $\mathcal{H}$.[12]

Physically, the transitions between the ground state and excited state of the qubit (σ_x- and σ_y-terms in (2.68)) are due to the Rabi oscillations induced by the resonant perturbation with frequency ω_{01}, Eq. (2.66). In the absence of a time-dependent control current $\Delta I(t)$ the rotating frame Hamiltonian (2.68) becomes zero, which simply means that the Bloch vector of the system in the rotating frame remains stationary, and in the laboratory frame it trivially rotates around the z-axis with frequency ω_{01}, due to the unperturbed part of the Hamiltonian (2.67).

A direct way of confirming the quantum behaviour of a qubit is by observing its Rabi oscillations. As we know, in order to do this it is necessary to initialize the qubit in its ground state, apply a resonant field at a frequency ω_{01} for a duration τ, then read out the state of the qubit, and repeat these measurements many times to determine the probability of finding it in the excited state, $p_1(\tau)$. This probability should then oscillate as a function of $\Omega\tau$, with the Rabi frequency Ω in resonance proportional to the amplitude of the signal, i.e., to the square root of its power (see § 1.4).

There are two ways of reading out the phase qubit, both based on the fact that an escape from a local minimum of the Josephson potential is accompanied by a rise of the finite voltage across the junction, Fig. 2.3. First, one can apply at the moment τ an *rf readout pulse* at the frequency ω_{12}, of amplitude and duration calculated to realize the transition $|1\rangle \to |2\rangle$. Then, since the tunnelling escape rate out of the state $|2\rangle$, $\Gamma_{|2\rangle}$ is significant, if within $\sim 1/\Gamma_{|2\rangle}$ from the end of the readout pulse there appears a voltage pulse, then the system was in the state $|1\rangle$ before the readout. Second, one can instead apply a *dc readout pulse*, to make the well shallower and, thus, increase the tunnelling rate out of the state $|1\rangle$. Now the appearance of a finite

---

[12] In order to prevent leakage from the two-dimensional subspace the pulse fronts must actually be smooth on the scale $1/\Delta\nu$.

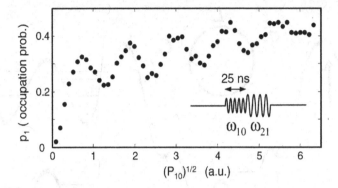

Fig. 2.5. Rabi oscillations in a phase qubit. The probability p_1 of finding the qubit in the excited state oscillates with the phase $\Omega\tau \propto P_{10}^{1/2}\tau$, where $\tau = 25$ ns and P_{10} (in pW range) are the duration and the power of the control pulse (with frequency $\omega_{01}/2\pi = 6.9$ GHz). The frequency of the readout pulse is $\omega_{12}/2\pi = 6.28$ GHz. The junction parameters are $I_c \approx 21$ μA, $C \approx 6$ pF; the average voltage in the resistive state ≈ 1 mV. (Reprinted with permission from Martinis et al., 2002, © 2002 American Physical Society.)

voltage within $\sim 1/\Gamma_{|1\rangle}$ signifies that at the moment τ the system was in the excited state. After repeating the cycle many times, the probability $p_1(\tau)$ can be directly established.

An experimental demonstration of the coherent manipulation of a phase qubit is shown in Fig. 2.5. Martinis et al. (2002) varied the power of the control signal while fixing $\tau = 25$ ns, and applied the first readout method. This allowed observation of the Rabi oscillations in spite of the relatively short decoherence time of their early device, ~ 10 ns. The decoherence time in phase qubits is currently more than ~ 80 ns (Steffen et al., 2006), which makes possible a number of complex manipulations with their states. Some of these results will be discussed later in the context of scaling up qubit-based structures.

2.2.3 *rf SQUID flux qubit*

Instead of using an external current source, one can bias a Josephson junction by the supercurrent induced in the closed loop by the magnetic flux through it, that is, using an rf SQUID configuration as in Fig. 2.1b. The Josephson potential is, in this case (Fig. 2.6)

$$U(\phi) = \frac{\Phi_{\text{tot}}^2}{2L} - E_J\cos\phi = \frac{\left(\Phi_0\phi/2\pi - \tilde{\Phi}\right)^2}{2L} - E_J\cos\phi, \qquad (2.71)$$

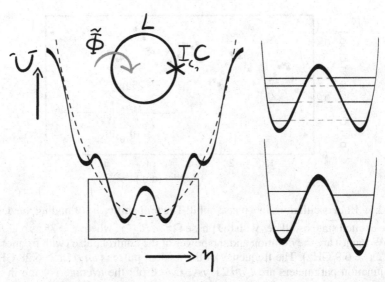

Fig. 2.6. rf SQUID qubit. The height of the potential barrier between the minima is proportional to the critical current of the Josephson junction, I_c, and can be tuned if the junction is replaced by a dc SQUID.

where $\widetilde{\Phi}$ is the external magnetic flux through the SQUID loop. Indeed, the minimum of this expression yields

$$\phi = 2\pi \frac{\widetilde{\Phi} - \frac{LI_c}{c}\sin\phi}{\Phi_0} \tag{2.72}$$

(plus an integer multiple of 2π), that is, the flux quantization condition (2.28); the second term in the numerator is the flux produced by the screening superconducting current in the loop.

The phase qubit states are, in this case, situated in one of the pockets on the sides of the large parabolic well. This method has serious advantages (for example, the resistive run of the system and the amount of decoherence it produces are restricted by the bottom of the potential well), and its refinements are now more or less standard. But the rf SQUID configuration also provides an opportunity to realize an altogether different type of qubit, a *flux qubit*. Note that if the external flux is half the flux quantum, the potential (2.71) is symmetric vs. ϕ, *and* its ground state is twice degenerate, if the critical current and the self-inductance of the SQUID loop are large enough. Introducing $\eta = \phi - \pi$, we see that

$$U(\eta)|_{\widetilde{\Phi}=\Phi_0/2} = \frac{\Phi_0^2}{4\pi^2 L}\frac{\eta^2}{2} + E_J\cos\eta \approx -E_J\left(1 - \frac{\pi}{\beta_L}\right)\frac{\eta^2}{2} + E_J\frac{\eta^4}{24} + \text{const.} \tag{2.73}$$

Here the parameter

$$\beta_L = 2\pi L I_c / c \Phi_0 \tag{2.74}$$

is the maximal extra phase shift that can be produced by the self-inductance flux in the SQUID loop. The quartic potential (2.73) contains degenerate quantized levels in the left and right potential wells. The degeneracy is lifted by tunnelling through the potential barrier in the middle, forming the bonding–antibonding type $((|L\rangle \pm |R\rangle)/\sqrt{2})$ states, where $|L\rangle$ ($|R\rangle$) correspond to the supercurrent flowing clockwise or anticlockwise in the rf SQUID loop. Taking the system away from the degeneracy point by $\delta\widetilde{\Phi} = \widetilde{\Phi} - \Phi_0/2$ skews the potential, adding to it

$$\delta U(\eta) = -\frac{\Phi_0 \delta\widetilde{\Phi}}{2\pi L}\eta + \frac{(\delta\widetilde{\Phi})^2}{2L} \tag{2.75}$$

and killing the tunnelling by taking the levels out of resonance. The anharmonicity of the potential allows us to separate the two bottom levels, and switching them on and off resonance, together with the naturally occurring tunnelling, provides a realization for the qubit dynamics in literal agreement with the Hamiltonian (1.85). The parameters of the double-well potential (the interwell distance $\Delta\eta = \sqrt{24(1 - \pi/\beta_L)}$ and the height of the potential barrier $\Delta U = \frac{3}{2}(1 - \pi/\beta_L)^2$) depend through β_L on the loop inductance and the critical current of the Josephson junction. If we replace a single junction by a small dc SQUID, then β_L and, therefore, the parameters of the rf SQUID can be controlled by the magnetic flux through the dc SQUID (see Eq. (2.32)).

Unfortunately, it follows from (2.73) that, in any case, we need $\beta_L > \pi$, and thus the inductance of the SQUID loop cannot be too small. This makes the device sensitive to the external magnetic noise, coupled to it through L. The tunnelling rate Δ in (1.85) also depends exponentially on both $\Delta\eta$ and ΔU, which makes it necessary to use a tunable junction in the rf SQUID loop. The sensitivity to noise and control make the rf SQUID qubits somewhat less popular, though these problems can be, in principle, overcome (see, e.g., Harris et al., 2007). Characteristically, one of the first ever experimental demonstrations of a macroscopic "Schrödinger's cat" state (Friedman et al., 2000) was achieved using an rf SQUID qubit, but instead of tuning down the potential barrier the quantum tunnelling was observed between the *excited* states near its top (more precisely, the fourth excited state in the left well and the tenth excited state in the right well; in the experiment, $\delta\widetilde{\Phi}$ varied within 1–1.5 % of Φ_0). The niobium SQUID loop contained a tunable Josephson junction (two Nb-AlOx-Nb junctions in a dc SQUID configuration) with $E_{J,\max} = 76$ K. The inductive energy was $\Phi_0^2/2L = 645$ K. The charging energy $E_C = (2e)^2/2C = 36.0$ mK was much less, thus the system was in the phase (flux) regime. The corresponding plasma frequency $\omega_J = 1.5$–1.8×10^{11} s^{-1}. The

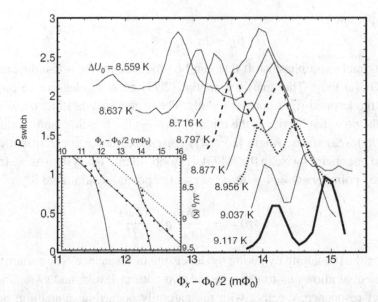

Fig. 2.7. Probability of tunnelling to the right potential well of the rf SQUID potential, P_{switch}, as a function of the external flux through the loop, Φ_x (in our notation, $\Phi_x - \Phi_0/2 \equiv \delta\widetilde{\Phi}$) and the tunnelling barrier height, ΔU_0 (in our notation, ΔU). The observed tunnelling splitting $\Delta \approx 0.1$ K. Inset: positions of the observed switching probability peaks in the $\Delta U_0 - \Phi_x$ plane. (Reprinted by permission from *Nature*, Friedman et al., 2000, © 2000 Macmillan Publishers Ltd.)

experiments were conducted at about 40 mK. The transition from the left to the right well was accompanied by a change of the magnetic flux of the system by about $\Phi_0/4$ (which was easily picked up by another dc SQUID used as a magnetometer), and the current in the loop by 2–3 μA. After multiple measurements one could restore the transition probability as a function of the barrier height ΔU and bias δU, controlled by the magnetic fluxes through the controlled junction loop and the rf SQUID loop, respectively. The results shown in Fig. 2.7 clearly demonstrate the transition probability peaks, which trace the energies of the corresponding states – with a clearly visible anticrossing due to the coherent tunnelling and formation of the "Schrödinger's cat" states, with the tunnelling splitting $\Delta \approx 0.1$ K. This is not a huge Schrödinger's cat, but its two states differed by about 10^{10} Bohr magnetons – not quite macro-, but a decently mesoscopic magnetic moment.

2.3 Circuit analysis for quantum coherent structures. More flux qubits

2.3.1 Lagrangian formalism for non-dissipative circuits

Before proceeding to other types of superconducting qubit, we will need some additional equipment. It was obvious enough how to choose a set of variables for a single Josephson junction. The situation becomes less straightforward if there are several Josephson junctions in the system, connected galvanically, electrostatically and inductively to each other and other elements of the circuit, and affected by externally applied gate potentials and magnetic fluxes. Then the choice of variables and the determination of the appropriate form of the Hamiltonian becomes nontrivial, and a general formalism is required.

Such a formalism is based on the Lagrange approach (Devoret, 1997; Burkard, 2005). Let us consider an arbitrary classical, dissipationless, lumped-element circuit. It can be described by a graph $\{\mathcal{G}, \mathcal{K}\}$, with nodes $\mathcal{G}(a)$ and edges $\mathcal{K}(a, b)$, where a, b label the nodes. Each edge contains a single lumped element: a capacitance, $C_\mathcal{K}$, an inductance, $L_\mathcal{K}$, or a Josephson junction (which we will consider later), see Fig. 2.8a. Any state of the circuit is defined by specifying the instantaneous voltages on each node, $V_a(t)$. One node voltage must be fixed as a reference point (the "zero node"). Of course, it does not mean that it must be physically grounded: the choice of the zero node is arbitrary. The rest of the nodes are "active".[13]

The current along the edge $\mathcal{K}(a, b)$, $I_{\mathcal{K}(a,b)}(t)$, will be determined from

$$i_{\mathcal{K}(a,b)}^{(\text{ind})} = -\frac{c(V_a - V_b)}{L_{ab}} \tag{2.76}$$

if the edge contains an inductor, or from

$$I_{\mathcal{K}(a,b)}^{(\text{cap})} = C_{ab}(\dot{V}_a - \dot{V}_b), \tag{2.77}$$

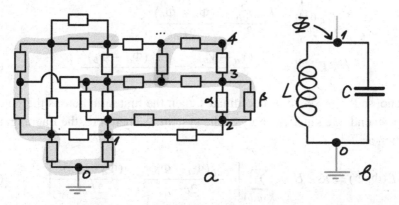

Fig. 2.8. (a) Graph $(\mathcal{G}, \mathcal{K})$ representing a lumped-elements circuit. A spanning tree $\mathcal{T}$ of the graph $\mathcal{G}$ is highlighted. Edges $\mathcal{K}(1, 2)$, $\mathcal{K}_\alpha(2, 3)$, $\mathcal{K}(3, 4)$, etc. are closure edges. The choice of the zero node is arbitrary and *does not* require it to be physically grounded. (b) An example: a parallel LC circuit (with zero external magnetic flux).

---

[13] The "dangling" nodes, which are connected to the rest of the circuit by a single capacitive or inductive edge, are not active. Their instantaneous voltages are fixed by the external "gate voltages" or "current sources".

if it contains a capacitor. Let us introduce "node phases",

$$\varphi_a(t) = \frac{2e}{\hbar} \int^t dt' V_a(t'), \tag{2.78}$$

and "node fluxes",

$$\Phi_a(t) = \Phi_0 \frac{\varphi_a(t)}{2\pi} \equiv c \int^t dt' V_a(t'). \tag{2.79}$$

Then the currents in inductive/capacitive edges are given by

$$I_{ab}^{(ind)} = \frac{c^2 \hbar}{2eL_{ab}} (\varphi_a(t) - \varphi_b(t)) = \frac{c\Phi_0}{L_{ab}} \frac{\varphi_a(t) - \varphi_b(t)}{2\pi} \equiv \frac{c(\Phi_a - \Phi_b)}{L_{ab}}; \tag{2.80}$$

$$I_{ab}^{(cap)} = \frac{\hbar C_{ab}}{2e} (\ddot{\varphi}_a(t) - \ddot{\varphi}_b(t)) = \frac{C_{ab}\Phi_0}{c} \frac{\ddot{\varphi}_a(t) - \ddot{\varphi}_b(t)}{2\pi} \equiv \frac{C_{ab}(\ddot{\Phi}_a - \ddot{\Phi}_b)}{c}. \tag{2.81}$$

We have assumed for simplicity that the inductors and capacitors in the circuit are conventional, linear lumped elements. This simplification is not crucial and, if necessary, nonlinear circuit elements can be taken into account in the same way. We will see in a moment how it works, e.g., with Josephson junctions. We have also dropped the edge label $\mathcal{K}$, with the understanding that it will be restored if there are, e.g., several inductive edges in parallel between the nodes $\mathcal{G}(a)$ and $\mathcal{G}(b)$.

The corresponding energies are

$$E_{ab}^{(ind)} = \frac{L_{ab} I_{ab}^2}{2c^2} = \frac{(\Phi_a - \Phi_b)^2}{2L_{ab}}; \tag{2.82}$$

$$E_{ab}^{(cap)} = \frac{C_{ab}(V_a - V_b)^2}{2} = \frac{C_{ab}(\dot{\Phi}_a - \dot{\Phi}_b)^2}{2c^2}. \tag{2.83}$$

If we choose the fluxes Φ as coordinates, then the first term here is the potential, and the second, the kinetic energy of the system.[14] Therefore, the Lagrangian can be written as

$$\mathcal{L}(\Phi, \dot{\Phi}) = K - U = \sum_{\mathcal{K}(a,b)} \left[\frac{C_{ab}(\dot{\Phi}_a - \dot{\Phi}_b)^2}{2c^2} - \frac{(\Phi_a - \Phi_b)^2}{2L_{ab}} \right]. \tag{2.84}$$

---

[14] This is contrary to the usual view of capacitances as "spring constants" and inverse inductances as "masses", but is in line with the quantum treatment of the Josephson junction, Eq (2.36, 2.37), where the charging term is analogous to kinetic energy. The difference simply reflects the fact that in a more conventional approach (e.g., Wells, 1967, §15.2.A) the electric charge corresponds to the position, while in our case, as we shall see, it corresponds to the canonical momentum. Both choices are equivalent. Nevertheless, when the circuit contains Josephson junctions, in our case we have non-quadratic potential, rather than kinetic, energy, which is somewhat more convenient to deal with.

Then the Lagrange equations are

$$\frac{d}{dt}\frac{\partial \mathcal{L}}{\partial \dot{\Phi}_a} - \frac{\partial \mathcal{L}}{\partial \Phi_a} = 0 = \sum_b \left[\frac{C_{ab}(\ddot{\Phi}_a - \ddot{\Phi}_b)}{c^2} + \frac{(\Phi_a - \Phi_b)}{L_{ab}} \right]. \tag{2.85}$$

Consider, for example, the parallel LC circuit of Fig. 2.8b. After "grounding" the node 0 we are left with a single variable, $\Phi_1 \equiv \Phi$, the Lagrangian

$$\mathcal{L} = \frac{C\dot{\Phi}^2}{2c^2} - \frac{\Phi^2}{2L}, \tag{2.86}$$

and the Lagrange equations

$$\frac{C\ddot{\Phi}}{c^2} + \frac{\Phi}{L} = 0.$$

Recalling the definition (2.79) and taking one more time derivative, we obtain the standard equation,

$$\ddot{V} = -\frac{c^2 V}{LC}.$$

The expression (2.84) and set of equations (2.85) do not include the effects of the external magnetic fluxes and electric potentials. To do so, let us choose a *spanning tree* $\mathcal{T}$, a subset of the graph edges, which connect each of the nodes to the zero (ground) node by one and only one path through $\mathcal{T}$ (Fig. 2.8a). The edges, which do *not* belong to the spanning tree, are called *closure edges*, since they close the loops in the original circuit $\mathcal{G}$. The choice of both $\mathcal{T}$ and the zero node is not unique and is guided by convenience (e.g., symmetries of the circuit). Now, while for the inductive edges belonging to $\mathcal{T}$ we can keep the term $(-(\Phi_a - \Phi_b)^2)/(2L_{ab})$ in the Lagrangian, for a closure edge $\mathcal{K}'(a, b)$ the term must be replaced by

$$-\frac{(\Phi_a - \Phi_b + \tilde{\Phi}_{ab})^2}{2L_{\mathcal{K}'(a,b)}}, \tag{2.87}$$

where $\tilde{\Phi}_{ab}$ is the external magnetic flux through the *irreducible loop*, the smallest loop formed by the edge $\mathcal{K}'(a, b)$ and other inductive edges. A dangling inductive edge (with inductance L_a) must be connected to a current source (otherwise it is irrelevant), which means that the node flux on its other end must be fixed to

$$\Phi_x = \frac{L_a I_x}{c}, \tag{2.88}$$

where I_x is the bias current. This adds to the Lagrangian the term

$$-\frac{(\Phi_a - \Phi_x)^2}{2L_a} \rightarrow \frac{\Phi_a I_x}{c} - \frac{\Phi_a^2}{2L_a} \rightarrow \frac{\Phi_a I_x}{c} \tag{2.89}$$

as $L_a \to \infty$ (i.e., if the node is directly connected to the current source; a constant term $L_a I_x^2/2c^2$ is, of course, dropped, since it does not contribute to the equations of motion).

The influence of the external potentials is as simple to take into account: they are applied to the corresponding nodes through "gate capacitances" (gates), C_a^g, and produce in the Lagrangian the terms

$$\frac{C_a^g \left(\dot{\Phi}_a - c\widetilde{V} \right)^2}{2c^2}. \tag{2.90}$$

In other words, an external potential corresponds to a fixed value of $\dot{\Phi}$ on the far end of a dangling capacitive edge.

Now it is straightforward to take into account Josephson junctions. As we have seen, each junction can be considered as a parallel circuit of a Josephson junction proper and a capacitor (putting aside for now the resistor of the RSJ model). The corresponding contribution to the Lagrangian will be

$$\frac{C_J \dot{\Phi}^2}{2c^2} + E_J \cos\left(2\pi \frac{\Phi}{\Phi_0} \right). \tag{2.91}$$

With our choice of variables, Eqs (2.78, 2.79), the Josephson relation (2.19) between the phase and voltage across the junction is automatically satisfied.

The simplest example is provided by an rf SQUID. There is only one active node here, and the Lagrangian becomes

$$\mathcal{L}_{\text{rf SQUID}} = \frac{C_J \dot{\Phi}^2}{2c^2} - \frac{(\Phi - \widetilde{\Phi})^2}{2L} + E_J \cos 2\pi \frac{\Phi}{\Phi_0}. \tag{2.92}$$

We see that the Josephson potential energy term here is precisely the one of Eq. (2.71).[15]

2.3.2 Dissipative elements in a circuit – Lagrange approach

We can easily incorporate in our formalism dissipation through the dissipative function, as long as it is determined only by the first-order time derivatives of the fluxes, $\dot{\Phi}$. If, for example, the circuit has conductance, $G_{ab} = 1/R_{ab}$, including the shunts of the RSJ model, the dissipative function (half the power dissipated in the system)

$$\mathcal{Q}(\dot{\Phi}) = \frac{1}{2}\frac{dE}{dt} = \sum_{ab} \frac{G_{ab}(\dot{\Phi}_a - \dot{\Phi}_b)^2}{2c^2}, \tag{2.93}$$

---

[15] If instead of a standard tunnelling Josephson junction, with its $\sin\phi$ current-phase dependence, the system includes, e.g., an SNS junction or a high-Tc junction, all we need to do is replace the Josephson energy with the one satisfying Eq. (2.21) with the appropriate $I_J(\phi)$ (e.g., Eqs (2.164, 2.165)).

and the Lagrange equations will be modified (Landau and Lifshitz, 1976, § 111):

$$\frac{d}{dt}\frac{\partial \mathcal{L}}{\partial \dot{\Phi}} - \frac{\partial \mathcal{L}}{\partial \Phi} = -\frac{\partial \mathcal{Q}}{\partial \dot{\Phi}}, \tag{2.94}$$

that is (including, for good measure, the Josephson terms and the bias fluxes)

$$\sum_b \left[\frac{C_{ab}(\ddot{\Phi}_a - \ddot{\Phi}_b)}{c^2} + \frac{(\Phi_a - \Phi_b + \tilde{\Phi}_{ab})}{L_{ab}} + \frac{C_{ab,J}(\ddot{\Phi}_a - \ddot{\Phi}_b)}{c^2} \right.$$

$$\left. + \frac{2\pi E_{ab,J}}{\Phi_0} \sin 2\pi \frac{\Phi_a - \Phi_b}{\Phi_0} \right] = -\sum_b \frac{G_{ab}}{c^2}\left(\dot{\Phi}_a - \dot{\Phi}_b \right). \tag{2.95}$$

2.3.3 Hamilton and Routh functions for a circuit

In non-relativistic quantum mechanics it is usually more convenient to deal with Hamiltonians rather than Lagrangians. The canonical momentum conjugate to the variable Φ_a is (Landau and Lifshitz, 1976, § 40)

$$\Pi_a = \frac{\partial \mathcal{L}}{\partial \dot{\Phi}_a}, \tag{2.96}$$

and the Hamilton function

$$\mathcal{H}(\Pi, \Phi) = \sum_a \Pi_a \dot{\Phi}_a - \mathcal{L}(\Phi, \dot{\Phi}). \tag{2.97}$$

The canonical momenta conjugate to the node fluxes or phases are sometimes called *node charges*. For the LC circuit (2.86)

$$\Pi = \frac{C\dot{\Phi}}{c^2}, \quad \mathcal{H}(\Pi, \Phi) = \frac{c^2\Pi^2}{2C} + \frac{\Phi^2}{2L}, \tag{2.98}$$

the momentum $\Pi = CV/c = Q/c$ is indeed, up to the factor $1/c$, the electrical charge on the capacitor. In the general case, when the ith node is electrostatically connected to several other active nodes and gates (voltage sources), the corresponding term in the Lagrangian will be

$$\sum_j \frac{C_{ij}}{2c^2}(\dot{\Phi}_i - \dot{\Phi}_j)^2 + \sum_k \frac{C_{g,ik}}{2c^2}(\dot{\Phi}_i - cV_{g,k})^2, \tag{2.99}$$

yielding the canonical momentum

$$\Pi_i = \sum_j \frac{C_{ij}}{c^2}(\dot{\Phi}_i - \dot{\Phi}_j) + \sum_k \frac{C_{g,ik}}{c^2}(\dot{\Phi}_i - cV_{g,k}), \qquad (2.100)$$

that is, up to the factor $1/c$, the total electric charge in the node. If instead of fluxes we used as the coordinates the node phases, then we would get instead the canonical momentum

$$\Pi_\phi = \frac{\partial \mathcal{L}}{\partial \dot{\phi}}, \qquad (2.101)$$

which is again the electric charge, but with a different numerical factor:

$$\Pi_\phi = \left(\frac{\hbar}{2e}\right)^2 C\dot{\phi} = \frac{\hbar}{2e}CV = \frac{\hbar}{2e}Q. \qquad (2.102)$$

In either case, in order to quantize the system, we directly replace the classical momentum with an appropriate operator, obtain the Hamiltonian and continue as before when considering a small biased Josephson junction or a phase qubit. The currents and voltages in the rest of the circuit can be expressed through the Bose creation/annihilation operators.

While one can treat the whole coherent circuit in a fully quantum-mechanical fashion (at least as long as the dissipative elements can be ignored), it is sometimes convenient to treat only part of it quantum mechanically, and the rest classically. This happens, for example, when an external circuit is used to control the qubit and transfer classical signals to it. Another case is similar to the quasiclassical theory of atom–field interactions, when the atom is treated as a quantum, and the electromagnetic field as a classical object. Such a situation arises, e.g., when a qubit is placed in a transmission line, where the amplitude of the electromagnetic wave is not too small.

For the analysis of a classical network the Lagrangian approach is often more convenient, while the quantum-mechanical part is easier to deal with using a Hamiltonian. This situation – when some degrees of freedom are better treated via the Hamiltonian, and some via the Lagrangian approach – arises also in classical mechanics, and so it is convenient to introduce the *Routh function* (e.g., Landau and Lifshitz, 1976, § 41). If Ψ's are the fluxes, which we want to deal with using the Hamiltonian dynamics, and Π_Ψ's their conjugate canonical momenta, $\Pi_{\Psi_a} = (\partial \mathcal{L}/\partial \dot{\Psi}_a)$, then the Routh function is obtained from the Lagrangian by a partial Legendre transformation:

$$\mathcal{R}(\Pi_\Psi, \Psi; \Phi, \dot{\Phi}) = \sum_a \Pi_{\Psi_a}\dot{\Psi}_a - \mathcal{L}(\Psi, \Phi; \dot{\Psi}, \dot{\Phi}). \qquad (2.103)$$

The Routh function ("Routhian") satisfies the Lagrange equations for the $(\Phi, \dot{\Phi})$-variables, and the Hamilton equations for the (Ψ, Π_Ψ)-ones:

$$\frac{d}{dt}\frac{\partial \mathcal{R}}{\partial \dot{\Phi}_b} - \frac{\partial \mathcal{R}}{\partial \Phi_b} = +\frac{\partial \mathcal{Q}}{\partial \dot{\Phi}_b}; \tag{2.104}$$

$$\frac{d}{dt}\Pi_{\Psi_a} = -\frac{\partial \mathcal{R}}{\partial \Psi_a}; \quad \frac{d}{dt}\Psi_a = \frac{\partial \mathcal{R}}{\partial \Pi_{\Psi_a}} \tag{2.105}$$

(we assume that dissipation is restricted to the "Lagrangian" degrees of freedom; note the change of sign in (2.104) compared to Eq. (2.94), because of the definition (2.103) of the Routhian).

2.3.4 Second quantization formalism for circuits

The transition to the quantum-mechanical description of a circuit is usually achieved by replacing the momenta Π with operators, $(\hbar/i)(\partial/\partial\Phi)$, in the Hamilton function as we did in Section 2.2. Sometimes it is more convenient instead of this "position representation" to express both momentum (i.e., nodal charge) and flux (phase) operators through Bose creation/annihilation operators, e.g.,

$$\hat{\Phi} = \frac{a + a^\dagger}{2}\Lambda, \quad \hat{\Pi} = \hbar\frac{a - a^\dagger}{i\Lambda}. \tag{2.106}$$

Here Λ is an arbitrary constant. Obviously, since $[a_j, a_j^\dagger] = 1$, the defined flux and charge operators satisfy the commutation relations,

$$[\hat{\Phi}, \hat{\Pi}] = i\hbar. \tag{2.107}$$

What physical meaning would such a formal representation have? Consider, for example, a nonlinear oscillator, with

$$\mathcal{L} = \frac{C\dot{\Phi}^2}{2c^2} - U(\Phi) = \frac{C\dot{\Phi}^2}{2c^2} - \frac{\Phi^2}{2L} - \delta U(\Phi),$$

$$\mathcal{H} = \frac{c^2\Pi^2}{2C} + \frac{\Phi^2}{2L} + \delta U(\Phi) = \frac{c^2\Pi^2}{2C} + \frac{C\omega_0^2\Phi^2}{2c^2} + \delta U(\Phi). \tag{2.108}$$

Here $\omega_0 = c/\sqrt{LC}$, and $\delta U(\Phi)$ describes the anharmonicity. Then, choosing $\Lambda = \sqrt{2\hbar/C\omega_0}$, we find the Hamiltonian

$$H = \hbar\omega_0\left(a^\dagger a + \frac{1}{2}\right) + \delta U(\Phi(a, a^\dagger)). \tag{2.109}$$

The operators $a, a^\dagger$, therefore, operate in the Fock space of a harmonic oscillator with frequency ω_0, which approximates our anharmonic oscillator, and provide straightforward expressions for different observables as matrices in this Fock basis. For example, Eq. (2.109) gives a shorter way to find the energy levels of a cubic oscillator, than the standard approach, the result of which we used when discussing the phase qubit.

2.3.5 Persistent current flux qubit

A relative disadvantage of the rf SQUID flux qubit is the large value of $\beta_L > \pi$, as well as the need to use a tunable Josephson junction. In order to minimize the dangerous coupling of the system to the external sources of noise, it would be advantageous to use a device with small self-inductance. Such a device, a *persistent current qubit*, was proposed in Mooij et al. (1999) and Orlando et al. (1999) and successfully used since then. The idea of the device is to interrupt a loop of negligible inductance ($L \to 0$) with three or more Josephson junctions (Fig. 2.9). The flux quantization condition (2.28)

$$\varphi_1 + \varphi_2 + \varphi_3 + 2\pi \frac{\widetilde{\Phi}}{\Phi_0} = 2\pi q \tag{2.110}$$

in this case fixes one of the three phase differences across the Josephson junctions, leaving two independent variables, e.g., φ_1 and φ_2. The Josephson energy is accordingly a function of two variables, not of one (as for an rf SQUID), and

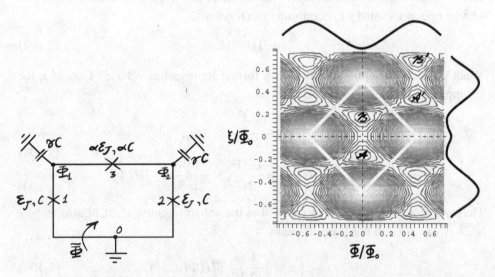

Fig. 2.9. (Left) Circuit diagram of the persistent current qubit. Self-inductance of the qubit loop is ignored. (Right) Contour plot of the Josephson energy $U(\Phi, \xi)$, Eq. (2.113), with $\alpha = 1$ and $\Phi = \Phi_0/2$. The pattern is periodic, with the unit cell indicated by the white square. The minima (A and B) are degenerate.

one can hopefully find a combination of critical currents and external flux, which creates two sufficiently close minima in this potential to be used for the qubit states.

The circuit diagram of this qubit is shown in Fig. 2.9 (as usual, the zero node does not have to be physically grounded). Note that we have neglected the self-inductance, which leaves us with only two active nodes and two independent node fluxes, Φ_1 and Φ_2. We will assume that the system is symmetric, with the equivalent junctions 1 and 2 (each with the Josephson energy E_J, Josephson capacitance C and stray capacitance γC), and junction 3 with the Josephson energy αE_J and Josephson capacitance αC. The Lagrangian of this circuit is

$$\mathcal{L}(\Phi_1, \Phi_2; \dot{\Phi}_1, \dot{\Phi}_2) = \frac{C(1+\gamma)}{2c^2}\left(\dot{\Phi}_1^2 + \dot{\Phi}_2^2\right) + \frac{\alpha C}{2c^2}\left(\dot{\Phi}_1 - \dot{\Phi}_2\right)^2$$
$$+ E\left(\cos 2\pi\frac{\Phi_1}{\Phi_0} + \cos 2\pi\frac{\Phi_2}{\Phi_0} + \alpha\cos 2\pi\frac{\Phi_1 - \Phi_2 + \tilde{\Phi}}{\Phi_0}\right). \tag{2.111}$$

It is convenient to use the symmetry and introduce new variables, $\Phi = (\Phi_1 + \Phi_2)/2$ and $\xi = (\Phi_1 - \Phi_2)/2$. Then the Lagrangian becomes

$$\mathcal{L}(\Phi, \xi; \dot{\Phi}, \dot{\xi}) = \frac{C(1+\gamma)}{c^2}\dot{\Phi}^2 + \frac{\alpha C(1+\gamma+2\alpha)}{c^2}\dot{\xi}^2 + U(\Phi, \xi), \tag{2.112}$$

where

$$U(\Phi, \xi) = -E\left(2\cos 2\pi\frac{\Phi}{\Phi_0}\cos 2\pi\frac{\xi}{\Phi_0} + \alpha\cos 2\pi\frac{2\xi + \tilde{\Phi}}{\Phi_0}\right). \tag{2.113}$$

The Hamilton function,

$$\mathcal{H}(\Pi_\Phi, \Pi_\xi; \Phi, \xi) = \frac{1}{2}\frac{\Pi_\Phi^2 c^2}{2C(1+\gamma)} + \frac{1}{2}\frac{\Pi_\xi^2 c^2}{2C(1+\gamma+2\alpha)} + U(\Phi, \xi) \tag{2.114}$$

is the Hamilton function of a particle with anisotropic mass in the periodic potential landscape shown in Fig. 2.9. When the bias flux is close to a half-flux quantum, the two minima are almost degenerate; the corresponding lowest-energy states are characterized by the clockwise or anticlockwise flow of supercurrent around the loop and can be used as the working states of a qubit. Their hybridization, due to tunnelling through a small separating barrier, produces the σ_x-term in the qubit Hamiltonian Eq. (1.85), while the σ_z-term is directly controlled by the bias flux.

The periodicity of $U(\Phi, \xi)$ reflects the periodicity of the Josephson energy as a function of the superconducting phase difference. This would seem to indicate that one should simply identify the points shifted by Φ_0 in the Φ- or ξ-direction, and obtain a discrete energy spectrum. Actually, this is not necessarily so: if the change of phase difference by 2π leads to observable physical changes (e.g., work performed by the external sources, which produce the external constant bias fluxes and currents or gate voltages, to keep their values fixed), the potential (2.113) must be treated as a truly periodic landscape familiar from solid-state physics (Ziman, 1979, Chapter 1; Kittel, 1987, Chapter 9; Kittel, 2004, Chapter 7) with energy bands and eigenstates characterized by *quasi*momenta instead of momenta. (In our formalism, momenta correspond to electric charges, so we will encounter *quasicharges*.) For example, this is the case for a current-biased junction, where the phase change by 2π moves the system one step up or down the washboard potential and, thus, changes its energy for an arbitrarily small bias current. We will consider this question in some detail in § 2.4.5. For a persistent current qubit this subtlety is irrelevant, since with the proper choice of parameters the lowest energy states practically do not tunnel out of the two-well area, and the would-be bands are squeezed to essentially discrete levels.

A detailed analysis of the system (2.114) is given in Mooij et al., 1999; Orlando et al., 1999. In the quasiclassical approximation, the tunnelling matrix element between two minima is (Landau and Lifshitz, 2003, § 50)

$$t \approx \frac{\hbar\omega_{att}}{\pi} \exp[-S/\hbar], \tag{2.115}$$

where the attempt frequency of escape $\omega_{att} \sim \omega_0$ (the frequency of small oscillations in the well), and $S \sim \hbar\sqrt{E_J/E_C}$ is the action between the initial and final points. For the choice of parameters ($E_J/E_C = 25, \alpha = 0.8, \gamma = 0.02$) the tunnelling rate between the lowest states in the minima in the same unit cell (A and B in Fig. 2.9) is about four orders of magnitude larger than that between the minima in adjacent unit cells (e.g., B and A′); it remains dominant for E_J/E_C as low as 20, which is close to the typical value. It follows from Eq. (2.115) that the tunnelling rate – the parameter Δ in the qubit Hamiltonian – is exponentially sensitive to the material parameters of the system, which requires either very precise fabrication, or tunability.

The usable local minima in the potential (2.113) exist only in a narrow range of bias fluxes, between 0.5 and 0.485 Φ_0. Using the latter as the operating point, and choosing the critical current of the bigger junctions $I_c = 400$ nA ($E_J = 200$ GHz) and the loop diameter $d = 10$ μm ($L \approx 10$ pH), the circulating current $I_p \approx 0.7I_c$ should produce the magnetic flux $\pm10^{-3}\Phi_0$, which is much less than the $\sim \Phi_0/4$-difference in an rf SQUID qubit, but is still observable by, e.g., a dc SQUID detector.[16]

---

[16] It is, of course, necessary for the measurement to bias the system away from the exact degeneracy point, $\tilde{\Phi} = 0.5\Phi_0$, because there the quantum expectation value of the persistent current-generated flux, proportional to the expectation value of the operator σ_z, will be exactly zero in either the ground or excited state of the system, $(|L\rangle \pm |R\rangle)/\sqrt{2}$.

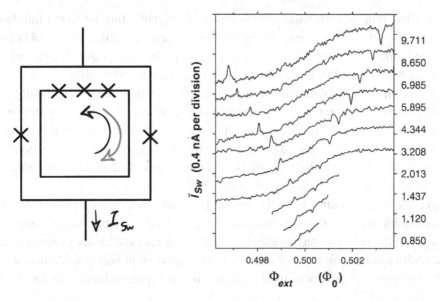

Fig. 2.10. Observation of coherent tunnelling in a persistent current qubit. (Left) A scheme of the measurement used by van der Wal et al. (2000). The qubit is placed inside a dc SQUID, which acts as a detector. (Right) (From van der Wal et al., 2000. Reprinted with permission from AAAS.) The dependence of the current bias, I_{Sw}, at which the dc SQUID switches to the resistive state, on the flux through the qubit loop. (The plotted quantity $\tilde{I}_{Sw}$ is averaged over a series of measurements, and has the background subtracted.) The smooth step reflects the change of direction of the average persistent current in the qubit loop as the qubit flux bias passes through $\Phi_0/2$. The spikes appear in the presence of a microwave signal (frequency, in GHz, shown on the right axis) and indicate the resonant transitions from the ground to the excited state, which precipitate the switching of the ds SQUID to the resistive state.

A measurement by van der Wal et al. (2000) was conducted using a dc SQUID detector (Fig. 2.10).[17] Its critical current (2.32) was sensitive to the small variations of the magnetic flux through the dc SQUID loop, produced by the persistent current in the qubit. By measuring the current I_{Sw} at which the dc SQUID switched to the resistive state, one could find the qubit-induced flux and therefore measure its quantum state. The smooth step in I_{Sw} as a function of the bias flux (Fig. 2.10) shows the behaviour of the average persistent current. In the presence of a harmonic signal with frequency ω, the qubit can undergo a resonant transition from the ground to

---

[17] The device was fabricated of aluminium and had the following (quite typical) parameters: $I_c = 570 \pm 60$ nA, $C = 2.6 \pm 0.4$ fF for the largest junctions in the loop; $\alpha = 0.82 \pm 0.1$; $E_J/E_C = (38 \pm 8)/4$; $I_p = 450 \pm 50$ nA; the self-inductance of the inner loop $L = 11 \pm 1$ pH, and the mutual inductance between the qubit and the dc SQUID $M = 7 \pm 1$ pH. To decrease the decoherence effects of the detector SQUID, it was made underdamped, with the resulting broad distribution of the switching current ($\sim 10 \cdot 10^{-3} \Phi_0$ compared to the expected qubit flux $\sim 3 \cdot 10^{-3} \Phi_0$). This made it necessary to average over multiple measurements of the switching current.

the excited state at such values of the external magnetic flux when the interlevel distance $E_{01}(\widetilde{\Phi}) = \hbar\omega$. These transitions would produce spikes in the switching current, which are indeed clearly seen in the experimental graph. The tunnelling splitting extrapolated from these data to $\widetilde{\Phi} = 0.5\Phi_0$ was ≈ 0.3 GHz.

A controlled manipulation of the quantum state of a persistent current qubit was demonstrated in a number of subsequent experiments. In particular, a modification of the basic design of Fig. 2.9 was used to control the tunnel splitting Δ (Paauw et al., 2009). This was realized by replacing the smaller junction ("α-junction") in the qubit loop by an "α-loop", a dc SQUID was employed as a tunable junction (see (2.32)), with the results shown in Fig. 2.11.

The additional feature of this design was the gradiometer structure (figure-eight-shaped qubit loop). It played a dual role. First, it made the qubit bias insensitive to changes in the external magnetic field, as long as it remained homogeneous (since any additional flux through both halves of the figure eight will contribute equally, but in an opposite direction, to any change in the superconducting phases in the

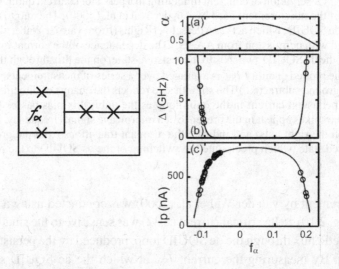

Fig. 2.11. (Left) Circuit diagram of a tunable persistent current qubit of Paauw et al. (2009). The α-junction of Fig. 2.9 is replaced by an Figure of "α-loop" (a dc SQUID used as a tunable junction). The role of the qubit loop is played by an eight-shaped gradiometer, with the magnetic flux trapped inside. The state of the qubit is read out by an inductively coupled, asymmetrically placed switching SQUID. The $12 \times 12\ \mu m^2$ qubit was fabricated of aluminium and could be controlled slowly by the uniform external magnetic field and quickly by ac and dc current pulses through the control lines adjacent to the qubit (not shown), allowing, e.g., Rabi oscillations to be produced. (Right) Qubit parameters (coefficient α, tunnel splitting Δ and persistent current I_p) measured as a function of "frustration" $f_\alpha = \widetilde{\Phi}_\alpha/\Phi_0$ in the α-loop (Reprinted with permission from *Phys. Rev. Lett.*, Paauw et al., 2009, © American Physical Society 2009).

loop). Second, the solid (i.e., containing no Josephson junctions) outer loop of the gradiometer pre-trapped the magnetic flux (Majer et al., 2002) and, therefore, eliminated the need for the external bias $\tilde{\Phi} \approx \Phi_0/2$, necessary for the persistent current qubit to realize the two-well Josephson potential. It would seem that the flux quantization condition, Eq. (2.13), will completely insulate both the α-loop and the qubit from the external magnetic field and so defeat its purpose, but this is not so. We derived Eq. (2.13) for massive superconductors – at least, ones much thicker than the penetration length of the magnetic field into the superconductor, typically $\sim 500 \text{Å}$. Then we could enclose the hole with a contour, running well inside the superconductor, where there was no current and, therefore, no superconducting phase gradient, which would otherwise contribute to the phase of the macroscopic wave function (see Eq. (2.12)). When, as in experiments by Majer et al. (2002) and Paauw et al. (2009), this condition is not satisfied – the loop is too thin! – one cannot drop the left-hand side of Eq. (2.12), and instead of flux quantization, Eq. (2.13), one has the *fluxoid quantization* in a superconducting ring:

$$2\pi \frac{\Phi_{\text{tot}}}{\Phi_0} + 2\pi \frac{L_k I}{\Phi_0} = 2\pi q. \tag{2.116}$$

The second term on the left is due to the term with $\mathbf{v}_s$ in (2.12), L_k is the kinetic inductance of the ring (which we mentioned in our discussion of the rf SQUID), and Φ_{tot} includes the contribution from the (geometric) self-inductance. Eq. (2.116) is an analogue of the quantization condition in a contour with a Josephson junction, Eq. (2.28). In either case it is the sum of the contributions from the magnetic flux and from the supercurrent that is quantized, only in the latter case all this supercurrent-related phase change can be considered as happening across the narrow tunnel barrier instead of being distributed along the superconductor. The change in the fluxoid number, q, requires crossing a prohibitively high energy barrier (estimated in Majer et al. (2002) as 10 000 K). It is easy to see that if the fluxoid number in the gradiometer loop is odd, the qubit of Fig. 2.11 will be pre-biased to its working point, corresponding to $\tilde{\Phi} = \Phi_0/2$.

2.4 Charge qubits

2.4.1 Charge regime: Normal conductors

So far we have encountered situations when Josephson energy exceeded the charging energy of the junction, $E_C = (2e)^2/2C$. What happens in the opposite limit, when even a single Cooper pair plays a crucial role?[18] First consider a small *normal* island connected to the external leads by tunnelling barriers and capacitors (a device

[18] See, e.g., Likharev (1986), Ch 16; Tinkham (2004), § 7.4; Zagoskin (1998), § 4.6.

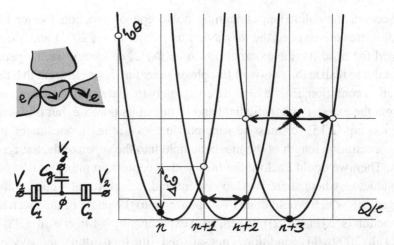

Fig. 2.12. Coulomb blockade of normal tunnelling. Left: an electron tunnelling through a small normal island and the circuit diagram of a SET. Right: The electrostatic energy of the island as a function of its charge Q can take only discrete values, corresponding to an integer number of electrons. At low temperatures ($k_B T \ll k_B T_Q \sim e^2/C_\Sigma$) only the lowest states are occupied (indicated by filled circles). Coulomb blockade: The tunnelling through the island is suppressed as long as the bias between the terminals is less than the energy necessary to add one electron, $eV < \Delta E$ (left curve). When the induced charge $Q^* = (n + 1/2)e$, the states with n and $n + 1$ electrons on the island become degenerate (e-degeneracy), and the blockade is lifted by resonant tunnelling (central curve). The $2e$-degeneracy (between the states with n and $n + 2$ electrons), when $Q^* = (n + 1)e$, is not realized in normal systems, since then the state with n electrons on the island has the smaller energy (right curve).

known as a single-electron transistor, SET), see Fig. 2.12. Each of the barriers also has a certain capacitance, and the electrostatic contribution to the classical energy of the system is obtained in the standard way as

$$E_Q(Q, Q^*) = \Pi_\Phi \dot{\Phi} - \mathcal{L} = \frac{1}{2c^2} \left\{ \sum_j C_j (\dot{\Phi} - cV_j)^2 + 2 \sum_j C_j (\dot{\Phi} - cV_j) cV_j \right\}$$

$$= \frac{1}{2C_\Sigma} (Q - Q^*)^2 - \frac{1}{2} \sum_j C_j V_j^2. \tag{2.117}$$

Here the charge on the island $Q = c\Pi_\Phi$, and the effective (induced) charge due to the external voltages $Q^* = -\sum_j C_j V_j$.

For a normal conductor we cannot switch to the quantum picture by simply replacing the charge Q by $-2ie\partial_\phi$, since there is no condensate, no bosonic Cooper pairs and no quantum-mechanical phase to differentiate over. The proper treatment of the problem is, as a result, too involved to be presented here and can be found in Averin and Likharev (1991) (with a simpler presentation in Likharev (1986)). Nevertheless, some important conclusions can easily be made.

The induced charge Q^* can vary continuously, since it reflects the effect of the redistribution of charge between the terminal capacitors and tunnel junctions, and connected to them infinitely large charge reservoirs, which maintain fixed values of the electrostatic potential. However, the charge Q on the finite island must be a multiple of e. Therefore, the allowed states of the system can have only discrete values of electrostatic energy, as shown in Fig. 2.12. This leads to the *Coulomb blockade* of tunnelling: neglecting thermal fluctuations, there can be no current through the island between terminals as long as the bias between them is less than the charging energy ΔE of the island necessary to add or remove an extra electron. This energy, $\Delta E_{n,n\pm1} = E_Q((n \pm 1)e, Q^*) - E_Q(ne, Q^*)$, can be tuned by the gate voltages (the potentials on the capacitors, connected to the island), to the point of lifting the blockade: when $Q^* = (n + 1/2)e$, $\Delta E_{n,n+1} = 0$, allowing resonance tunnelling through the island. This effect (single-electron oscillations) was observed as a periodic dependence of the conductance through the island on the gate voltage (see Tinkham, 2004, § 7.5.2, Fig. 7.7). Note that while the states of the island differing by *two* electron charges can also be made degenerate: $\Delta E_{n,n+2}$ for $Q^* = (n + 1)e$, this does not lead to any resonant transitions, since the non-degenerate state of the island with charge $Q = ne$ will have the lowest energy.

If the tunnel junctions were replaced by resistors, the blockade would still be there as long as the discrete charge states can be resolved, that is, as long as the characteristic interlevel distance, $\sim e^2/C_\Sigma$, exceeds the linewidth, $\sim \hbar/\tau$. The time τ needed for the extra charge on the island to dissipate can be estimated as C_Σ/G_Σ, where G_Σ is the effective leakage conductance. This leads to the criterion:

$$G_\Sigma < \frac{e^2}{\pi\hbar} \equiv G_Q. \tag{2.118}$$

We included the factor of $1/\pi$ (which does not matter anyway in the order-of-magnitude estimates) in order to make clear the connection between the observability of charging effects and the *conductance quantum* $G_Q \approx (12.9\,\text{k}\Omega)^{-1}$, a fundamental quantity known from the quantum Hall effect (Stone, 1992).

2.4.2 Charge regime: Superconductors

If the island is superconducting, another energy scale, the superconducting gap Δ, comes into play: see Fig. 2.13. When we discussed large superconductors, neither the electron number nor its parity mattered, since the relative effect of a single electron quasiparticle or an extra Cooper pair was zero. For a small island this is no

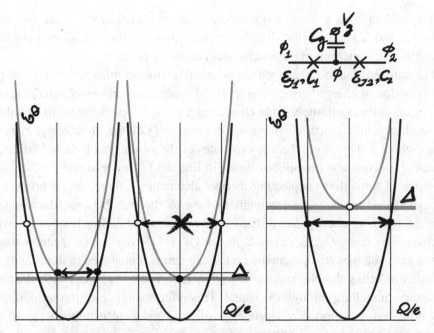

Fig. 2.13. Charging and parity effect in a superconducting island. Two separate energy curves (for even and odd numbers of electrons in the island) are shifted by the gap energy, Δ. If $\Delta < e^2/2C_\Sigma$, only e-degeneracy is possible (left), while $2e$-degeneracy is excluded (centre). For $\Delta > e^2/2C_\Sigma$ $2e$-degeneracy and resonant Josephson tunnelling are possible (right). Inset: Circuit diagram of SSET.

longer so, and we must take into account *both* the charging energy and the parity term, i.e. (at zero temperature)

$$E_Q(Q, Q^*) = \frac{(Q - Q^*)^2}{2C_\Sigma} + p\left(\frac{Q}{2e}\right)\Delta, \tag{2.119}$$

where $p(n)$ is zero if n is even and one if it is odd. The last term reflects the fact that an odd electron cannot form a Cooper pair and will therefore be forced to occupy a state above the superconducting gap (Fig. 2.13). This makes possible resonant tunnelling involving Cooper pairs, as long as the superconducting gap exceeds the single-electron charging energy $e^2/2C_\Sigma$.

For a superconducting island between two superconducting electrodes, the Hamiltonian is thus

$$H = E_C\left(\frac{1}{i}\frac{\partial}{\partial\phi} - \frac{Q^*}{2e} + \frac{p}{2}\right)^2 + p\Delta - E_{J1}\cos(\phi - \phi_1) - E_{J2}\cos(\phi - \phi_2) \tag{2.120}$$

(it is more convenient to use here "phase" rather than "flux" variables). We have assumed that there is no voltage drop between the superconducting banks and the island, therefore $Q^* = -C_g V_g$. We also assume that the Josephson couplings are the same, and – without loss of generality – take $\phi_1 = -\phi_2 = \phi_0/2$. Then the Josephson current through the island at zero temperature is

$$I(\phi_0) = \frac{2e}{\hbar} \frac{\partial}{\partial \phi_0} \langle H \rangle = \frac{2e}{\hbar} \frac{\partial}{\partial \phi_0} E_0(\phi_0), \tag{2.121}$$

where E_0 is the ground state energy of the Hamiltonian (2.120).

As long as Δ exceeds the charging energy, we can disregard the odd-parity states on the island, which reduces (2.120) to

$$H(V_g, \phi_0) = E_C \left(\frac{1}{i} \frac{\partial}{\partial \phi} + \frac{C_g V_g}{2e} \right)^2 - 2E_J \cos(\phi_0/2) \cos(\phi). \tag{2.122}$$

We need only consider the subspace near the degeneracy between the states with n and $n+2$ electrons (N and $N+1$ Cooper pairs) on the island. The matrix element of $\exp(\pm i\phi)$ is (recalling Eq. (2.52))

$$\langle N | e^{\pm i\phi} | N' \rangle = \int \frac{d\phi}{2\pi} e^{-iN\phi} e^{\pm i\phi} e^{iN'\phi} = \delta_{N'-N\pm 1}. \tag{2.123}$$

Therefore the Hamiltonian in the subspace $\{|N\rangle, |N+1\rangle\}$ is

$$
\begin{aligned}
H &= \begin{pmatrix} E_Q(N, V_g) & -E_J \cos(\phi_0/2) \\ -E_J \cos(\phi_0/2) & E_Q(N+1, V_g) \end{pmatrix} \\
&= \frac{E_Q(N, V_g) + E_Q(N+1, V_g)}{2} + \frac{E_Q(N, V_g) - E_Q(N+1, V_g)}{2} \sigma_z \\
&\quad - E_J \cos(\phi_0/2) \sigma_x,
\end{aligned}
\tag{2.124}
$$

where

$$E_Q(N, V_g) = E_C \left(N + \frac{C_g V_g}{2e} \right)^2. \tag{2.125}$$

The ground and excited state energies and the Josephson current are then

$$
\begin{aligned}
E_{0,1}(V_g, \phi_0) &= \frac{E_Q(N, V_g) + E_Q(N+1, V_g)}{2} \\
&\quad \mp \left[\left(\frac{E_Q(N, V_g) - E_Q(N+1, V_g)}{2} \right)^2 + E_J^2 \cos^2(\phi_0/2) \right]^{1/2}
\end{aligned}
\tag{2.126}
$$

and

$$I(\phi_0) = \frac{2e}{\hbar} \frac{E_J/4}{\left[\left(\frac{E_Q(N,V_g) - E_Q(N+1,V_g)}{2E_J} \right)^2 + \cos^2(\phi_0/2) \right]^{1/2}} \sin\phi_0. \tag{2.127}$$

The Josephson current is blocked unless the states with N and $N+2$ Cooper pairs are within $\sim E_J$ of each other. The physical mechanism of this blockade can be understood as due to the quantum uncertainty relation between number and phase, Eq. (2.47). The electrostatic interaction tends to fix the number of Cooper pairs on the island, and, therefore, make its phase indefinite. The phase fluctuations, in turn, suppress the observed value of the Josephson current between the island and either of the massive superconductors, averaging it to zero. In resonance, charge can freely flow, the number of Cooper pairs is no longer fixed, and the Josephson current is restored.[19] The detailed theory (Matveev et al., 1993) predicts that as the superconducting gap is suppressed below the charging energy (by applying a magnetic field, since we would not want to smear the energy levels by increasing the temperature), the critical current sharply drops, and its dependence on the gate voltage becomes e/C_Σ-, rather than $2e/C_\Sigma$-periodic, which has been confirmed by experiment (Joyez et al., 1994).

One could in principle use a small superconducting island (known as a Cooper pair box (CPB), or superconducting single-electron transistor (SSET)) as a tunable Josephson junction, controlled by voltage (dual to the magnetic flux-controlled dc SQUID, Eq. (2.32)), but this would not be practical. Instead, SSETs are exceptionally well suited for the role of qubits.

2.4.3 Charge qubit

The Hamiltonian of an SSET, Eq. (2.124), is, up to a constant term, the same as the qubit Hamiltonian (1.85). The "left-right" states are the states with definite charge, $|L\rangle \equiv |N\rangle$ and $|R\rangle \equiv |N+1\rangle$, with energies (2.126) dependent on the gate voltage. Due to the Josephson coupling, these states are hybridized, and a finite gap opens between the true $|0\rangle$ and $|1\rangle$-states at the degeneracy point, where otherwise we would have $E_0(N, V_g) = E_1(N+1, V_g)$.

Actually the charge qubit can be even simpler: one Josephson junction is enough (Fig. 2.14, top left). Its Hamiltonian is then

$$H(n^*) = E_C(\hat{N} - n^*/2)^2 - E_J \cos\phi. \qquad (2.128)$$

Here $\hat{N} = -i\partial_\phi$ is the Cooper pair number operator, and n^* characterizes the offset charge, $Q^* = en^*$. Ideally, this charge only includes the gate-induced contribution, $C_g V_g$, but in reality it is also influenced by impurities, external noise, etc. At any rate, the Hamiltonian (2.128) is readily reduced to the qubit form in the subspace

---

[19] Note also the deviation of $I(\phi_0)$ given by (2.127) from the simple sinusoid; in exact resonance, $I(\phi_0) \propto \sin(\phi_0/2)$! This does not mean that the current becomes 4π-periodic; simply as ϕ_0 passes $\pm\pi$, it suddenly switches sign.

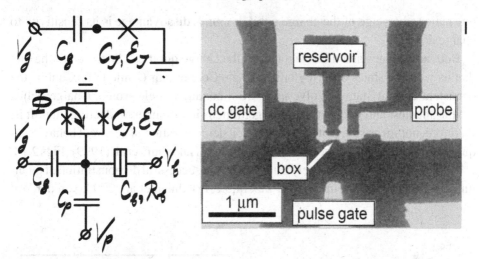

Fig. 2.14. (Left) Circuit diagram of a simple charge qubit (top) and of a charge qubit with tunable Josephson energy (bottom). (Right) Tunable charge qubit (reprinted by permission from *Nature* (Nakamura et al., 1999) © 1999 Macmillan Publishers Ltd). The aluminium "box" ($700 \times 50 \times 15$ nm^3) contains $\sim 10^8$ electrons. The (single-electron) charging energy, $e^2/2C_\Sigma = 117 \pm 3$ μeV, and the superconducting gap, $\Delta = 230 \pm 10$ μeV, greatly exceed the temperature at which the experiments were conducted, $k_B T = 3$ μeV(30 mK). The tunable Josephson junctions have the total normal resistance 10 kΩ; the resistance of the probe junction is 30 MΩ. There are two separate gate electrodes, to apply constant bias V_g and short (160 ps) pulses, $V_p(t)$.

spanned by the states $|L\rangle$, $|R\rangle$:

$$H(n^*) = -\frac{1}{2}\left(E_C(1-n^*)\sigma_z + E_J\sigma_x\right). \tag{2.129}$$

Is it possible to simplify the device even further, e.g., by using just one small Josephson junction? The answer is no, because the charge states of such a junction will be smeared out. Static impedance of a small Josephson junction – its tunnelling resistance – can, of course, be much larger than the quantum resistance, 12.9 kΩ, required by the criterion (2.118), but at the high frequencies (GHz) dictated by the small capacitance of the junction, it will be shunted by the small impedance of the connecting leads (superconducting banks) (e.g., Tinkham, 2004, § 7.1). It is, therefore, necessary to have an island, separated from the leads on all sides.

In practice, the single Josephson junction in an SSET is often replaced by two junctions in parallel, in a tunable dc SQUID configuration. In addition, the tunnelling term in the Hamiltonian (2.129) depends linearly on the parameters of the device (E_C and E_J), and not exponentially, like in flux qubits. This constitutes one major advantage of the charge qubit design. A disadvantage is its sensitivity to charge noise.

Both advantages and disadvantages of the charge qubit can be traced to the fact that its working states differ by only a single Cooper pair (while in flux qubits the persistent current states involve up to 10^{10} conduction electrons). Charge qubits are the smallest and liveliest Schrödinger kittens in the superconducting litter. It is, therefore, not surprising that the first explicit demonstration of coherent behaviour – quantum beats – was obtained in a charge qubit by Nakamura et al. (1999) (Figs 2.14, 2.15). These results, immediately followed by the successful demonstration of quantum behaviour in different types of flux qubit (Friedman et al., 2000; van der Wal

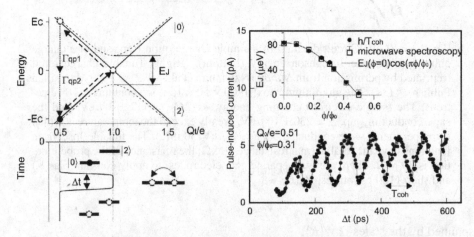

Fig. 2.15. Quantum beats in a superconducting charge qubit (reprinted by permission from *Nature* (Nakamura et al., 1999) © 1999 Macmillan Publishers Ltd). (Left) The scheme of the experiment. The two states of the qubit are denoted here by $|0\rangle$ and $|2\rangle$, according to the number of extra *electrons* on the island. The qubit was initialized in the ground state ($|0\rangle$ for the chosen static bias) by simply waiting long enough. Then it was brought to the degeneracy point by applying a short and sharp (160 ps duration, 30–40 ps rise/fall time) voltage pulse, $V_p(t)$, to a separate gate. The position of energy levels is schematically plotted vs. $Q_t = C_g V_g + C_b V_b + C_p V_p(t)$, the total charge induced on the island. After the pulse is lifted, the qubit is left in a superposition of states $|0\rangle$ and $|2\rangle$, with the probability of finding the qubit in state $|2\rangle$, $P(\Delta t) \propto \sin(E_J \Delta t/\hbar)$. The measurement was realized by the decay of state $|2\rangle$ via inelastic tunnelling of two quasiparticles (electrons) from the island through the high-resistivity tunnelling junction into the probe electrode, detected as a current pulse. The 160 ps pulses were repeated every 16 ns, and the average probe current, therefore, provided a direct measure of $P(\Delta t)$. (Right) Pulse-induced probe current vs. pulse duration for $Q_0/e = 0.51$. Inset shows the Josephson energy of the qubit junction as a function of the tuning magnetic flux. It can be determined from the period of quantum beats, $T_{\text{coh}} = 2\pi\hbar/E_J$, or using microwave spectroscopy. Both approaches agree with each other and with the formula (2.32).

et al., 2000), opened the floodgates, and have, now made possible almost a routine observation and manipulation of coherent quantum states in superconducting systems – as well as the writing of this book. They also indicated that the decoherence time of a simple charge qubit is short, of the order of 2 ns (at longer pulse durations oscillations were no longer seen). It turned out that this can be helped by a bit of (quantum) engineering.

2.4.4 Quantronium

The sensitivity of charge qubits to external electrical noise is caused by the sharp dependence of their energy levels, Eq. (2.126) or (2.129), on the induced charge $n^*e = -C_g V_g + Q^*$ near the degeneracy point,[20] $n^* = 1$:

$$E_{0,1}(n^*) = \mp \frac{E_C}{2}\sqrt{(1 - n^*)^2 + (E_J/E_C)^2}. \qquad (2.130)$$

The interlevel spacing, $\hbar\omega_{01} = E_1 - E_0$, sharply increases from E_J at the degeneracy point to $\approx E_C \gg E_J$ outside the narrow, $\sim E_J/E_C \ll 1$, region around it. Therefore, a minor change in local electrical potential will drastically affect the behaviour of the qubit. Charge qubits with a larger ratio E_J/E_C would be less sensitive, but reducing E_C would make the charge readout more difficult.

The design of *quantronium* (see Fig. 2.16) was specifically developed to decrease the sensitivity of the qubit to noise and to improve the readout (Cottet et al., 2002; Vion et al., 2002). From Eq. (2.130) it follows that small deviations from $\psi = 0$ and $n^* = 1$ change the energy only in the second order. This point in the parameter space (the *optimal point*, or more colloquially, the "sweet spot") is, thus, protected from the magnetic and, what is especially important, electric noise. In the standard SSET it was necessary to leave this spot, since it was the charge on the island that was read out. The advantage of quantronium is that one instead reads out the phase variable, staying at $n^* = 1$. The efficiency of the phase readout also allows us to make the ratio $E_J/E_C \sim 1$, further reducing the sensitivity to charge fluctuations.

The device is essentially an SSET with a tunable junction, but with the dc SQUID loop closed by another, large Josephson junction, and with E_J and E_C of the same order of magnitude (in the experiment by Vion et al. (2002) $E_{J0} \approx 20E_J$, and $E_J/E_C = 1.27$). The Hamiltonian of the SSET part of the quantronium is thus

$$H = E_C(\hat{N} - 2n^*)^2 - E_J \cos\frac{\psi}{2}\cos\theta, \qquad (2.131)$$

where $\psi = \phi_2 - \phi_3$ is the total phase drop across the two smaller junctions of the SSET, and $\theta = (\phi_2 - \phi_1) - (\phi_1 - \phi_3)$ is conjugate to the Cooper pair number

---

[20] Here Q^* is the charge induced on the SSET island by everything else except the gate voltage.

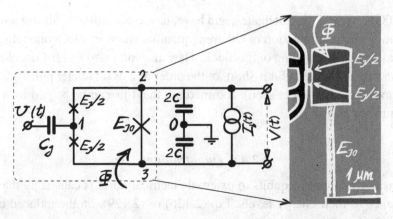

Fig. 2.16. Circuit diagram of quantronium (inside the dashed area) and readout circuit (after Vion et al., 2002). Charging energy of the SSET (node 1) and the Josephson energies $E_J/2$ of the two junctions connecting it to the loop are comparable. In Vion et al. (2002) $E_J/E_C = 1.27$; $E_J/k_B = 0.865$ K, and neither ϕ_1, nor its conjugate, $\hat{N}_1$ (the number of Cooper pairs on the SSET) are good quantum numbers. Instead of directly reading out the charge on the SSET (cf. Fig. 2.14), the state of the qubit is entangled with the phase difference, $\phi_2 - \phi_3$, across the large Josephson junction, $E_{J0} \approx 20E_J$, which is in turn measured through the latter's transition into the resistive state (like in a phase qubit, Fig. 2.4).

operator $\hat{N}$ on the SSET. The supercurrent through the SSET depends on the latter's state $|s\rangle$ and equals (cf. (2.127))

$$I_{\text{SSET},0(1)} = \frac{2\pi c}{\Phi_0} \frac{\partial E_{0(1)}(n^*)}{\partial \psi} = \pm \frac{2e}{\hbar} \frac{E_J^2}{8E_C} \frac{\sin\psi}{\sqrt{(1-n^*)^2 + (E_J/E_C)^2 \cos^2(\psi/2)}}.$$

The SSET quantum state can, therefore, be read out even in the charge degeneracy point,

$$I_{0(1)} = \pm \frac{2e}{\hbar} \frac{E_J}{4} \sin\frac{\psi}{2}, \tag{2.132}$$

since it determines the direction of the current.

Neglecting the self-inductance of the dc SQUID loop, we obtain from the flux quantization condition

$$\gamma = -\psi - 2\pi \frac{\widetilde{\Phi}}{\Phi_0}, \tag{2.133}$$

where γ is the phase drop on the large junction and $\widetilde{\Phi}$ is the external flux through the dc SQUID loop. This fixes the phase difference ψ. Now by tuning the flux $\widetilde{\Phi}$ and the bias current I_b one can send such an additional pulse of current (or flux) that the readout junction will switch to the resistive state with significantly larger probability if, e.g., the quantronium is in state $|0\rangle$, rather than in state $|1\rangle$.

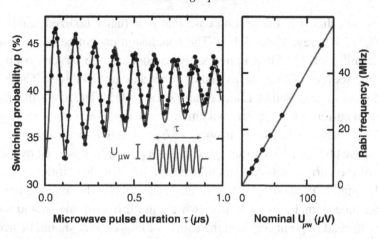

Fig. 2.17. Rabi oscillations in a quantronium (from Vion et al., 2002, reprinted with permission from AAAS). The measurements were performed at the base temperature 15 mK. The interlevel spacing, $\nu_{01} = E_{01}/h$, could be tuned in the range $\approx$ 12–18 GHz, and in the optimal spot the spectroscopically determined value $\nu_{01} = 16.46365$ GHz, with FWHM (full width at half-maximum) of 0.8 MHz only.

This method was used in the experiment by Vion et al. (2002). The manipulation of the qubit was realized through microwave pulses of the gate voltage, $V_g(t)$. In particular, Rabi oscillations were detected (Fig. 2.17). The relaxation and dephasing time extracted from this experiment were as high as $1.8\,\mu s$ and $0.5\,\mu s$, respectively.

We defer until Section 5.4 the theoretical description of the decoherence suppression at the optimal point. There we will also discuss another development of the charge qubit, the "transmon", the design of which maximizes this advantage.

2.4.5 *Charge and quasicharge. Bloch oscillations*

We have extensively used the coincidence between the Hamiltonian of a Josephson junction and the Hamiltonian of a particle in a potential. Putting the bias current I to zero in Eq. (2.55), we place our "particle" in a periodic sinusoidal potential. Since the periodicity is due to the superconducting phase defined modulo 2π, it would seem that after a 2π-shift the system will be indistinguishable from its initial state. In this case we would have a quantum pendulum, with a discrete set of levels. If, however, the 2π-shift produces observable changes in the system, we would have the same situation as in the band theory of solids (Kittel, 1987, 2004; Ziman, 1979) and will obtain energy bands, even though the periodic pattern arises, not from any actual periodic arrangement of atoms in space, but from the fact that the superconducting phase is defined only modulo 2π.

It turns out, that it is the second case that is actually realized (Likharev and Zorin, 1985; Likharev, 1986, § 1.4). The conclusion cannot be made based only on the Hamiltonian (2.55): it is necessary to consider the larger system, which includes the sources of bias fluxes and currents and gate voltages. There are two lines of argument proposed by Likharev and Zorin (1985). The first is that if even an infinitesimal current δI is applied to a junction, the potential term $-(\delta I \Phi_0 \phi / 2\pi c)$ in Eq. (2.33) leads to a difference in energy $\delta E = -(\delta I \Phi_0 / c)$ accompanying every 2π-shift of the phase ϕ. This difference can be, in principle, observed by measuring the state of the current source, which generates δI. The description, valid for an arbitrarily small bias, should work in its absence as well. The other argument does not invoke an external current source (which, in all rigour, one should somehow describe). Instead, it points out that the source of bias current should be irrelevant; one could, e.g., include the junction in a superconducting loop with a very large inductance, $L \gg c\Phi_0 / \delta I$. Then a 2π-shift of the phase would correspond to a change of the flux through the loop by Φ_0, which would also be observable.

Now, after the physics is established, one notes that the stationary Schrödinger equation with energy E for the Hamiltonian (2.55) is, mathematically, the Mathieu equation (Likharev and Zorin, 1985; Likharev, 1986):

$$\Psi(z)'' + (a - 2q \cos 2z)\Psi(z) = 0 \qquad (2.134)$$

where $z = \phi/2$, $a = 4E/E_C$, $q = -2E_J/E_C$. The general solution is found as a superposition of *Bloch states*

$$\Psi_k^{(n)}(z) = u_k^{(n)}(2z)e^{2ikz}, \quad u_k^{(n)}(2z + 2\pi) = u_k^{(n)}(2z), \qquad (2.135)$$

with $k \in (-\infty, \infty)$ and the eigenvalues periodic in k: $E^{(n)}(k+1) = E^{(n)}(k)$, forming energy bands (see Abramowitz and Stegun, 1964). The first Brillouin zone is, accordingly, $k \in [0, 1)$ (or $k \in [-1/2, 1/2)$, etc.). One can introduce the *quasicharge*,

$$q^* = 2ek, \qquad (2.136)$$

related to the charge on the junction in the same way as the quasimomentum of an electron is related to its momentum in band theory (Ziman, 1979, Chapter 1; Kittel, 1987, Chapter 9; Kittel, 2004, Chapter 7). The quasicharge is conserved modulo $2e$, with the extra charge taken or supplied by the superconducting condensate of the banks, like the extra momentum of a band electron is taken or supplied by the lattice as a whole in an *Umklapp process* (when the momentum of a particle in the periodic lattice changes by a vector of the reciprocal lattice times $\hbar$).

It is down to the physics, and not the mathematics, as to which solutions of Eq. (2.134) apply to the actual system described by it. That the choice of Bloch functions was right, was confirmed by the experimental observation (Kuzmin and

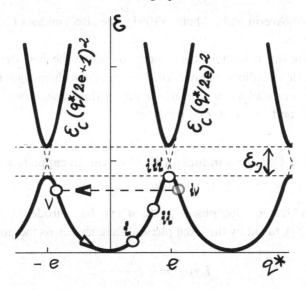

Fig. 2.18. Bloch oscillations (after Likharev and Zorin, 1985; Likharev, 1986).
The band structure $E^{(n)}(q^*)$ is sketched for the case $E_C \gg E_J$ (the counterpart
to the first correction to the "empty lattice" approximation (see, e.g., Kittel, 2004,
Chapter 7) in 1D). The bandgaps are opened by the Josephson coupling. A cycle
of Bloch oscillations is indicated by numbered circles. The quasicharge on the
Josephson junction grows ((i), (ii)) until it reaches e (iii). Then the quasicharge
returns to the first Brillouin zone, as points (iv) and (v) are equivalent. (Their
infinitesimal displacements from $q^* = \pm e$ are exaggerated.) Physically, the switch
from (iv) to (v) corresponds to recharging the Josephson junction (from $+e$ to
$-e$), as a single Cooper pair passes through. The frequency of Bloch cycles is
determined by the applied current via Eq. (2.137).

Haviland, 1991; Kuzmin, 1993) of theoretically predicted *Bloch oscillations* (see
Fig. 2.18). If the quasicharge were increased by an external current source, as soon
as it reaches the edge of the Brillouin zone, the quasicharge changes by $2e$, and
the process repeats. This leads to the oscillations with the bias *current*-dependent
Bloch frequency

$$f_B = I/2e, \tag{2.137}$$

"dual" to the ac Josephson effect, where the frequency of oscillations

$$f_J = eV/h \tag{2.138}$$

is determined by the bias *voltage*. Detailed discussion of the theory of Bloch oscillations and related effects in superconducting and non-superconducting devices (such
as SETs), as well as metrological and other applications and experimental evidence,
falls beyond the scope of this book and is best found in Likharev and Zorin (1985),

Likharev (1986), Averin and Likharev (1991); see also Tinkham (2004, § 7.6) and references therein.

For our purposes, it is more convenient to choose the first Brillouin zone as $q^* \in [0, 2e)$, which we did implicitly. The operation of a charge qubit involves only a narrow region around $q^* = e$, in the centre of this zone, where the difference between charge and quasicharge is irrelevant.

2.5 Quantum inductance and quantum capacitance

2.5.1 Quantum inductance

We have seen that the Josephson junction can be considered as a nonlinear inductance (2.27), tuned by the fixed phase difference across the junction,

$$L_J(\phi) = \frac{\hbar c^2}{2e I_c \cos \phi}.$$

We have seen also that the phase difference itself can be considered as a quantum variable, and it therefore makes sense to ask, what change does it make to the "classical" formula (2.27)? For example, if the junction is included in the loop of an rf SQUID qubit near the degeneracy point, the system will be in a superposition of states with different phase differences and, therefore, conceivably, different inductances.

Let us consider such a Josephson junction connected to an LC circuit and biased by an external flux $\widetilde{\Phi}$ (Fig. 2.19a). The Lagrangian of the system is

$$\mathcal{L} = \frac{(C + C_J)\dot{\Phi}^2}{2c^2} - \frac{(\Phi + \widetilde{\Phi})^2}{2L} + E_J \cos 2\pi \frac{\Phi}{\Phi_0}, \tag{2.139}$$

and the Hamilton function

$$\mathcal{H} = \frac{c^2 \Pi_\Phi^2}{2(C + C_J)} - E_J \cos 2\pi \frac{\Phi}{\Phi_0} + \frac{(\Phi + \widetilde{\Phi})^2}{2L}. \tag{2.140}$$

The quantum-mechanical Hamiltonian is obtained from the latter by exchanging $c\Pi_\Phi$ for $2e\hat{N} = -i\hbar c \partial_\Phi$. For small deviations $\delta\Phi$ from the equilibrium value of Φ (that is, $\Phi = \Phi_{\min} + \delta\Phi$, and $\Phi_{\min}$ is the minimum of the potential energy term in (2.140)) we obtain, in the same way as in deriving Eq. (2.58):

$$\delta\ddot{\Phi} = -\frac{c^2}{C + C_J} \frac{\partial^2}{\partial\Phi^2} \left\langle -E_J \cos 2\pi \frac{\Phi}{\Phi_0} + \frac{(\Phi + \widetilde{\Phi})^2}{2L} \right\rangle_{\Phi = \Phi_{\min}} \delta\Phi$$

$$= -\frac{c^2}{C + C_J} \frac{\partial^2}{\partial\Phi^2} \langle H \rangle \Big|_{\Phi = \Phi_{\min}} \delta\Phi. \tag{2.141}$$

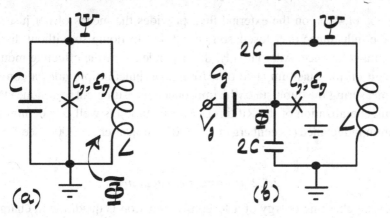

Fig. 2.19. Quantum inductance (a) and quantum capacitance (b) in an LC circuit.

(In this form this expression is valid for a general, not necessarily sinusoidal, functional form of the Josephson energy.) We can now identify the (inverse) effective inductance,

$$L_{\text{eff}}^{-1} = \frac{\partial^2}{\partial \Phi^2} \langle H \rangle \bigg|_{\Phi = \Phi_{\min}}, \tag{2.142}$$

where the average is taken over the quantum state of the system.

For example, if the Josephson junction is in its ground (excited) state, with energy $\epsilon_{0(1)}(\Phi, \tilde{\Phi})$ tuned by the external flux,

$$L_{\text{eff},0(1)}^{-1} = \frac{1}{L} + \frac{\partial^2 \epsilon_{0(1)}}{\partial \Phi^2} \bigg|_{\Phi = \Phi_{\min}} = \frac{1}{L} + \frac{\partial^2 \epsilon_{0(1)}}{\partial \tilde{\Phi}^2} \bigg|_{\Phi = \Phi_{\min}}. \tag{2.143}$$

The second term is the (inverse) *quantum inductance*, determined by the curvature of the given energy level as a function of flux (or of the external flux). In the classical limit it becomes simply $\partial_{\Phi,\Phi}^2 U_J(\Phi = \Phi_{\min})$, sensitive to the bias flux.

Indeed, it enters the expression as a contribution from an inductor connected in parallel with the LC circuit's own inductance, and one can readily relate this term to an additional voltage produced in the branch when an ac current flows through it. The particular way in which the junction is connected to the rest of the circuit is irrelevant. We could, e.g., consider an inductive coupling between a flux qubit and the LC circuit. This, all the same, would make the effective inductance of the circuit dependent on the qubit state. In the classical limit, this setup is used in the Rifkin–Deaver method (Rifkin and Deaver, 1976) of measuring the magnetic flux in an rf SQUID through the variation of impedance in an LC circuit inductively coupled to it.

Near the degeneracy point the contribution to quantum inductance from the ground and excited states, obviously, have opposite signs. This, together with the dependence of $\Phi_{\min}$ on the external flux, provides the means of not just manipulating the inductance of a Josephson circuit, but of doing so without disrupting its quantum coherence. Alternatively, the dependence of the effective inductance of a circuit on the quantum state of a Josephson junction provides a convenient way of measuring the qubit state (the impedance measurement technique (IMT), a quantum generalization of the Rifkin–Deaver method) as well as a demonstration of quantum inductance (Greenberg et al., 2002; Grajcar et al., 2004; see § 5.5.3).

2.5.2 Quantum capacitance

The classical charging energy of a Josephson junction is quadratic in charge, that is, it is a linear capacitor in the classical limit. It is only due to the quantum effects of order E_J/E_C that the nonlinearity appears; voltage-dependent Josephson capacitance – unlike Josephson inductance – is, therefore, an essentially quantum phenomenon and can only be seen when charge states are well defined. For the reasons explained at the end of the paragraph on charge qubits, a single Josephson junction will not work; an SSET is required. As before, for the analysis we include the quantum capacitor in an LC-circuit (Fig. 2.19b), with the Lagrangian

$$\mathcal{L} = \frac{C_g}{2c^2}\left(\dot{\Phi} - cV_g\right)^2 + \frac{C}{c^2}\left(\dot{\Phi} - \dot{\Psi}\right)^2 + \frac{C}{c^2}\dot{\Phi}^2 + \frac{C_J}{2c^2}\dot{\Phi}^2 + E_J\cos 2\pi\frac{\Phi}{\Phi_0} - \frac{\Psi^2}{2L}$$
(2.144)

and the Hamilton function

$$\mathcal{H} = \frac{1}{2}\frac{\left(c\Pi_\Phi + c\Pi_\Psi + C_gV_g\right)^2}{C_g + 2C + C_J} + \frac{c^2}{2}\frac{\Pi_\Psi^2}{2C} - E_J\cos 2\pi\frac{\Phi}{\Phi_0} + \frac{\Psi^2}{2L}.$$
(2.145)

Here, for a change, instead of the Heisenberg equations of motion we will use the Hamilton equations – the results will be the same. From the Hamilton equations

$$\dot{\Psi} = \frac{\partial\mathcal{H}}{\partial\Pi_\Psi}, \quad \dot{\Pi}_\Psi = -\frac{\partial\mathcal{H}}{\partial\Psi}$$
(2.146)

it follows that, since there are no terms in the Hamilton function that contain simultaneously Ψ and Π_Ψ, and $\mathcal{H}$ is not an explicit function of time,

$$\ddot{\Psi} = \frac{\partial^2\mathcal{H}}{\partial\Pi_\Psi\partial\Psi}\dot{\Psi} + \frac{\partial^2\mathcal{H}}{\partial\Pi_\Psi^2}\dot{\Pi}_\Psi = -\frac{\partial^2\mathcal{H}}{\partial\Pi_\Psi^2}\frac{\partial\mathcal{H}}{\partial\Psi} = -\left[\frac{\partial^2\mathcal{H}}{\partial\Pi_\Psi^2}\frac{\partial^2\mathcal{H}}{\partial\Psi^2}\right]\Psi.$$
(2.147)

The last equality holds exactly if the Hamilton function is quadratic in Ψ (as in this case). The expression in the brackets is the squared frequency of small oscillations near equilibrium, and we identify the quantity

$$C_{\text{eff}}^{-1} = \frac{1}{c^2} \frac{\partial^2 H}{\partial \Pi_\psi^2} \tag{2.148}$$

as the (inverse) effective capacitance of the circuit, similarly to the effective inductance (2.142). One can arrive at (2.148) or an equivalent expression in other ways, e.g., directly through the relation between induced charge and applied voltage (Averin and Bruder, 2003; Sillanpää et al., 2005; Duty et al., 2005; Johansson et al., 2006a).

In the quantum description, as usual, we replace the canonical momentum with the corresponding operator. In the absence of the SSET Josephson junction and neglecting single-electron resonance tunnelling, the device reduces to a classical LC circuit,

$$\mathcal{H} = \frac{c^2 \Pi_\psi^2}{2C} + \frac{1}{2} \frac{\left(c \Pi_\psi + C_g V_g\right)^2}{2C + C_g} + \frac{\psi^2}{2L}, \tag{2.149}$$

so that, as expected,

$$\ddot{\Psi} = c^2 \dot{\Pi}_\psi \left(\frac{1}{2C} + \frac{1}{2C + C_g} \right) = -\frac{c^2}{L C_{\text{eff}}} \psi. \tag{2.150}$$

The effective capacitance

$$C_{\text{eff}}^{-1} = (2C)^{-1} + (2C + C_g)^{-1} = \frac{1}{c^2} \frac{\partial^2 \mathcal{H}}{\partial \Pi_\psi^2}, \tag{2.151}$$

where $c \Pi_\psi$ is the electric charge in the node. Of course, the system is linear and the effective capacitance and resonance frequency do not depend on the gate voltage.

In the presence of a Josephson junction, while still treating the node flux Ψ as a classical variable, we will consider the superconducting flux variable, Φ, on the SSET quantum-mechanically and get rid of it by taking the expectation value of the Hamilton equations. Near the degeneracy point of the SSET its energy levels are given by Eq. (2.130) with $n^* = -(c\Pi_\psi/e + C_g V_g/e) \equiv -c\Pi_\psi/e + n_g$. Therefore, the expectation value of H when the SSET is in state $|s\rangle$ $(s = 0, 1)$ is, up to an additive constant,

$$H_s = \epsilon_s(n_g, \Pi_\psi) + \frac{c^2 \Pi_\psi^2}{4C} + \frac{\psi^2}{2L}, \tag{2.152}$$

where the energy of the SSET

$$\epsilon_{0(1)}(n_g, \Pi_\Psi) = \mp \frac{E_C}{2} \sqrt{\left(1 - n_g + \frac{c\Pi_\Psi}{e}\right)^2 + \left(\frac{E_J}{E_C}\right)^2}, \tag{2.153}$$

and $E_C = 2e^2/(C_g + 2C + C_J)$. The effective capacitance now depends on the SSET quantum state and can be tuned by the gate voltage:

$$C^{-1}_{\text{eff},0(1)}(n_g) = \frac{1}{c^2} \frac{\partial^2 \langle H \rangle}{\partial \Pi_\Psi^2} = \frac{1}{2C} \mp \frac{E_C}{2c^2} \frac{\partial^2}{\partial \Pi_\Psi^2} \sqrt{\left(1 - n_g + \frac{c\Pi_\Psi}{e}\right)^2 + \left(\frac{E_J}{E_C}\right)^2} \tag{2.154}$$

$$\approx \frac{1}{2C} \mp \frac{1}{C_g + 2C + C_J} \frac{\partial^2}{\partial n_g^2} \sqrt{\left(1 - n_g\right)^2 + \left(\frac{E_J}{E_C}\right)^2}.$$

The second term in this expression, which is the quantum correction to geometric inverse capacitance, is negative in the ground state. Therefore, the quantum contribution to the capacitance itself is *positive* in the ground state and negative in the excited state of the SSET (Sillanpää et al., 2005; Duty et al., 2005; Johansson et al., 2006a).

Expressions (2.153) and (2.154) apply not too far from the SSET degeneracy point, where only the two adjacent charge states can be taken into account. For an arbitrary SSET offset, one still can write for the effective capacitance

$$C^{-1}_{\text{eff},s}(n_g) = \frac{1}{2C} + \frac{c^2}{e^2} \frac{\partial^2}{\partial n_g^2} \langle \epsilon_s(n_g, \Pi_\Psi) \rangle \bigg|_{c\Pi_\Psi = Q_{\min}}, \tag{2.155}$$

where the exact expression for $\epsilon_s(n_g, \Pi_\Psi)$ is used; $Q_{\min}$ is the equilibrium value of the nodal charge, and the quantum statistical average was taken. Quantum capacitance was independently demonstrated in aluminium SSETs at temperatures below ~ 100 mK by Sillanpää et al. (2005) and Duty et al. (2005).

2.6 *Superconductivity effects in normal conductors

2.6.1 *Andreev reflection and proximity effect

An interesting situation arises when one considers (at zero temperature) the flow of the electric current, driven by an infinitesimally small bias, $\delta\mu = e\delta V$, through a clean interface between a normal conductor and a superconductor (the NS-boundary). In a normal conductor it is carried by quasiparticles – electronic excitations with energies within $\delta\mu$ of the Fermi energy (see, e.g., Kittel, 1987,

2004), while in a superconductor there are no quasiparticles with energies less than Δ, and the current is carried by the condensate of Cooper pairs. Therefore, some conversion mechanism is required.

This mechanism is provided by *Andreev reflection* (Andreev, 1964, see also, e.g., Zagoskin, 1998, § 4.5). The process is sketched in Fig. 2.20. An electron incident from the normal side with energy ϵ above the Fermi-level, μ, cannot penetrate the superconductor. Instead, it picks up another electron with the same energy,

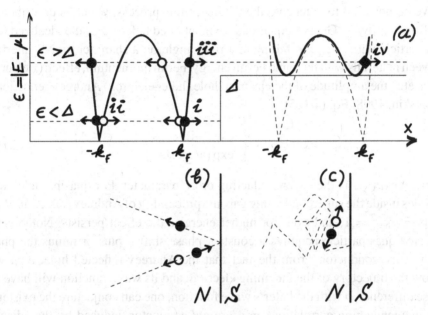

Fig. 2.20. Andreev reflection from a sharp NS boundary. Energies of quasiparticles are measured from the common Fermi level, μ, and we therefore distinguish electrons ($\epsilon = E - \mu > 0$, $k > k_F$, black circles) and holes ($\epsilon = \mu - E > 0$, $k < k_F$, white circles). In a superconductor, the quasiparticles are *bogolons*, the coherent superposition of electron- and hole-like excitations, and can only exist outside the energy gap (see, e.g., Zagoskin, 1998, Chapter 4). (a) An incoming electron with energy $\mu + \epsilon$, $\epsilon < \Delta$ (i) cannot penetrate the superconductor and must be reflected either as an electron (with momentum $\sim -k_F$) or a hole (with momentum $\sim k_F$) *with the same energy* ϵ. The latter process (Andreev reflection) is favoured if there is no potential barrier at the NS interface, which could provide the $\sim 2k_F$-change in the electron momentum required for the normal reflection. An incoming hole (ii) can undergo the same process. An electron (or hole) with energy above the gap (iii) can penetrate the superconductor, as an electron(hole)-like bogolon (iv, diamond), but can still be normally or Andreev-reflected. (b) Unlike normal reflection, Andreev reflection changes *all* the components of the velocity of the reflected particle, and the reflected hole (electron) will retrace the path of the incident electron (hole). (c) The same remains true in the presence of elastic scattering in the normal conductor.

almost the same momentum and opposite spin, forms a Cooper pair and joins the condensate on the superconducting side, thus, carrying the net momentum and the electric current of *two* electrons across the NS boundary. This leaves a *hole* on the normal side of the boundary. The hole's group velocity, $\mathbf{v}(\mathbf{p}) = \partial_{\mathbf{p}}\epsilon(\mathbf{p})$, is *opposite* to its momentum, so the hole moves away from the boundary (and retraces the path of the incoming electron). The whole process can, therefore, be considered as an Andreev reflection of an electron, when an electron impinging on the NS boundary is reflected back as a hole. The opposite process – that of the Andreev reflection of a hole – realizes the hole–electron conversion.

We do not need to enter into the details of this process, which is described by the Bogoliubov–de Gennes equations we mentioned before. For the idealized case of a particle with energy ϵ falling at a right angle on a sharp, clean NS interface between materials with the same Fermi energy, Fermi momentum, effective electron mass, etc., the amplitude of the electron–hole (hole–electron) Andreev reflection is (Zagoskin, 1998, Eq. (4.165))

$$r_{eh(he)} = \exp(\mp\phi) \times \begin{cases} \exp\left(-i \arccos\frac{\epsilon}{\Delta}\right), & \epsilon \le \Delta; \\ \exp\left(\operatorname{arccosh}\frac{\epsilon}{\Delta}\right), & \epsilon > \Delta, \end{cases} \qquad (2.156)$$

where $\Delta \exp(i\phi)$ is the superconducting order parameter. For quasiparticles with energies inside the superconducting gap, the probability of Andreev reflection $R_A = |r_{eh}|^2 = |r_{he}|^2 = 1$, and even for higher energies the effect persists. Notably, the reflected quasiparticle acquires a constant phase shift – plus or minus the phase of the superconductor. From the fact that the Andreev-reflected hole, e.g., will follow the trajectory of the incoming electron, and its wave function will have the phase difference ϕ with the latter's wave function, one can conjecture the existence of *superconducting* correlations in a *normal* conductor, induced by the adjacent superconductor – the *proximity effect*. This is indeed the case (see, e.g., Tinkham, 2004; Zagoskin, 1998). These correlations exist only in a finite layer near the boundary. The reason is simple. It is clear from Fig. (2.20) that the momenta of the incident electron (hole) and the Andreev-reflected hole (electron) are slightly different:

$$k_e - k_h \approx 2\frac{\partial k}{\partial E}\bigg|_{E=\mu} = \frac{2\hbar v_F}{\epsilon}. \qquad (2.157)$$

Therefore, the phase difference between the electron and Andreev-reflected hole will grow with distance from the reflection point, and at length $l \sim \hbar v_F/2\epsilon$ becomes significant. At finite temperatures, the characteristic energy spread of incoming particles $\epsilon \sim k_B T$, and therefore the *normal metal coherence length* in the clean case is

$$l_{T,\text{clean}} = \frac{\hbar v_F}{k_B T}. \qquad (2.158)$$

The clean case means that there is no impurity scattering in the normal layer and, therefore, the electron trajectories are straight lines (Fig. 2.20b). In the opposite limit there is so much elastic scattering (by impurities or defects) in the normal layer that electrons *diffuse* there, with some diffusion coefficient D. The loss of phase coherence between an electron and Andreev-reflected hole (or vice versa) is still determined by the distance along the trajectory (Fig. 2.20c), but it is the displacement from the NS boundary that can be directly measured. For diffusion, the displacement is $\langle x^2 \rangle = Dt$, and since between collisions the particle moves with speed $\approx v_F$, then

$$l_{T,\text{dirty}} = \sqrt{Dt} = \sqrt{\frac{Dl_{T,\text{clean}}}{v_F}} = \sqrt{\frac{\hbar D}{k_B T}}. \tag{2.159}$$

In either case, we see that superconducting correlations can exist in normal conductors. This allows us to use a normal layer between two superconductors to form a Josephson junction (SNS junction). One interesting property of such junctions is that the Josephson current there does not follow the $\sin\phi$-rule of Eq. (2.18).

2.6.2 *Andreev levels and Josephson current in SNS junctions*

If a normal layer is sandwiched between two superconductors, quasiparticles with energies below Δ will be trapped in a kind of potential well. As a result, there appear quantized *Andreev levels* (Kulik, 1970). Consider the one-dimensional case (Fig. 2.21). The simplest way to discuss them is with the quasiclassical quantization condition (Landau and Lifshitz, 2003, § 48),

$$\oint p(\epsilon)dq = 2\pi n\hbar, \tag{2.160}$$

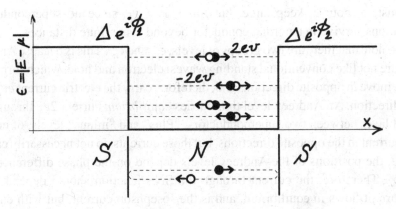

Fig. 2.21. Andreev levels in an SNS junction.

where p and q are the conjugate momentum and coordinate. For a quasiparticle in the normal layer, which undergoes two Andreev reflections before returning to the initial point, for the quantized level number n (assuming that Eq. (2.156) holds)

$$-2\arccos\frac{\epsilon_n^{\pm}}{\Delta} \pm (\phi_1 - \phi_2) + (k_e(\epsilon_n^{\pm}) - k_h(\epsilon_n^{\pm}))L, \qquad (2.161)$$

where $k_{e(h)}(\epsilon) = \sqrt{2m^*(\mu \pm \epsilon)}/\hbar$ is the wave vector of electron (hole) with effective mass m^*, and L is the length of the normal conductor. Eq. (2.161), generally has only numerical solutions, but in two limiting cases, $L = 0$ and $L \to \infty$, it gives explicit answers. If $L = 0$, (2.161) reduces to

$$\arccos\frac{\epsilon_n^{\pm}}{\Delta} = \frac{1}{2}(\pm\phi - 2\pi n),$$

and, therefore, there are only two degenerate levels:

$$\epsilon^+ = \epsilon^- = \Delta\cos\frac{\phi}{2}. \qquad (2.162)$$

If $L \to \infty$, there will be many levels in the well, and for the majority of them $\epsilon \ll \Delta$. We can, therefore, approximate the momentum difference in (2.161), $k_e(\epsilon_n^{\pm}) - k_h(\epsilon_n^{\pm}) \approx k_F\epsilon_n^{\pm}/\mu$, and set $\arccos(\epsilon_n^{\pm}/\Delta) \approx \pi/2$, so that

$$\epsilon_n^{\pm} = \frac{\hbar v_F}{2L}[\pi(2n+1) \pm \phi] \qquad (2.163)$$

(we must, of course, keep large, but finite, $L < l_T$, since no superconducting correlations survive in a normal conductor beyond some finite distance).

First, note that there are two groups of levels, ϵ_n^+ and ϵ_n^-. This is because Andreev levels are not like conventional standing waves: electron and hole, which form such a level, move in opposite directions and, therefore, carry the electric current ev in the same direction. An Andreev level thus carries *equilibrium* current $2ev$ through the normal layer between two superconductors. "Plus" and "minus" levels, of course, carry current in the opposite directions, but these currents do not necessarily cancel. Second, the positions of the Andreev levels depend on the phase difference $\phi = \phi_1 - \phi_2$. Therefore, the current through the SNS junction should depend on ϕ. This current flows in equilibrium, and is the Josephson current, but with current-phase dependence drastically different from the simple formula (2.18). At zero

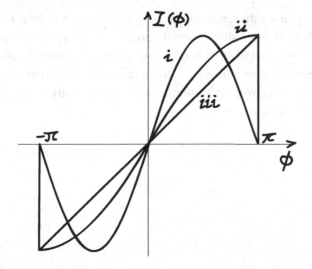

Fig. 2.22. Josephson current dependence on the phase difference at zero temperature: (i) tunnel junction, (ii) short clean SNS junction, (iii) long clean SNS junction.

temperature, in a short SNS junction (Kulik and Omelyanchuk, 1977; see also Zagoskin 1998, § 4.5.3): see Fig 2.22

$$I(\phi) = \frac{\pi \Delta}{e R_N} \sin \frac{\phi}{2}. \tag{2.164}$$

In a long junction instead (Ishii, 1970; Bardeen and Johnson, 1972; see also Zagoskin, 1998, § 4.5.4): see Fig. 2.22

$$I(\phi) = \frac{e v_F}{L} \left[\frac{2}{\pi} \sum_{s=1}^{\infty} (-1)^{s+1} \frac{\sin s\phi}{s} \right]. \tag{2.165}$$

Eq. (2.164), incidentally, gives the same $\sin \phi/2$-dependence as Eq. (2.127) gave for the resonant Josephson current through an SSET. In the long junction the current is a nice sawtooth.[21] At finite temperatures or with impurity scattering off the sharp edges, by suppressing first the high harmonics of (2.164, 2.165), the current will eventually be reduced to a basic sinusoidal shape.

The results we obtained in the previous sections will not depend on the shape of $I(\phi)$. The relation between the Josephson potential and Josephson current, Eq. (2.21), stands, and all one has to do is to consider a slightly different potential well. The presence of a normal layer is unpleasant from a different point of

[21] It is easy to check that the expression in square brackets in Eq. (2.165) converges to a 2π-periodic sawtooth of unit amplitude.

view: when the phase difference depends on time, there will be a Josephson voltage, which will tend to produce normal currents, with the accompanying loss of quantum coherence. On the other hand, some kinds of normal conductor may have sufficiently useful properties to outweigh this particular disadvantage – as, possibly, a two-dimensional electron gas in semiconductor heterostructures, which we will discuss in the next chapter.

3

Quantum devices based on two-dimensional electron gas

> ... Men by various ways arrive at the same end.
>
> M. de Montaigne, *Essays*, translated by Charles Cotton

3.1 Quantum transport in two dimensions

3.1.1 Formation of two-dimensional electron gas in heterojunction devices

Given all the advantages of superconducting structures, with their tunability, intrinsic protection against decoherence, well-understood physics and well-developed fabrication and experimental techniques, it would seem superfluous even to consider other possibilities for quantum engineering. Nevertheless, it would be short-sighted to neglect other possibilities, especially such rich ones as provided by devices based on a two-dimensional electron gas (2DEG). Here one has a *normal* electron system, which, nevertheless, maintains quantum coherence over comparatively large distances, and can literally be shaped into the desirable form (two-, one- or zero-dimensional) during an experiment by a simple turn of a knob. It can use both charge and spin degrees of freedom, and serve as a basis for qubits, sensitive quantum detectors, quantum interferometers or other interesting devices. It is edifying, showing that one does not necessarily need a macroscopic quantum state (like superconductivity) to observe macroscopic quantum coherence.

As a bonus, there are interesting and useful effects that can be realized in hybrid, superconductor-2DEG structures. Probably, if we had discussed these devices first, we would have asked, who needs superconductors?

The 2DEG layers can be formed, e.g., in MOSFETs (metal-oxide-semiconductor field effect transistors) (see, e.g., Smith, 1996). Nevertheless, the devices predominantly used in the research of interest to us are based on heterojunctions, mostly GaAs/AlGaAs (the so-called HEMT – high electron mobility transistor – devices). Unlike MOSFETs, here the 2DEG layer is produced, not by the pulling of electrons

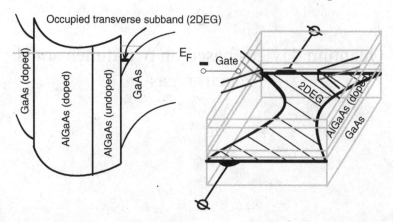

Fig. 3.1. The formation of a 2DEG in a heterojunction device. (Left) The band structure of an GaAs/AlGaAs heterostructure. (Right) The formation of a 2DEG quantum point contact using the split-gate technique.

from the doped semiconductor (silicon) under the positively charged gate, but due to the natural band mismatch between the two sides of the heterojunction. The main advantage of these devices is the high mobility of the electron gas. Other systems with a high-mobility 2DEG include Si/SiGe and InAs/AlSb heterostructures. For our discussion, which, after all, concentrates on the theory side, we will use GaAs/AlGaAs as the representative example.

A schematic structure of a system is shown in Fig. 3.1. The layers of doped and undoped GaAs and $Al_xGa_{1-x}As$ are placed on top of each other using the molecular beam epitaxy (MBE) technique. Due to the fact that the lattice constants of AlAs (5.660 Å) and GaAs (5.653 Å) are close, while the bandgaps are different, one can vary the concentration of aluminium to form the desired band structure.[1] As a result, an approximately triangular potential well is formed at the undoped GaAs/AlGaAs interface. The 2DEG layer is created by the electrons leaking there from the n-type donors, which, using the modulation doping, are placed across a spacer layer from the 2DEG. Therefore, the donors' electrostatic influence on 2DEG electrons is reduced to a spatially smooth scattering potential (additionally smoothened and weakened by the screening by the 2DEG itself, which will be discussed a little later). As a result, the mobility of electrons in a 2DEG, $\mu \sim 10^7 \, cm^2V^{-1}s^{-1}$, and their elastic scattering length, $l_e \sim 10 \, \mu m$, are remarkably high at moderately low temperatures ~ 1 K (Smith, 1996). The inelastic electron scattering length l_i (which is mainly due to electron–phonon interactions) in this case greatly exceeds l_e. When l_e and l_i are greater than the characteristic size

---

[1] Smith (1996) quotes from Casey and Panish (1978) the approximate formula for the bandgap in $Al_xGa_{1-x}As$:

$$E_{gap} \, (eV) = 1.424 + 1.247x.$$

of the device, the electronic transport is both quantum coherent and *ballistic*, with obviously important implications for our engineering applications. Other important properties of the system are a relatively high electron density, $\sim 10^{11-12}$ cm^{-2}, and low effective mass compared to the mass of a free electron, $m^* = 0.067m_0$.

The electron–electron scattering length in the layer, $l_{e-e} \sim 100$ μm, is even larger, and electrons can be considered in most cases as non-interacting. A free electron in the layer satisfies the stationary Schrödinger equation:

$$\left[-\frac{\hbar^2}{2m^*}\left(\frac{\partial^2}{\partial x^2} + \frac{\partial^2}{\partial y^2} + \frac{\partial^2}{\partial z^2}\right) + U(z) + V(x,y,z) \right]\psi(x,y,z) = E\psi(x,y,z).$$
(3.1)

Here $U(z)$ is the (approximately triangular) confining potential created by the band bending at the interface and the averaged electric field of the charged donors, while $V(x,y,z)$ describes the local fluctuations of their potential, contributions from other impurities, gate electrodes, etc. Expanding the wave function over the discrete eigenstates of the transverse Hamiltonian,

$$\left[-\frac{\hbar^2}{2m^*}\frac{\partial^2}{\partial z^2} + U(z) \right]\phi_j(z) = E_j\phi_j(z),$$
(3.2)

we obtain a set of coupled 2D Schrödinger equations:

$$-\frac{\hbar^2}{2m^*}\left(\frac{\partial^2}{\partial x^2} + \frac{\partial^2}{\partial y^2}\right)\psi_j(x,y) + \sum_k V_{jk}(x,y)\psi_k(x,y) = (E - E_j)\psi_j(x,y).$$
(3.3)

The transverse energies are approximately given by (Smith, 1996)

$$E_j = \left(\frac{\hbar^2}{2m^*}\right)^{1/2}\left(\frac{3\pi}{2}\left(j + \frac{3}{4}\right)eF\right)^{2/3},$$
(3.4)

where the effective transverse field per unit area $F = F_0 + \frac{en}{2\epsilon_0\epsilon}$, F_0 is the field of the ionized donors, n is the 2D density of electrons, and $\epsilon = 12.9$ is the dielectric constant of GaAs. The energy separation $E_1 - E_0 \sim 30$ meV, and for densities below $\sim 10^{12}$ cm^{-2} only the lower band is occupied. Moreover, the interband scattering potential given by

$$V_{jk}(x,y) \equiv \langle \phi_j | V | \phi_k \rangle = \int \phi_j^*(z)V(x,y,z)\phi_k(z)\,dz, \quad j \neq k$$
(3.5)

is negligible. Therefore, we are left with a single 2D equation

$$-\frac{\hbar^2}{2m^*}\left(\frac{\partial^2}{\partial x^2} + \frac{\partial^2}{\partial y^2}\right)\psi(x,y) + V(x,y)\psi(x,y) = E\psi(x,y),$$
(3.6)

where we have suppressed the now unnecessary subscripts, and denoted $V(x, y) = V_{00}(x, y) + E_0$.

The electron density is, of course,

$$n = 2\frac{\pi k_F^2}{4\pi^2} = \frac{k_F^2}{2\pi} = \frac{2\pi}{\lambda_F^2}, \tag{3.7}$$

which for a typical experimental value $n = 3 \times 10^{11}$ cm^{-2} corresponds to a long Fermi wavelength $\lambda_F \sim 100$ nm. (Typically in experiments $\lambda_F \approx 60$ nm.) The transport in such a 2DEG will reveal the wave nature of the electron as soon as the characteristic scale of the device nears λ_F.

Note that for 2D electrons the density of states (per spin) is constant

$$N^{(2D)}(E) = \frac{\partial}{\partial E} \frac{\pi k^2(E)}{4\pi^2} = \left(\frac{\partial}{\partial E} \frac{\hbar^2 k^2(E)}{2m^*}\right) \frac{2m^*}{4\pi \hbar^2} = \frac{m^*}{2\pi \hbar^2}, \tag{3.8}$$

equal to $(1/2) \times 2.8 \times 10^{10}$ cm^{-2}/meV for a GaAs/n-GaAs structure (Hiyamizu, 1990, § III.2). For $n = 3 \times 10^{11}$ cm^{-2} this yields $E_F = n/(2N^{(2D)}) \approx 10$ MeV $\sim$ 100 K. In a 1D case we would get instead

$$N^{(1D)}(E) = \frac{\partial}{\partial E} \frac{k(E)}{2\pi} = \left(\frac{\partial}{\partial E} \hbar k(E)\right) \frac{1}{2\pi \hbar} = \frac{1}{2\pi \hbar v(E)}, \tag{3.9}$$

where $v(E) = (\partial \hbar k / \partial E)^{-1} = \hbar k / m^*$ is the electron velocity. We will soon make use of this formula.

3.1.2 Conductance quantization in a point contact

If for a moment we ignore the not-so-important contribution of charged donors, the potential $V(x, y)$ in Eq. (3.6) would be due to the gate electrodes deposited on top of the structure (Fig. 3.1a). This *split-gate* technique was greatly improved over the years and now allows fabrication of pretty sophisticated structures, which can be maintained at different voltages and tuned independently. Therefore, the experimentally realizable shapes of $V(x, y)$ are quite versatile. The depleted 2DEG regions under the gates form a two-dimensional structure, which can be easily (for a theorist) tuned during an experiment by turning a few knobs. These electrostatically defined structures are inevitably smooth, which is an advantage, since any sharp corners would strongly scatter the electronic wave, reducing the efficiency of transport.

The simplest such structure is a point contact, a short constriction between two wide 2DEG regions ("source" and "drain", Fig. 3.1b). By increasing the negative voltage on the gate, this bottleneck can be completely pinched off. One can apply a

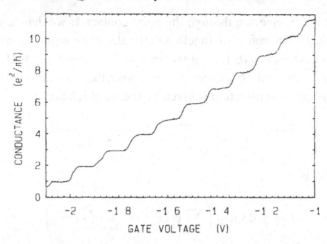

Fig. 3.2. Conductance quantization in a 2DEG point contact (reprinted with permission from van Wees et al., 1988. © 1988 American Physical Society).

bias eV between the source and drain and measure the conductance, $G = I/V$ (or, more precisely, *differential conductance*, dI/dV at $V = 0$). Experimental results (Wharam et al., 1988; van Wees et al., 1988) were surprising: within the accuracy of $\sim 1\%$ the conductance was quantized (Fig. 3.2),

$$G = n\frac{e^2}{\pi\hbar} \equiv nG_Q, \qquad (3.10)$$

in the same conductance quanta of $\approx (12.9 \text{ k}\Omega)^{-1}$, which we encountered in Chapter 2 when discussing superconducting charge qubits. While the accuracy of this quantization is a far cry from what is achieved for the quantum Hall effect, it is a remarkable phenomenon.

In order to understand what is going on, let us recall that under the conditions of the experiment the size of the bottleneck is small compared to the electron scattering length (both elastic and inelastic). Any electron arriving at the point contact will be scattered only by the gate potentials (i.e., by the boundary of the 2DEG region). We can, therefore, consider the electrons independently of each other.

The point contact eventually connects two electronic reservoirs, the "left" and "right" banks, or "source" and "drain", which are maintained at different chemical potentials:

$$\mu_L - \mu_R = eV. \qquad (3.11)$$

The interaction of electrons with phonons, impurities and each other in these reservoirs is strong enough to establish equilibrium at these chemical potentials, and they can be considered infinitely large, so that there is no effect on them whatsoever from electrons leaving or entering through the point contact. If we denote the properly normalized single-electron wave function of the electron coming from the left/right by $\psi_{k,q}^{L(R)}(x, y)$, where $k = 0, 1, \ldots$ is the transverse state in the 2DEG region (3.4), and q labels the states in the corresponding reservoir, then the corresponding electric current density per spin direction is given by the usual relation

$$\mathbf{j}^{L(R)}(x, y) = -\frac{e\hbar}{m^*}\,\mathrm{Im}\left[\psi_{k,q}^{L(R)}\nabla\psi_{k,q}^{L(R)*}\right]. \tag{3.12}$$

The net current is thus

$$I = \int \mathrm{d}S\,\mathbf{n}_S \cdot \left(\sum_q^{\mu_L}\mathbf{j}^L(x, y) - \sum_q^{\mu_R}\mathbf{j}^R(x, y)\right), \tag{3.13}$$

where the current density is integrated along an arbitrary surface (actually, line) S across the junction (Fig. 3.3). The details of the reservoirs, their connection to the 2DEG region, etc. are irrelevant as long as they supply impinging electrons with equilibrium distributions at respective chemical potentials. Therefore, the problem is essentially reduced to a scattering problem. We can consider the point contact as a scatterer placed in the middle of an infinitely large 2DEG layer, and employ as functions $\psi^{L(R)}$ the scattering states $\chi_{\mathbf{k}}^{L(R)}$, which are labelled by their wave vectors:

$$\chi_{\mathbf{k}}^L(x \to -\infty, y) \sim \mathrm{e}^{\mathrm{i}\mathbf{k}\,\mathbf{r}}; \quad \chi_{\mathbf{k}}^R(x \to \infty, y) \sim \mathrm{e}^{\mathrm{i}\mathbf{k}\,\mathbf{r}}. \tag{3.14}$$

These scattering states are mutually orthogonal and form a complete basis. We have thus reduced a transport problem to a scattering problem, and can write the

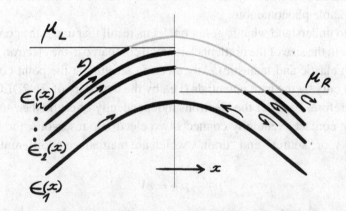

Fig. 3.3. Current through a ballistic point contact is carried by transverse subbands, which carry over the maximum of the corresponding effective transverse potential $\epsilon_n(x)$. Position-dependent subband occupation is shown in black.

net current as

$$I = \int dE \, (N_L(E)f(E - \mu_L)I_L(E) - N_R(E)f(E - \mu_R)I_R(E)), \qquad (3.15)$$

where $N_{L(R)}(E)$ is the corresponding density of states, $f(E - \mu)$ the Fermi distribution function, and $I_{L(R)}(E)$ the average current to the right (left) carried by the impinging electrons with energy E (and wave vectors $\mathbf{k}(E)$). The latter is determined by the properties of the bottleneck, and we will now find it.[2]

Let us return to the Schrödinger equation (3.3) and, following Glazman et al. (1988), use the adiabatic separation of variables. We seek a solution in the form

$$\psi_{nE}(x, y) = \varphi_n(x, y)\phi_{nE}(x), \qquad (3.16)$$

where the normalized function $\varphi_n(x, y)$, $\int_{-\infty}^{\infty} |\varphi_n(x, y)|^2 dy = 1$, is the eigenfunction of the transverse Hamiltonian:

$$\left(-\frac{\hbar^2}{2m^*} \frac{\partial^2}{\partial y^2} + V(x, y) \right) \varphi_n(x, y) = \epsilon_n(x)\varphi_n(x, y). \qquad (3.17)$$

Without loss of generality, we can assume that the width of the constriction (along the y-direction) is always finite, though it can be arbitrarily large. Then the transverse energies $\epsilon_n(x)$ belong to a discrete spectrum.

Now for the longitudinal function $\phi_{nE}(x)$ we obtain the equation

$$-\frac{\hbar^2}{2m^*} \frac{d^2\phi_{nE}(x)}{dx^2} = (E - \epsilon_n(x)) \phi_{nE}(x, y)$$

$$+ \frac{\hbar^2}{2m^*} \left(2\frac{d\phi_{nE}(x)}{dx} \cdot \frac{1}{\varphi_n(x, y)} \frac{\partial \varphi_n(x, y)}{\partial x} + \phi_{nE}(x) \cdot \frac{1}{\varphi_n(x, y)} \frac{\partial^2 \varphi_n(x, y)}{\partial x^2} \right).$$

$$(3.18)$$

The last term on the right, which contains the x-derivatives of φ_n, can be neglected, if we recall that the potential $V(x, y)$, and with it the eigenmodes of the transverse Hamiltonian, varies slowly in the x-direction. Therefore, we are left with a 1D Schrödinger equation with an effective potential barrier created by the "transverse" potential $\epsilon_n(x)$ (see Fig. 3.3). Electrons with energies exceeding the maximal value of $\epsilon_n(x)$ propagate across the barrier, and the rest are reflected back. The approximate WKB-type solution to (3.18) with a given energy E for a transverse mode n is

$$\phi_{nE}^{(s)}(x) = e^{is \int_{-s\infty}^{x} k_{nE}(x') \, dx'} \qquad (3.19)$$

---

[2] The rigorous treatment of quantum conductivity requires the introduction of nonequilibrium Green's functions, which goes beyond the scope of our discussion. See Zagoskin (1998, § 3.6) and references therein.

with $s = \pm$ determining the direction of propagation. The longitudinal wave vector

$$k_{nE}(x) = \frac{1}{\hbar}\sqrt{2m^*(E - \epsilon_n(x))} \tag{3.20}$$

is everywhere real for the propagating solutions ($E > \max \epsilon_n(x)$). The corresponding electric current density

$$j_x^{(s)}(x,y) = \frac{ie\hbar}{m^*}\left[\varphi_n(x,y)\phi_{nE}^{(s)}(x)\right]\frac{\partial}{\partial x}[\varphi_n(x,y)\phi_{nE}(x)]^* = s|\varphi_n(x,y)|^2\frac{e\hbar k_{nE}(x)}{m^*}, \tag{3.21}$$

and the partial current through the junction at a given point

$$I^{(s)}(x) = \int j^{(s)}(x,y)\,dy = s\frac{e\hbar k_{nE}(x)}{m^*} \equiv sev_{nE}(x), \tag{3.22}$$

where $v_{nE}(x) = (\partial k_{nE}(x)/\partial E)^{-1}$ is the velocity in the nth transverse subband. The solutions for which $k_{nE}(x)$ somewhere becomes complex are classically reflected at the point where $k_{nE}(x) = 0$ and can only tunnel beyond this point and do not contribute to the current.

The total current through the junction can therefore be expressed as

$$I = \int_{-\infty}^{\infty}\left[\sum_{\max \epsilon_n(x)\leq E}(f(E-\mu_L) - f(E-\mu_R))N_n(E,x)\frac{e\hbar k_{nE}(x)}{m^*}\right]dE, \tag{3.23}$$

where $N_n(E,x)$ is the longitudinal density of states in the transverse mode n at point x:

$$N_n(E,x) = \frac{\partial}{\partial E}\frac{k_{nE}(x)}{2\pi} = \frac{1}{2\pi\hbar}\frac{m^*}{\hbar k_{nE}(x)}, \tag{3.24}$$

exactly like in Eq. (3.9), except for the dependence on x. But this dependence does not actually matter, since in the expression for the current $N_n(E,x)$ enters in the combination $N_n(E,x)v_{nE}(x) = 1/2\pi\hbar$, which is position-independent. This was to be expected, since the total current through the contact cannot depend on where we calculate it. Finally, taking into account the twofold spin degeneracy, we find

$$I = 2\frac{e}{2\pi\hbar}\int_{-\infty}^{\infty}\left[\sum_{\max \epsilon_n(x)\leq E}(f(E-\mu_L) - f(E-\mu_R))\right]dE. \tag{3.25}$$

The differential conductance $G = dI/dV|_{V=0}$, where $\mu_L - \mu_R = eV$, can be directly obtained from (3.25). At zero temperatures, differentiation of a Fermi distribution yields a delta function, and eventually we obtain the quantized conductance

$$G = \frac{e^2}{\pi \hbar} \sum_{\max \epsilon_n(x) \le \mu} 1 = \frac{e^2}{\pi \hbar} n_{\max}. \tag{3.26}$$

Each transverse mode constitutes a quantum conductance channel with the conductance $e^2/\pi \hbar$, and the total conductance is their sum.

The result does not depend on the details of the transverse potential, except for the dependence of the number of open channels $n_{\max}$ on the gate voltage. If, e.g., we assume a (completely unrealistic) rectangular potential well of width $W(x)$ with infinitely high walls, then, obviously,

$$n_{\max} < N_F = \frac{k_F W_{\min}}{\pi} = \frac{W_{\min}}{\lambda_F/2}, \tag{3.27}$$

where $k_F = \sqrt{2m^*\mu}/\hbar$. This is usually a good rule-of-thumb estimate, and N_F has a transparent physical meaning as the number of Fermi half-wavelengths that can be packed across the narrowest part of the channel.

The corrections to Eq. (3.26), as long as we maintain the assumption of an adiabatically smooth potential profile, come from two sources. First, there is the tunnelling in the non-propagating modes through their effective barriers. This leakage is exponentially small: it is shown by Glazman et al. (1988) that the correction is

$$\Delta G_n = \frac{e^2}{\pi \hbar} e^{-\pi^2(N_F-n)\sqrt{2R/W_{\min}}}, \tag{3.28}$$

where $R \gg W_{\min}$ is the radius of curvature of the constriction near its narrowest point. Second, there is the *above-barrier reflection* in the propagating modes,[3] but it is also exponentially small due to the smoothness of the potential.

We can take into account deviations from the infinitely sharp step by introducing the form factor $0 \le T_n(E) \le 1$, so that

$$I = 2\frac{e}{2\pi \hbar} \int_{-\infty}^{\infty} \left[\sum_{\max \epsilon_n(x) \le E} T_n(E)(f(E - \mu_L) - f(E - \mu_R)) \right] dE; \tag{3.29}$$

$$G = \frac{e^2}{\pi \hbar} \sum_{\max \epsilon_n(x) \le \mu} T_n(\mu). \tag{3.30}$$

---

[3] This is a specifically quantum-mechanical phenomenon, which is in a sense a counterpart to quantum tunnelling. In tunnelling, a particle with energy E can leak through a potential barrier of magnitude $U > E$. In above-barrier reflection, instead, a particle can be reflected by a barrier of magnitude $U < E$ (Landau and Lifshitz, 2003, § 25).

If $\mu_L = \mu + \alpha eV$, $\mu_R = \mu - (1-\alpha)eV$, the differential conductance at a finite voltage V is given by

$$G(V) \equiv \frac{dI}{dV} = \frac{e^2}{\pi\hbar} \int_{-\infty}^{\infty} \left[\sum_{\max \epsilon_n(x) \leq E} T_n(E)(-\alpha \frac{d}{dE} f(E - \mu_L) \right.$$

$$\left. -(1-\alpha)\frac{d}{dE} f(E - \mu_R)) \right] dE$$

$$= \int_{-\infty}^{\infty} G(E)(-\alpha \frac{d}{dE} f(E - \mu_L) - (1-\alpha)\frac{d}{dE} f(E - \mu_R)) dE,$$

and the second derivative of the current–voltage characteristics

$$\frac{d^2 I}{dV^2} = -e \int_{-\infty}^{\infty} \frac{dG(E)}{dE} (\alpha^2 \frac{d}{dE} f(E - \mu_L) - (1-\alpha)^2 \frac{d}{dE} f(E - \mu_R)) dE.$$

At zero temperatures the above equations reduce to

$$G(V) = -\alpha G(\mu_L) - (1-\alpha) G(\mu_R) \tag{3.31}$$

and

$$\frac{d^2 I}{dV^2} = e \left[\alpha^2 \frac{dG(E)}{dE} \Big|_{E=\mu_L} - (1-\alpha)^2 \frac{dG(E)}{dE} \Big|_{E=\mu_R} \right]. \tag{3.32}$$

From here we can find the spacing between the bottoms of the transverse subbands from the dependence $d^2 I(V)/dV^2$ (Zagoskin, 1990), and infer the shape of the confining potential $V(x, y)$ in the y-direction. Indeed, the linear conductance $G(E)$ increases by $e^2/\pi\hbar$, and its derivative $dG(E)/dE$ contains sharp peaks when the next transverse channel opens, that is, when $E = \epsilon_n(0)$. Correspondingly, $d^2 I/dV^2$ will contain positive and negative peaks. If at zero voltage the chemical potential lies between $\epsilon_n(0)$ and $\epsilon_{n+1}(0)$, then the first positive peak will appear at $eV_+ = (\epsilon_{n+1}(0) - \mu)/\alpha$, and the first negative one at $eV_- = (\mu - \epsilon_n(0))/(1-\alpha)$. Assuming that the potential drop is symmetric, $\alpha = 1/2$, we see that

$$\epsilon_{n+1}(0) - \epsilon_n(0) = \frac{eV_+ + eV_-}{2} \tag{3.33}$$

(see Fig. 3.4).

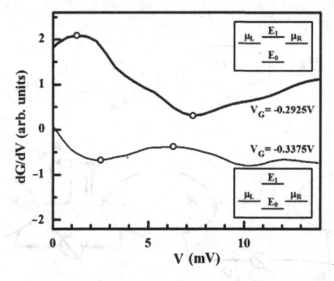

Fig. 3.4. Derivative of the differential conductance of a point contact in a Si/SiGe heterostructure as a function of applied bias voltage for two different gate voltages, V_G (reprinted with permission from Scappucci et al., 2006, © 2006 American Physical Society). The intersubband spacing can be estimated from the positions of the first two extrema using Eq. (3.33).

3.1.3 Quantum transport from scattering matrix: Landauer formalism. Landauer formula and its modifications

The electrical conductance through a point contact is the simplest example of nonlocal quantum transport. Here the processes which determine the conductivity (coherent scattering of the electron wave by the gate potential) are spatially separated from the processes that are responsible for decoherence, including energy and momentum relaxation – these processes take place in the thermal reservoirs and are, so to speak, taken to the outside of the parentheses. The elegance and power of this minimalist picture is because no explicit description of these non-unitary processes, which caused us so much trouble in Chapter 1, is necessary.

Clearly, this approach is not limited to the case of a point contact in a 2DEG. Any system through which a coherent propagation of the electron waves is possible, will do. Ballistic propagation is not actually important as long as the electron does not undergo inelastic scattering (e.g., by phonons, other electrons or fluctuating impurities), that is, the electron wave does not undergo random phase variations.

Since the non-unitarity does not need to be explicitly introduced, the general description of transport (Landauer, or Landauer–Büttiker, formalism) is based on the unitary scattering matrix of the system (see, e.g., Imry, 2002, §5.2).

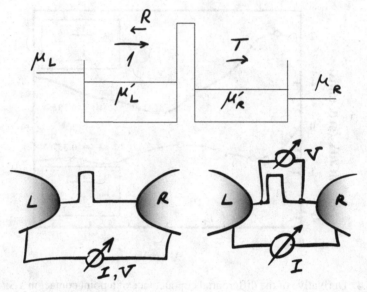

Fig. 3.5. Landauer formalism in the 1D case: two- and four-point conductance.

Originally Landauer (1957) considered the situation where an ideal 1D conductor with an internal barrier connected two thermal reservoirs with chemical potentials μ_L and $\mu_R = \mu_L - eV$ (Fig. 3.5). If the transmission coefficient of the barrier is $T(E)$, then the current through the wire (cf. Eqs (3.23, 3.24))

$$I = \frac{e}{\pi\hbar} \int_{-\infty}^{\infty} (f(E - \mu_L) - f(E - \mu_R))T(E)\mathrm{d}E, \qquad (3.34)$$

and in the limit of zero temperature, and assuming that the transmission coefficient is energy-independent, this yields

$$G = \frac{I}{V} = \frac{eI}{\mu_L - \mu_R} = \frac{e^2}{\pi\hbar}T. \qquad (3.35)$$

This result is precisely what we have obtained for a quantum point contact (QPC) with a single transmission channel, but not what Landauer himself originally derived, namely, not

$$G^{(4pt)} = \frac{e^2}{\pi\hbar}\frac{T}{1-T} \equiv \frac{e^2}{\pi\hbar}\frac{T}{R} \qquad (3.36)$$

(the meaning of $(4pt)$ will become clear in a moment). The discrepancy is due to the difference in the definition of conductance. Landauer was interested in the conductance of the barrier itself. Therefore he defined conductance as the ratio

$$G^{(4pt)} = \frac{eI}{\mu_L' - \mu_R'}, \qquad (3.37)$$

where the chemical potentials are taken not in the reservoirs, but in the 1D *leads*. To measure such a quantity we would have to use a *four-point* scheme: two electrodes would measure the current between the reservoirs, and another two the voltage between the leads. This is a different setup from the two-point scheme, whereby the conductance of Eq. (3.35) can be measured. Therefore, as the operational definitions of G and $G^{(4pt)}$ are different, there is nothing strange in the difference between (3.35) and (3.36).

Still, before measuring the difference between the chemical potentials of the leads in the four-point scheme, one can and should question how a chemical potential – an essentially equilibrium characteristic – can be ascribed to an essentially nonequilibrium system. Indeed, considering for the sake of simplicity the case of zero temperature, the left lead contains (a) right-moving electrons with energies from $-\infty$ to μ_L, incident from the left reservoir; (b) left-moving electrons with the same energies, reflected from the barrier (there will be fewer of them, since some will have gone through); and (c) left-moving electrons with energies from $-\infty$ to μ_R, incident from the right reservoir, which have passed through the barrier. In the right lead, the picture is similar, with obvious left–right index swapping. (Fortunately we don't need to bother with multiple reflections, since by assumption any electron arriving at the reservoir is instantaneously and completely absorbed.) The picture is far from equilibrium. In each point there are different "tribes" of electrons, with their own history-dependent chemical potentials.

One way to deal with the situation is to relate effective chemical potentials μ_L', μ_R' to the electronic densities in the leads. For example, in the left lead the density

$$n_L = \int N(E) \times \frac{1}{2}(f(E - \mu_L) + Rf(E - \mu_L) + Tf(E - \mu_R))\,dE \qquad (3.38)$$

$$= \int N(E)f(E - \mu_L')\,dE.$$

In the first line we explicitly account for the contributions (a), (b) and (c). At zero temperature and assuming constant $N(E)$, this definition of μ_L', and similarly for μ_R', leads to

$$\mu_L' = \frac{1}{2}[\mu_L(1 + R) + T\mu_R], \quad \mu_R' = \frac{1}{2}[\mu_R(1 + R) + T\mu_L], \qquad (3.39)$$

and

$$\mu_L' - \mu_R' = R[\mu_L - \mu_R]. \qquad (3.40)$$

Since the 1D density of states $N(E)$ is a slow function of energy, it will not change significantly on the scale of $\mu_L - \mu_R$, and our assumption of constant $N(E)$ is justified in arriving at this formula. Substituting (3.40) in (3.37), we recover the four-point Landauer conductance (3.36).

Comparing Eqs (3.36) and (3.35), we see that the two-point resistance of the system, $G^{-1} = (\pi\hbar/e^2)(1/T) = (\pi\hbar/e^2)(R/T + 1) = (G^{(4pt)})^{-1} + 2(\pi\hbar/2e^2)$, exceeds the four-point resistance by the so-called constriction, or Sharvin, resistance, $2(\pi\hbar/2e^2)$ (see Zagoskin, 1998, § 3.6). We should ascribe the latter to the points of contact between the reservoirs and the 1D leads, where the electrons' motion is drastically restricted. Each contact contributes a half quantum resistance unit. An analogy in classical statistical physics is given by the resistance of a small hole to a flow of rarefied gas (the Knudsen regime).[4] Without such a resistance it would have been impossible to maintain the two equilibrium reservoirs at their different chemical potentials.

Four-point conductance provides information about the transmission/reflection properties of the barrier, but two-point conductance is the more fundamental quantity: it characterizes the current and voltage between the equilibrium reservoirs, where the chemical potentials are well defined.

The one-dimensional leads are an especially useful abstraction, because the density of states there is inversely proportional to the velocity and, therefore, cancels the velocity in the definition of current. We saw this working in Eqs (3.23) and (3.34). We also saw that in a quantum point contact these leads act as transverse modes, each with its own longitudinal wave vector, velocity and 1D density of states. When deriving (3.29) and (3.30) we treated each transverse mode separately, characterizing it by its transmission coefficient T_n. This was justified by the smoothness of the QPC potential, though even in this situation, scattering between different transverse modes is possible (e.g., Zagoskin and Shekhter, 1994). For a general mesoscopic system, one has no reason to make such simplifications, and it is necessary to introduce the full scattering matrix,

$$S(E) = \begin{pmatrix} r(E) & \bar{t}(E) \\ t(E) & \bar{r}(E) \end{pmatrix}. \tag{3.41}$$

This is a $2N_\perp \times 2N_\perp$-matrix, where $N_\perp$ is the number of transverse channels in each lead. The $N_\perp \times N_\perp$ blocks $t(E), r(E)$ describe the electrons incident from the left reservoir at energy E. The transmission coefficient from the mode i on the left into mode j on the right is thus $T_{ij} = |t_{ij}|^2$. The unitarity of S ensures the conservation of the current (Fisher and Lee, 1981). This means, in particular, that

$$\sum_{i,j=1}^{N_\perp} \left(|t_{ij}|^2 + |r_{ij}|^2 \right) = N_\perp, \tag{3.42}$$

---

[4] See, e.g., Pitaevskii and Lifshitz (1981), § 15.

that is, the total current of electrons with a given energy from the left reservoir (proportional to the number of channels) is either transmitted to the right reservoir, or reflected back to the left one. The net current in the left lead is (taking into account spin degeneracy)

$$
\begin{aligned}
I &= \int \sum_{i=1}^{N_\perp} 2N_i(E) e v_i(E) \left[\left(1 - \sum_j |r_{ij}(E)|^2 \right) f(E - \mu_L) \right. \\
&\qquad \left. - \left(\sum_j |\bar{t}_{ij}(E)|^2 \right) f(E - \mu_R) \right] dE \\
&= \frac{e}{\pi \hbar} \int \sum_{i,j} |\bar{t}_{ij}(E)|^2 (f(E - \mu_L) - f(E - \mu_R)) dE \\
&\approx \frac{e(\mu_L - \mu_R)}{\pi \hbar} \int \sum_{i,j} |\bar{t}_{ij}(E)|^2 \left(-\frac{df(E)}{dE} \right) dE \\
&= \frac{e(\mu_L - \mu_R)}{\pi \hbar} \int \operatorname{tr} \left(\bar{t}(E) \bar{t}(E)^\dagger \right) \left(-\frac{df(E)}{dE} \right) dE.
\end{aligned}
\tag{3.43}
$$

At zero temperatures this yields the two-point conductance, expressed only through the transmission submatrices of the unitary scattering matrix at the Fermi level (Fisher and Lee, 1981)

$$
G = \frac{e^2}{\pi \hbar} \operatorname{tr} \left(\bar{t}(E_F) \bar{t}(E_F)^\dagger \right) = \frac{e^2}{\pi \hbar} \operatorname{tr} \left(t(E_F) t(E_F)^\dagger \right).
\tag{3.44}
$$

If the eigenvalues of $\bar{t}(E_F)\bar{t}(E_F)^\dagger$ (or $t(E_F)t(E_F)^\dagger$) are denoted by T_n, then (3.44) can be written as

$$
G = \frac{e^2}{\pi \hbar} \sum_{n=1}^{N_\perp} T_n
\tag{3.45}
$$

reminiscent of (3.30); but while there the T_n's denoted transparencies of separate, non-mixing modes, in (3.45) we cannot generally ascribe any T_n to a specific transverse mode.

The four-point conductance can also be determined in this case (see, e.g., Imry, 2002, §5.2.2). Introducing, as before, the effective chemical potentials in the leads via

$$n_L = \frac{1}{2} \sum_i \int N_i(E) \left(f(E - \mu_L)(1 + \sum_j |r_{ij}(E)|^2) \right.$$

$$\left. + f(E - \mu_R) \sum_j |\bar{t}_{ij}(E)|^2 \right) dE$$

$$= \sum_i \int N_i(E) f(E - \mu'_L) dE \qquad (3.46)$$

and similarly for the electron density n_R in the right lead, we find at zero temperature

$$G^{(4pt)} = \frac{e^2}{\pi \hbar} \frac{2 \sum_{ij} |t_{ij}|^2 \sum_i v_{F,i}^{-1}}{\sum_i v_{F,i}^{-1}(1 + \sum_j (|r_{ij}|^2 - |t_{ij}|^2))}. \qquad (3.47)$$

If all longitudinal Fermi velocities, $v_{F,i}$, were the same, this would simplify to become $N_\perp$ times the four-point conductance of Eq. (3.36).

3.1.4 Quantum point contact as a quantum detector

The importance of a quantum point contact is not limited to the role of a nice entry-level example of a 2DEG-based quantum device. This is a working device, useful as a *quantum detector* of electric charge. The idea behind the QPC detector is simple. If the gate voltages are tuned to just pinch off the QPC, a small variation of the electrostatic potential in its vicinity will change its conductance by a finite value, the conductance quantum, which is quite detectable (Fig. 3.6). This is a natural way of measuring the charge state of a gate connected to a quantum system. It is enough to determine whether the QPC conducts electricity. In accordance with general quantum-mechanical ideology, the measurement should lead to the quantum collapse of the state of our system:

$$\rho_i = |\psi_i\rangle\langle\psi_i| \equiv \left(\sum c_i |i\rangle\right) \left(\sum c_j^* \langle j|\right) \to \rho_f \equiv \sum |c_j|^2 |j\rangle\langle j|. \qquad (3.48)$$

In the case of a QPC detector this comes about in a rather direct way.

We have seen that quantum transport through a QPC is naturally described within the Landauer formalism in terms of the scattering matrix S, without referring to the details of the scattering potential, which makes it a very convenient tool. In our case, the S-matrix would depend on the state $|j\rangle$ of the measured system: $S[|j\rangle] \equiv S_j$,

Fig. 3.6. Scheme of a 2DEG quantum point contact charge detector. The transparency of the effective barrier depends on the gate potential, and therefore on the quantum state $|j\rangle$ of the measured system.

with the corresponding change in transport. As a matter of fact, any mesoscopic quantum coherent system can be described in the same way and would constitute a *mesoscopic scattering detector* (Clerk et al., 2003). A QPC with its easily manipulated conductance is one of its most convenient implementations.

Let us first consider the effect of measurement on the system following the transparent heuristic argument of Averin and Sukhorukov (2005). Describing the QPC by a scattering matrix

$$S_j = \begin{pmatrix} r_j & \bar{t}_j \\ t_j & \bar{r}_j \end{pmatrix} \tag{3.49}$$

we can describe the "measured system plus detector" before and after the interaction by the wave functions

$$|in\rangle \otimes \sum c_j|j\rangle \rightarrow \sum c_j \left(r_j|L\rangle + t_j|R\rangle \right) \otimes |j\rangle. \tag{3.50}$$

Here $|in\rangle$ is the initial state of an individual electron impinging on the QPC from the left reservoirs, and $|L\rangle$, $|R\rangle$ correspond to the outgoing states of this electron in the corresponding reservoir. The reduction to the system's wave function is achieved by tracing out the electron states and obtaining for the initial, $\bar{\rho}_i$, and final, $\bar{\rho}_f$, the reduced density matrix:

$$\bar{\rho}_i = \text{tr}_e\{|in\rangle\langle in| \otimes |\psi_i\rangle\langle\psi_i|\} = \rho_i; \qquad (3.51)$$

$$\bar{\rho}_f = \text{tr}_e\left\{\left(\sum c_j\left(r_j|L\rangle + t_j|R\rangle\right) \otimes |j\rangle\right)\left(\sum c_k^*\left(r_k^*\langle L| + t_k^*\langle R|\right) \otimes \langle k|\right)\right\}$$

$$= \sum_{jk} c_j c_k^*(t_j t_k^* + r_j r_k^*)|j\rangle\langle k|. \qquad (3.52)$$

Since $|t_j t_k^* + r_j r_k^*| \leq 1$ (for $j \neq k$ there is equality only if $S_j = S_k$, that is, the QPC is insensitive to the state of the system and no measurement takes place), Eq. (3.52) indeed describes the dephasing: the off-diagonal elements of the density matrix of the measured system are suppressed.

The measurement can be thought of as due to the sequence of electrons passing through the QPC; the QPC conductance in the final analysis reflects the statistics of such passages and, therefore, of the state of the measured system. Electrons that impinge on the QPC can have any energy and arrive from any reservoir. Therefore, in addition to (3.52) we should consider the other possibilities and weight them with the appropriate Fermi distribution functions ($f_L(E) = n_F(E - \mu_L)$, $f_R(E) = n_F(E - \mu_R)$, $\mu_L - \mu_R = eV$, see Fig. 3.3). For example, the "event" described by Eq. (3.52) is possible if the state with energy E is occupied in the left and empty in the right reservoir, and it must, therefore, be weighted with the factor $f_L(1 - f_R)$. The average change in the density matrix element $\bar{\rho}_{jk}$ due to a one-electron "event" with energy E is thus

$$\bar{\rho}_{jk} \rightarrow \bar{\rho}_{jk} a_{jk},$$

$$a_{jk} = \left| f_L(E)(1 - f_R(E))(t_j t_k^* + r_j r_k^*) + f_R(E)(1 - f_L(E))(\bar{t}_j \bar{t}_k^* + \bar{r}_j \bar{r}_k^*) \right.$$

$$\left. + (1 - f_L(E))(1 - f_R(E)) + f_L(E)f_R(E)e^{i\alpha_{jk}} \right|. \qquad (3.53)$$

The first term in the modulus is the process (3.52); the second is the reverse process, when an electron impinges from the right reservoirs; the third is when there are no electrons at this energy, so nothing happens at all; and the last term is the case when both right and left states are occupied, then there is still no transport, but extra electrons may change the energy of the system and thus produce a phase shift in the density matrix. After many such events happening during a sufficiently long time interval t, we see that on average (assuming that the "events" with energy E happen at a characteristic frequency E/h, which is the only reasonable quantity with proper dimensionality)

$$\bar{\rho}_{jk}(t) = \bar{\rho}_{jk}(0)\langle\Pi a_{jk}\rangle \approx \bar{\rho}_{jk}(0)\exp\left[\left\langle\sum \ln a_{jk}\right\rangle\right]$$

$$\approx \bar{\rho}_{jk}(0) \exp\left[t \int \frac{dE}{2\pi\hbar} \Big| f_L(E)(1 - f_R(E))(t_j t_k^* + r_j r_k^*) \right.$$

$$+ f_R(E)(1 - f_L(E))(\bar{t}_j \bar{t}_k^* + \bar{r}_j \bar{r}_k^*)$$

$$\left. + (1 - f_L(E))(1 - f_R(E)) + f_L(E) f_R(E) e^{i\alpha_{jk}} \Big| \right]. \tag{3.54}$$

Therefore the off-diagonal density matrix elements indeed decay exponentially in the limit $t \to \infty$:

$$\bar{\rho}_{jk}(t) \approx \bar{\rho}_{jk}(0) e^{-\Gamma_{jk} t};$$

$$\Gamma_{jk} = -\int \frac{dE}{2\pi\hbar} \Big| f_L(E)(1 - f_R(E))(t_j t_k^* + r_j r_k^*) + f_R(E)(1 - f_L(E))$$

$$(\bar{t}_j \bar{t}_k^* + \bar{r}_j \bar{r}_k^*) + (1 - f_L(E))(1 - f_R(E)) + f_L(E) f_R(E) e^{i\alpha_{jk}} \Big|. \tag{3.55}$$

We have derived the *back-action dephasing* by a QPC detector. If the system to be measured is a qubit, i.e., $|j\rangle = |0\rangle, |1\rangle$, then there is only one dephasing rate, $\Gamma_{01} \equiv \Gamma$, which at zero temperature becomes

$$\Gamma = -\frac{eV}{2\pi\hbar} \ln |t_1 t_2^* + r_1 r_2^*|. \tag{3.56}$$

3.1.5 *Back-action dephasing by a QPC detector: a more rigorous approach*

The way we obtained the key Eq. (3.55) is just a hand-waving demonstration. The actual derivation starts with the Liouville equation for the total density matrix, $\rho(t)$, with the Hamiltonian $H = H_s + H_e[|j\rangle]$, where H_s describes the evolution of the measured system, and $H_e[|j\rangle] \equiv H_0 + H_j$ is the Hamiltonian of the quantum point contact, including the scattering potential, which depends on the state $|j\rangle$ of the system. Assuming, as usual, that the initial density matrix was factorized, $\rho(0) = \rho_s(0) \otimes \rho_e(0)$, and also that $H_s = 0$ (that is, the system does not evolve in the process of measurement, except due to the measurement process itself), and using the cyclic invariance of trace, we easily find for the reduced density matrix of the measured system, $\bar{\rho}(t) = \mathrm{tr}_e \rho(t)$, the expression

$$\bar{\rho}_{jk}(t) = \bar{\rho}_{jk}(0) \langle e^{iH_k t} e^{-iH_j t} \rangle; \tag{3.57}$$

the average is taken over the unperturbed state of the electron system in the point contact. Averin and Sukhorukov (2005) proceed using the original approach from *full counting statistics* (Levitov et al., 1996). Here we will instead apply an approach of Schönhammer (2007) based on Klich's trace formula (Klich, 2003):

$$\mathrm{tr}_{\mathrm{Fock}}\left(e^{\tilde{A}} e^{\tilde{B}}\right) = \left[\det\left(\hat{1} - \xi e^A e^B\right)\right]^{-\xi}. \tag{3.58}$$

Here $\tilde{A}$ is the second-quantized representation of a one-particle operator A, $\hat{1}$ is a unit operator, and $\xi = 1$ for bosons and -1 for fermions. The trace on the left is taken over the Hilbert space of a system of identical fermions (bosons), that is, a Fock space, while the determinant on the right is taken over the Hilbert space of single-particle operators, which is certainly a simpler task. This equation is readily generalized to a product of an arbitrary number of operator exponents using the Baker–Hausdorff formula (for a proof see, e.g., Gardiner and Zoller, 2004, Appendix 4A), which allows us to switch from $e^{\tilde{A}} e^{\tilde{B}}$ to $e^{\tilde{C}}$ and back. This formula,

$$e^{\hat{a}+\hat{b}} = e^{\hat{a}} e^{\hat{b}} e^{-\frac{1}{2}[\hat{a},\hat{b}]} = e^{\hat{b}} e^{\hat{a}} e^{\frac{1}{2}[\hat{a},\hat{b}]}, \tag{3.59}$$

holds for any two operators $\hat{a}, \hat{b}$, which both commute with their commutator: $[\hat{a}, [\hat{a}, \hat{b}]] = [\hat{b}, [\hat{a}, \hat{b}]] = 0$.

We will also use the well-known formula

$$\ln \det M = \operatorname{tr} \ln M. \tag{3.60}$$

As usual, the logarithm of a matrix, or an operator, should be understood as the sum of the corresponding series.[5]

The average of the exponents in (3.57) has the right form for Klich's formula:

$$\langle e^{iH_k t} e^{-iH_j t} \rangle = \operatorname{tr}_{\text{Fock}} \left(e^{iH_k t} e^{-iH_j t} \rho_e(0) \right) = \operatorname{tr}_{\text{Fock}} \left(e^{\frac{\Omega}{k_B T}} e^{-\frac{H_0'}{k_B T}} e^{iH_k t} e^{-iH_j t} \right). \tag{3.61}$$

Since what is fixed is the chemical potentials in the left/right reservoirs, and not the number of electrons in them, we use the grand canonical ensemble; in (3.61) $H_0' = H_0 - \mu N$, and Ω is the grand potential. The operators in this equation are second-quantized with respect to the Fock space of the electrons coming from the left and right reservoirs. It is therefore convenient to consider them as two-by-two matrices in the space spanned by "left–right" electron states. For example,

$$H_0' = \begin{pmatrix} H_0 - \mu_L N_L & 0 \\ 0 & H_0 - \mu_R N_R \end{pmatrix}. \tag{3.62}$$

---

[5] The formula (3.60) is easy to verify for matrices of finite dimensionality. Any square matrix M with complex elements allows a *Schur decomposition*:

$$M = U M_S U^\dagger,$$

where U is a unitary matrix, and M_S is an upper diagonal one (Schur form of matrix M). For an arbitrary function F of a matrix,

$$U F(M) U^\dagger = U \left(\sum f_k M^k \right) U^\dagger = \sum f_k (U M U^\dagger)^k = F(U M U^\dagger).$$

The trace is unitarily invariant. Therefore, we can calculate both sides of (3.60) directly for the Schur form of M. Eq. (3.60) obviously holds: $\operatorname{tr} \ln M_S = \sum (\ln \mu_i)$, and $\ln \det M_S = \ln (\prod \mu_i)$, where μ_i are the eigenvalues of M_S (and M).

Using the Klich formula with $\xi = -1$,

$$\mathrm{tr}_{\mathrm{Fock}}\left(\rho_e(0)e^{iH_kt}e^{-iH_jt}\right) = e^{\frac{\Omega}{k_BT}}\,\mathrm{tr}_{\mathrm{Fock}}\left(e^{-\frac{H_0'}{k_BT}}e^{iH_kt}e^{-iH_jt}\right)$$

$$= e^{\frac{\Omega}{k_BT}}\det\left(\hat{1}+e^{-\frac{H_0'}{k_BT}}e^{iH_kt}e^{-iH_jt}\right)$$

$$= \det\left(\hat{1}+e^{-\frac{H_0'}{k_BT}}\right)^{-1}\det\left(\hat{1}+e^{-\frac{H_0'}{k_BT}}e^{iH_kt}e^{-iH_jt}\right)$$

$$= \det\left(\hat{1}+\hat{n}_F\left(e^{iH_kt}e^{-iH_jt}-\hat{1}\right)\right), \qquad (3.63)$$

where $\hat{n}_F = \left(\hat{1}+e^{-\frac{H_0'}{k_BT}}\right)^{-1}e^{-\frac{H_0'}{k_BT}}$ is the Fermi occupation number operator.

With the help of (3.60), we now obtain

$$\det\left(\hat{1}+\hat{n}_F\left(e^{iH_kt}e^{-iH_jt}-\hat{1}\right)\right) = \exp\left(\mathrm{tr}\ln\left(\hat{1}+\hat{n}_F\left(e^{iH_kt}e^{-iH_jt}-\hat{1}\right)\right)\right),$$

$$(3.64)$$

where the trace is to be taken over the one-electron states (labelled, e.g., by their momenta and "left–right" origin). In the limit $t \to \infty$ the exponents become stationary scattering matrices for electrons with a given energy, $e^{-iH_jt/\hbar} \to S_j$, $e^{iH_kt/\hbar} \to S_k^\dagger$. The expression for $\langle e^{iH_kt}e^{-iH_jt}\rangle$ becomes

$$\langle e^{iH_kt}e^{-iH_jt}\rangle = \exp\left\{t\int\frac{dE}{2\pi\hbar}\ln\det\left[\hat{1}+\hat{n}_F\left(S_k^\dagger S_j-\hat{1}\right)\right]\right\} \qquad (3.65)$$

(the determinant is in the "left-right" space, so that $\hat{n}_F = \mathrm{diag}(f_L(E), f_R(E))$, and S_j, S_k are given by (3.49)). We have indeed arrived at Eq. (3.55), with $e^{i\alpha_{jk}}$ replaced by $\det S_k^\dagger S_j$ (Averin and Sukhorukov, 2005).

The measurement described here is an example of *weak continuous measurement* (see § 5.5.1): each of the electron-scattering events only weakly disturbs the quantum state of the dot, and the measurement itself requires averaging over many such events, which produces the average current and QPC conductance. We note here that a quantum point contact can be tuned to be a *quantum limited* detector. By definition, the decoherence rate Γ_{jk} induced in the measured system by such a detector coincides with the *measurement rate* (see § 5.5.4). In other words, the loss of quantum coherence in the system, leading to the reduction of its state to an incoherent sum of different measurement outcomes (cf. §§ 1.2.2, 1.3.3), is in this case solely due to the measurement.

3.2 2DEG quantum dots

3.2.1 *Linear and nonlinear transport through a double quantum dot*

Two quantum point contacts in series produce a 2DEG quantum dot, if the distance between them is comparable to the Fermi wavelength (that is, in the submicron range). This structure contains a large number of electrons, typically 10^3–10^9, but only the highest occupied states contribute to transport. The energy spectrum of electrons is discrete, since their motion is limited in all directions. Therefore this zero-dimensional object can be considered as an artificial atom. This distinguishes 2DEG quantum dots from superconducting islands, which we considered in the previous chapter: the Fermi wavelength in a 2DEG is comparable to the size of the dot. Moreover, the electron energy spectrum can now be tuned through reshaping the 2DEG region by changing the gate voltages. We have seen that the energy of a sufficiently small system is sensitive to the addition or subtraction of even a single electron. This holds, of course, for 2DEG quantum dots. Their state can be directly measured by a QPC detector, which is readily incorporated into the design (see Fig. 3.7).

A large electron wavelength makes 2DEG quantum dots really similar to artificial atoms. Electrons confined by the gate potential form shells, with certain electron numbers being "magic", that is, minimizing the total energy of the system. To determine the electronic configuration, it is necessary to take into account Coulomb repulsion between the electrons and the charge redistribution in the gates; the full many-body problem can be treated only numerically. Remarkably,

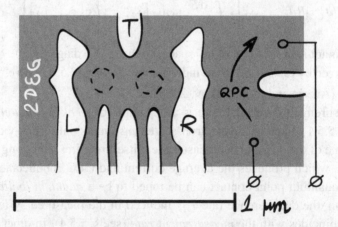

Fig. 3.7. A 2DEG double quantum dot (after Petta et al., 2005). Gates L and R control the number of electrons in the corresponding dots, T controls the interdot tunnelling rate. The quantum point contact to the right of gate R acts as a detector of charge on the right dot.

Kumar et al. (1990) found using a self-consistent Hartree approach that the potential of square gates produces an almost radially symmetric, parabolic effective confinement potential. This justifies the use of a 2D harmonic confinement potential as the standard approximation for the calculations of electronic states on a single 2DEG quantum dot.

For our goals this interesting topic, reviewed in detail by Reimann and Manninen (2002), is not very relevant. We already discussed the Coulomb blockade in both normal and superconducting systems when considering superconducting charge qubits (§ 2.4.1), and we did not need the details of the electronic structure of a dot. Here we will immediately proceed to the case of double 2DEG quantum dots (DQD) (reviewed by van der Wiel et al. (2002) and Hanson et al. (2007)), which is richer and more instructive and, on the other hand, is directly related to the 2DEG-based implementation of qubits. If a single quantum dot is an "atom", a DQD is a tunable "molecule", made to order.

We begin with the description of transport in a double-dot system (Fig. 3.8) using the classical level. The electrostatic energy of the system as a function of the numbers N_1, N_2 of extra electrons on the dots is given by

$$U(N_1, N_2) = \frac{N_1^2 E_{C1}}{2} + \frac{N_2^2 E_{C2}}{2} + N_1 N_2 E_{Cm} + f(N_{g1}, N_{g2}), \qquad (3.66)$$

$$f(N_{g1}, N_{g2}) = \frac{N_{g1}^2 E_{C1}}{2} + \frac{N_{g2}^2 E_{C2}}{2} + N_{g1} N_{g2} E_{C1}$$
$$- \left(N_{g1}(N_1 E_{C1} + N_2 E_{Cm}) + N_{g2}(N_1 E_{Cm} + N_2 E_{C2}) \right), \quad (3.67)$$

where $(j = 1, 2)$

$$N_{gj} = C_{gj} V_{gj} / |e| \qquad (3.68)$$

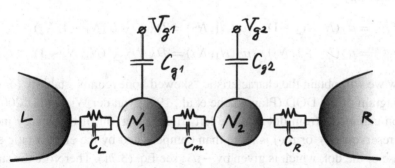

Fig. 3.8. Equivalent scheme for a double quantum dot. In the limit of large interdot capacitance C_m (small energy E_{Cm}) the system is equivalent to a single quantum dot.

are the effective numbers of electrons at the gates,

$$E_{Cj} = \frac{e^2}{C_j}\left(1 - \frac{C_m^2}{C_1 C_2}\right)^{-1}; \quad E_{Cm} = \frac{e^2}{C_m}\left(\frac{C_1 C_2}{C_m^2} - 1\right)^{-1}, \quad (3.69)$$

and the summary capacitances of the dots are

$$C_1 = C_L + C_{g1} + C_m; \quad C_2 = C_R + C_{g2} + C_m. \quad (3.70)$$

The chemical potential of a dot follows directly from its definition (the thermodynamic potential per particle):

$$\mu_1(N_1, N_2) = U(N_1, N_2) - U(N_1 - 1, N_2);$$
$$\mu_2(N_1, N_2) = U(N_1, N_2) - U(N_1, N_2 - 1). \quad (3.71)$$

The result is

$$\mu_1(N_1, N_2) = (N_1 - 1/2)E_{C1} + N_2 E_{Cm} - (N_{g1}E_{C1} + N_{g2}E_{Cm});$$
$$\mu_2(N_1, N_2) = (N_2 - 1/2)E_{C2} + N_1 E_{Cm} - (N_{g2}E_{C2} + N_{g1}E_{Cm}). \quad (3.72)$$

Now we can better understand the physical meaning of the parameters E_{Cj}, E_{Cm}. The *addition energy* of dot 1 is defined as the difference in its chemical potential due to one extra charge on it,

$$\mu_1(N_1 + 1, N_2) - \mu_1(N_1, N_2) = D_{N_1}^2 U(N_1, N_2) = E_{C1}. \quad (3.73)$$

Here $D^2 f(n) \equiv Df(n) - Df(n-1) \equiv f(n+1) - 2f(n) + f(n-1)$ is the second difference, which is an analogue of the second derivative for a function of a discrete variable (see, e.g., Bender and Orszag, 1999, Chapter 2). For the second dot the addition energy is $\mu_2(N_1, N_2 + 1) - \mu_2(N_1, N_2) = E_{C2}$. Finally, E_{Cm} is the cross addition energy:

$$E_{Cm} = \mu_1(N_1, N_2 + 1) - \mu_1(N_1, N_2) = D_{N_2} D_{N_1} U(N_1 - 1, N_2)$$
$$= \mu_2(N_1 + 1, N_2) - \mu_2(N_1, N_2) = D_{N_1} D_{N_2} U(N_1, N_2 - 1). \quad (3.74)$$

Now we will obtain the characteristic "skewed honeycomb" stability (or charging) diagram of the DQD (Pfannkuche et al., 1998; van der Wiel et al., 2002). An electron can move from dot 1 to the left (right) reservoir, if the energy increase of the reservoir (μ_L or μ_R) is more than compensated by the electrostatic energy decrease of the dot, which is given by $-\mu_1$; see Eq. (3.71). Therefore the stability conditions are

$$\mu_1(N_1, N_2; N_{g1}, N_{g2}) \le \mu_L, \mu_R; \quad \mu_2(N_1, N_2; N_{g1}, N_{g2}) \le \mu_L, \mu_R. \quad (3.75)$$

In the absence of bias between left and right reservoirs we can measure energies from the level $\mu = \mu_L = \mu_R$ $(= 0)$. Substituting (3.72) in (3.75), we see that the stability diagram follows from the equations in the (N_{g1}, N_{g2})-plane (see Fig. 3.9a):

$$N_{g1}(E_{C1}/E_{Cm}) + N_{g2} \geq (N_1 - 1/2)(E_{C1}/E_{Cm}) + N_2;$$
$$N_{g2}(E_{C2}/E_{Cm}) + N_{g1} \geq (N_2 - 1/2)(E_{C2}/E_{Cm}) + N_1. \qquad (3.76)$$

For zero cross-capacitance, $C_m = 0$ $(E_{Cm} = \infty)$, these equations decouple, and the diagram looks like the checkerboard. Any finite coupling splits the quadruple degenerate points, where four different charge configurations have the same energy, into triple degenerate ones. The solid lines in the picture correspond to such a choice of N_{g1}, N_{g2} (or, which is the same, the gate voltages) that the chemical potential of at least one dot is zero; along the dotted lines the chemical potentials of adjacent charge configurations coincide. Therefore, at the gate voltages, for which N_{g1}, N_{g2} lie in the triple points or on the dotted lines, a transfer of an electron through the system does not entail an energy penalty, and the electric current flows in the linear response regime (at infinitesimal bias $\mu_L - \mu_R$). This is the same mechanism as resonant conductance through a single quantum dot in the Coulomb blockade regime, § 2.4.1. Along the solid lines this penalty is minimal, and current flows at small biases. An experimental picture of current vs. gate voltages, shown in Fig. 3.9b, demonstrates a typical honeycomb. Comparing this picture to the diagram Fig. 3.9a, one can see how the parameters of the system can be directly extracted from the measurements.

So far we have not considered *quantum* transport through a double quantum dot. The quantization of electron levels is expected, since the dot size L is comparable to the Fermi wavelength in a 2DEG. The interlevel spacing $\Delta E \sim E_F(\lambda_F/L)^2 \sim$ 100 K $\times$ (60 nm/300 nm)$^2 \sim$ 4 K is large enough to take into account only the lowest level (the ground state) when dealing with linear transport. Therefore, the honeycomb diagram in this regime applies to the quantum case almost unchanged, with the exception of bending near the triple points (Fig. 3.9c), which appears due to interdot tunnelling. Indeed, in this case the electrons are no longer localized on a single dot and occupy bonding and antibonding "molecular orbitals", with the energy of the former being less, and the latter larger than the sum of corresponding single-dot energies by Δ, the tunnelling matrix element. Therefore, at a triple point the first extra electron can enter the double dot at smaller gate voltages; in terms of the effective gate electron numbers, $N_{gj} \to N_{gj} - C_{gj}\Delta/e^2$. But the second electron must occupy the antibonding state, which shifts the second triple point in the opposite direction: $N_{gj} \to N_{gj} + C_{gj}\Delta/e^2$. Remarkably, this picture was

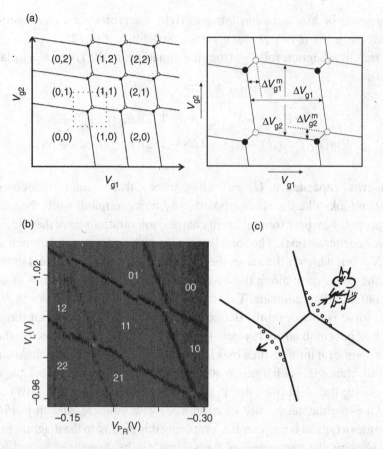

Fig. 3.9. (a) Classical stability (charging) diagram for a double quantum dot in the linear response regime (reprinted with permission from van der Wiel et al., 2002, © 2002 American Physical Society). Numbers in the cells denote the numbers of extra electrons (N_1, N_2) on the dots. In the triple points $\mu_L = \mu_1(N_1, N_2) = \mu_1(N_1 + 1, N_2) = \mu_2(N_1, N_2 + 1) = \mu_2(N_1, N_2) = \mu_R$ (black) or $\mu_L = \mu_1(N_1 + 1, N_2 + 1) = \mu_1(N_1 + 1, N_2) = \mu_2(N_1, N_2 + 1) = \mu_2(N_1 + 1, N_2 + 1) = \mu_R$ (white), which enables resonant transport of electrons (resp. holes) through the system. The dimensions of the hexagonal cell shown follow from (3.76): $\Delta V_{gi} = e/C_{gi}(i = 1, 2)$; $\Delta V_{g1}^m = eC_m/C_{g1}C_2$; $\Delta V_{g2}^m = eC_m/C_{g2}C_1$; here C_i is the summary capacitance of the ith dot. (b) Experimental stability diagram for a double quantum dot (reprinted with permission from Elzerman et al., 2003, © 2003 American Physical Society). Note the inverted voltage polarity. (c) Bending of transition lines near triple points due to quantum tunnelling.

observed in the experiment, even though real electrons have spin, the "orbitals" become degenerate, and there should be no need to occupy the antibonding state. The reason is, of course, strong Coulomb repulsion, which effectively localizes the electrons when the second electron is added (Hanson et al., 2007).

3.2.2 Coherent manipulation of charge in 2DEG quantum dots

An especially interesting application of a 2DEG quantum dot device is as a qubit. There are two degrees of freedom one can use: spin and charge. We start with charge, as it is more straightforward to both describe and measure.

A straightforward way of using charge states is to confine a single electron to a double quantum dot, and use the lowest-energy localized states on the left/right dot as qubit working states. Transitions between them are controlled by the gate voltage affecting the interdot tunnelling. Of course, the DQD should be electrically isolated by pinching off the quantum point contacts between it and the leads, or by taking the quantized electron states in the DQD off resonance with the leads. In Fig. 3.10a and Fig. 3.11a two experimental implementations of such a device are shown. In the second case the 2DEG approach is used to form quantum point contacts serving as charge detectors next to the corresponding quantum dots. In order to concentrate on the coherent behaviour of charge in the system, we will start by discussing the first, simpler device, where the readout is realized by measuring the current through the DQD itself, not through the QPC detectors.

At low enough temperatures we need to consider only the ground states of an extra electron on either dot. Therefore, we can immediately write the familiar Hamiltonian

$$H = -\frac{1}{2}\epsilon(t)\sigma_z - \frac{1}{2}\Delta\sigma_x, \qquad (3.77)$$

where the physical states are $|L\rangle$, $|R\rangle$, corresponding to the electron being localized on the left (right) dot, and the tunnelling matrix element Δ between the two is tuned by the voltage on gate G_C.

The parameters were chosen in such a way that when the chemical potentials of the left and right electrodes (source and drain) were zero, no current flows through the DQD, and the ground states of the dots are aligned. The initialization was performed by an application of a rectangular voltage pulse, changing the drain bias from zero to $-650\,\mu eV$ and back. Due to a fast inelastic tunnelling rate Γ in and from the DQD (compared to the inelastic tunnelling rate Γ_i between the dots) the system was prepared in the state $|L\rangle$. For the duration t_p it was allowed to undergo quantum beats (§ 1.4.2.), after which the bias was restored. Due to a high escape rate into the lead, and misalignment of the electron levels on the dot, the application of bias effectively stops the oscillations, and the electron escapes into the drain – if, of course, at the moment the bias was reapplied it was on the right dot. This is the same kind of readout as was used for superconducting phase qubits, § 2.2.2. In the experiment, the bias pulses of length $t_p \le 2$ ns and the rise time about 0.1 ns were repeated with the frequency $f_p = 100$ MHz, and the average pulse-induced current through the quantum dot was measured. The average number of electrons passing through the DQD due to the pulse sequence is $n_p(t_p) = I_p(t_p)/ef_p$. On the other hand, $n_p(t_p) = P_R(t_p)$, the probability for the electron to be found in state $|R\rangle$ at $t = t_p$, if at $t = 0$ it was in state $|L\rangle$. This probability should demonstrate oscillations with frequency $\Omega = \sqrt{\epsilon^2 + \Delta^2}$, which it duly does (Fig. 3.10c). From the fitting,

$$n_p(t_p) = A - \frac{1}{2}Be^{-t_p/T_2}\cos\Omega t_p - \Gamma_i t_p, \qquad (3.78)$$

the values for the tunnelling splitting at $\epsilon = 0$, $\Delta = \Omega \approx 2\pi \times 2.3$ GHz, and the dephasing rate, $T_2 \approx 1$ ns, were extracted. (The last term in (3.78) was associated with the suppression of the background inelastic tunnelling current during the pulse application.)

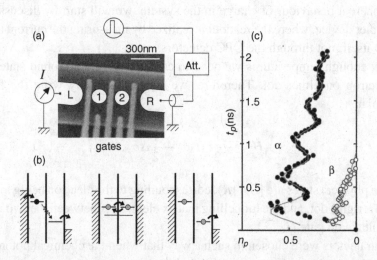

Fig. 3.10. (Reprinted with permission from Hayashi et al., 2003, © 2003 American Physical Society.) (a) Double-quantum-dot charge qubit formed in an GaAs/AlGaAs heterostructure. Each dot contained approximately 25 electrons, the charging energy was ~ 1.3 meV, and interdot coupling ~ 200 μeV. (b) Schematics of the initialization, manipulation, and measurement. A constant bias voltage is applied between the reservoirs. A bias voltage pulse of duration t_p brings the filled electron state on the left dot in resonance with the empty one on the right dot; the current through the DQD is blocked. If after the pulse, the electron ends up on the right dot, it will escape to the right reservoir, adding to the average current through the DQD. (c) Average number of electrons passing through the DQD per bias pulse, as a function of the delay time t_p. Oscillations of $n(t_p)$ reveal coherent quantum beats of an electron between the left and right quantum dots. Curve $\alpha(\beta)$ is probably associated with the resonance between the ground state of the left, and the first (second) excited state on the right dot. The measurements were conducted at 20 mK (though the electron temperature was higher, ~ 100 mK), and in the external magnetic field of 0.5 T to lift spin degeneracy.

This way of reading out the state of the DQD requires the actual transfer of an electron between the leads through the system. Therefore, the DQD is periodically connected to thermal reservoirs, leading to additional decoherence. Instead, we could use the QPC detectors to determine the charge on each quantum dot. Here all the decoherence can be, in principle, due to the measurement of the quantum state, that is, at the theoretical minimum. In addition, this method works well when there are only a few electrons in the DQD (which would make direct transport through the DQD difficult). It was implemented by Petta et al. (2004) when the double dot contained just a single electron.

The conductance of the QPC detector is sensitive to the average excess charge on the corresponding quantum dot. For a single electron on a double dot, from Eq. (3.77) it follows that in the ground state the probability of finding the electron on the left (right) dot is $\frac{1}{2}\left(1 \mp \frac{\epsilon}{\Omega}\right)$, and the other way around in the excited state. At finite temperatures of the system (that is, the effective temperature of the electron gas, since in the experiment it is usually different from the lattice temperature), the average occupation number on the left (right) dot is thus (DiCarlo et al., 2004)

$$\langle n_{L,R} \rangle = \frac{1}{2}\left(1 \pm \frac{\epsilon}{\Omega}\tanh\left(\frac{\Omega}{2k_B T}\right)\right). \tag{3.79}$$

Therefore, the conductance of the left (right) – or rather lower (upper) – QPC in Fig. 3.10b can be written as

$$G_{QPC} = G_{QPC}^{(0)}(\epsilon) \pm \delta G \frac{1}{2}\frac{\epsilon}{\Omega}\tanh\left(\frac{\Omega}{2k_B T}\right). \tag{3.80}$$

The first term is the background QPC conductance, including its dependence on the voltage applied to create the bias ϵ between the dots. Using this fit to the measurements of the QPC conductance, the number of electrons on the upper dot in the device of Fig. 3.11a was measured. In the presence of microwave radiation with frequency about 20 GHz, the occupation number curve, Fig. 3.11b, shows sharp resonances, which correspond to one- and two-photon transitions between the $(N_1 = 0, N_2 = 1)$-ground state and the $(N_1 = 1, N_2 = 0)$-excited state of the DQD "molecule".

The next step is to implement coherent "intermolecular" interactions, that is – in the context of quantum computing – two-qubit quantum gates (see the Appendix). This was achieved in a structure that incorporates two DQD qubits of Fig. 3.10a side by side (Shinkai et al., 2009), see Fig. 3.12a. The potential barriers between the two DQDs prevent tunnelling, so that the qubits are galvanically isolated and

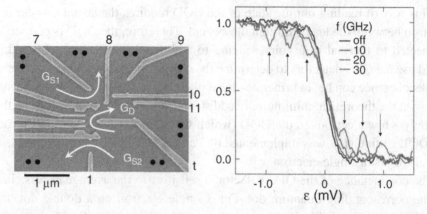

Fig. 3.11. (Reprinted with permission from Petta et al., 2004, © 2004 American Physical Society.) (Left) Double-quantum-dot charge qubit with QPC acting as charge detectors (S1 and S2). (Right) Average number of electrons on the upper dot as a function of bias ϵ, with and without external microwave radiation at 10, 20, and 30 MHz. Dips/peaks indicate the resonant photon-assisted transitions between the quantized electron levels.

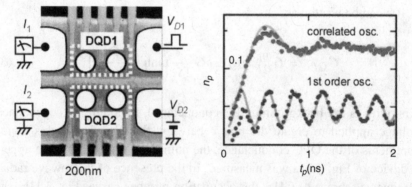

Fig. 3.12. (Reprinted with permission from Shinkai et al., 2009, © 2009 American Physical Society.) (Left) Two DQD charge qubits coupled electrostatically, but galvanically isolated. Each DQD contains a single electron. Source-drain currents I_1, I_2 can be measured independently. The measurement scheme is analogous to the one of Hayashi et al. (2003), Fig. 3.10. (Right) Average number of electrons passing through the first DQD per bias pulse, as a function of the delay time t_p. The lower curve corresponds to the single-qubit, and the upper one to the double-qubit quantum beats.

only interact electrostatically. The effective Hamiltonian can be written as

$$H = -\frac{1}{2} \sum_{j=1,2} \left(\epsilon_j \sigma_z^{(j)} + \Delta_j \sigma_x^{(j)} \right) + \frac{J}{4} \sigma_z^{(1)} \sigma_z^{(2)}. \tag{3.81}$$

The qubit–qubit coupling strength can be calculated as a function of mutual capacitances, etc., or – which is simpler – determined from the experimental data. What is important is that it is positive, that is, preferring "antiparallel" configurations $|LR\rangle \equiv |L\rangle_1 \otimes |R\rangle_2$ (where the electrons are on the left dot of the first DQD, and on the right dot of the second one), or $|RL\rangle$, over the "parallel" ones ($|LL\rangle$ and $|RR\rangle$). The source of this "antiferromagnetism" is, of course, direct Coulomb repulsion between the electrons on the DQDs.

By applying $\epsilon_1 = J/2$ in Eq. (3.81), the states $|LR\rangle$ and $|RR\rangle$ are made degenerate, thus enabling tunnelling between them; at $\epsilon_1 = -J/2$ resonant tunnelling and quantum beats become possible between $|LL\rangle$ and $|RL\rangle$. The length of the pulse allows us to coherently rotate the first qubit between its "left" and "right" states. Similarly, applying a bias to the second DQD, it is possible to change its state without disturbing the first one. Such an operation, called a CROT (conditional single-qubit rotation[6]), is an effect of the first order in Δ_1, Δ_2.

Electrons on both DQDs can also tunnel simultaneously, enabling, e.g., the SWAP operation, $|LR\rangle \leftrightarrow |RL\rangle$. This is a process of the second order in tunnelling matrix elements. For sufficiently small Δ's (as is the case), it can dominate only when the first-order effects are suppressed, $||\epsilon_{1,2}| \pm J/2| \gg \Delta_{1,2}$, and the DQD states are aligned. For example, when $\epsilon_1 = \epsilon_2 = \epsilon$, the states $|LR\rangle$ and $|RL\rangle$ would be degenerate in the absence of tunnelling. Tunnelling (with $\Delta_1 = \Delta_2 = \Delta$) forms the bonding and antibonding states, $\propto (|LR\rangle \pm |RL\rangle)$, with the energy splitting $\frac{1}{2}\Delta^2 J/|\epsilon^2 - J^2/4|$ between them, and a half-period of quantum beats will produce the desired SWAP operation. Similarly, when $\epsilon_1 = -\epsilon_2 = \epsilon$, the states $|LL\rangle$ and $|RR\rangle$ are mixed by tunnelling, and the transitions $|LL\rangle \leftrightarrow |RR\rangle$ are possible. The experiments (Shinkai et al., 2009) using the same readout and initialization technique as Hayashi et al. (2003) (see Fig. 3.10) demonstrated both single- and double-qubit quantum beats (Fig. 3.12).

3.2.3 Coherent manipulation of spin in 2DEG quantum dots

Now let us consider the spin of the electrons on quantum dots. The advantage of the spins as controllable quantum degrees of freedom is the same as their disadvantage: weak coupling to everything else. This makes them both relatively immune to

---

[6] That is, conditional on the state of the other qubit: for example, if we apply a bias $\epsilon_1 = J/2$, but the second DQD is in state $|L\rangle_2$, nothing happens.

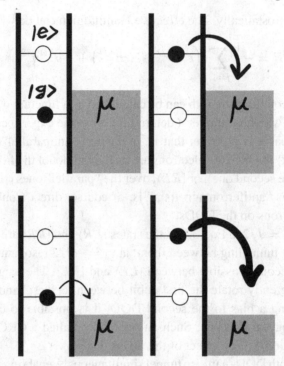

Fig. 3.13. Energy-selective (top) and tunnel rate-selective (bottom) spin-to-charge conversion in a quantum dot. In both cases the energy difference of spin states translates into charge current from the dot to the reservoir (with chemical potential μ).

decoherence and difficult to initialize, manipulate and read out. Quantum dots provide a nice opportunity to deal with the latter problem by coupling spins to charge degrees of freedom. (Of course, this immediately brings back the decoherence trouble, but that is another story.)

Of several proposed ways of spin-to-charge conversion, which are all essentially based on the Pauli principle, two have been realized experimentally: energy-selective and tunnel rate-selective readout (see Hanson et al., 2007, VI, and references therein). In both cases, the two spin states of an electron in a quantum dot ("up" and "down" for brevity) should have different energies (Fig. 3.13), which influence electron tunnelling off the dot. In the first case, the "up" state lies above, and the "down" state below the chemical potential of the lead, and only an electron in the "up" state can leave the dot. In the second case, both "up" and "down" states are energetically allowed to leave the dot, but the tunnelling rate of the "up" state is much higher, and for finite-time measurement again only an "up" electron will tunnel out. It is the charge transfer between the dot and the lead that is measured (e.g., by a QPC detector), and this allows us to realize a single-shot measurement (§ 5.5.1.)

The first approach was realized by Elzerman et al. (2004). The device consisted of a 2DEG quantum dot and a quantum point contact next to it. The 2DEG density in the GaAs/AlGaAs heterostructure was a typical $2.9 \cdot 10^{11}$ cm^{-2}. In the dot it was sufficiently depleted by the gate voltages to produce the single electron regime (that is, the dot could contain either one electron or none). The tunnelling barrier between the dot and the reservoir (a large-area 2DEG lead) was tuned to ensure the tunnel rate $\Gamma \approx 1/(0.05 \text{ ms})$. The magnetic field of 10 tesla applied in the plane of 2DEG produced Zeeman splitting $\Delta E_Z \approx 200$ μeV in the dot, which exceeded the thermal energy (25 μeV), but was less than the charging energy (2.5 meV). The dot was initially emptied by increasing the potential of both "up" and the "down" states in the dot above the chemical potential μ_0 of the reservoir and then waiting for $\sim 1/\Gamma$. Then the voltage was applied to put both states below the chemical potential and again waiting for $\sim 1/\Gamma$. Finally, the voltage was changed to the "readout" value, which places the "up" level above, and the "down" level below μ_0. Now the electron can escape into the reservoir only from the "up" state. This escape happens within $\Gamma_\downarrow^{-1}$ from this moment (here $\Gamma = \Gamma_\downarrow + \Gamma_\uparrow$ when both states are above μ_0, and all these rates are of the same order of magnitude). The escape reduces the charge on the dot and slightly opens the adjacent QPC (which is put in the tunnelling regime, to increase its sensitivity to slight variations of electrostatic potential), which enables additional current to flow through it, indicating that the electron on the dot was in the "up" state. Two QPC measurements (for the "up" and "down" states) from Elzerman et al. (2004) are shown in Fig. 3.14. This is a good example of a single-shot readout: after a single attempt, with the probability close to one, a superposition of "up"–"down" spin states on the dot is reduced to the mixture of these states with appropriate weights.

This experiment also allowed direct measurement of the spin relaxation time. Since it is the more energetic state that produces the observable electron transfer to the reservoir, the longer the interval between the initial filling of the quantum dot and the readout pulse (interval t_{hold} in Fig. 3.14), the more probable it is that the "up" state will relax into the "down" state by that time. The probability of observing the "up" state should exponentially decay with increased t_{hold} as $\exp(-t_{\text{hold}}/T_1)$. This dependence was indeed observed, with T_1 scaling from 0.85 ms at 8 tesla to 0.12 ms at 14 tesla. Note that even without special measures to counteract relaxation, T_1 is pretty large compared with the relaxation time in superconducting qubits. Of course, this does not say much until we compare this time with the characteristic time required for manipulating spins on a quantum dot.

Tunnel-rate selective readout has been realized for a *two*-electron quantum dot (Hanson et al., 2005) and essentially distinguished between the singlet and triplet state. In the triplet, one electron occupies the excited orbital state, which has a faster rate of tunnelling out through the confining potential of the dot. The advantages

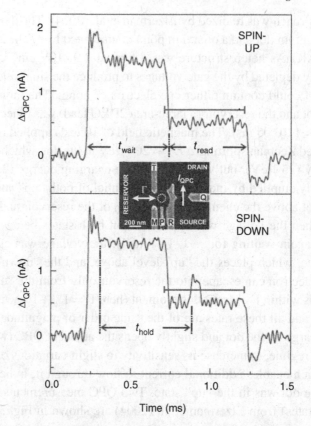

Fig. 3.14. Energy-selective electron spin readout in a 2DEG quantum dot (adapted by permission from *Nature*, Elzerman et al., 2004, © 2004 Macmillan Publishers Ltd). The voltage pulse applied to the central gate *P* shifts the quantum dot energy levels. The charge on the quantum dot is measured by the QPC detector (inset). The QPC current pulse at ~ 0.8–0.9 ms in the lower panel indicates the electron leaving the "down" state in the reservoir (current rise) and then another one filling the state from the reservoir (fall). The electron in the "up" state cannot initially leave the dot (upper panel).

of this approach were several, including no sensitivity to the absolute position of "up" and "down" (in this case, singlet and triplet) levels with respect to μ_0, which made the previous approach vulnerable to local electric potential fluctuations; and no need for a large energy difference between the resolved states, which could be taken down to less than $4k_B T \sim 0.1$ meV. The measured triplet-to-singlet relaxation time was also remarkably long ($T_1 = 2.58$ ms in near zero field).

Manipulation of electron spins in a quantum dot can be performed using the *exchange interaction* rather than magnetic fields. Let us recall that this electron–electron interaction,

$$H_{ex} = -\frac{1}{4} J \vec{\sigma}_1 \cdot \vec{\sigma}_2, \qquad (3.82)$$

where Pauli matrices for once denote actual spins, is not magnetic, but electrostatic: due to the Pauli principle electrons with the same spin being unable to have the same spatial wave functions, which leads to different electrostatic contributions to their energy reflected in (3.82) (White, 1983, §2.2.7). It is, therefore, rather strong, and the coupling strength J can have either sign, depending on the configuration of the system.

From our current perspective, this is a very good tool for coupling spins, since we can directly shape the potential barriers around adjacent quantum dots and tune J with a turn of a knob. It was used by Petta et al. (2005) to manipulate quantum states of spins on a double quantum dot, the same kind as shown in Fig. 3.11a. To a theorist, a pair of quantum dots with an electron with spin on each already looks like two·qubits, but the structure was still used as a single qubit – precisely to take advantage of the exchange interaction. The gate voltages were selected to allow either (0,2) or (1,1) charge configurations on the DQD. In the first case, the gate potential would push both electrons into the right dot, but only in a spin-singlet state, $S = 0$. (Due to the small size of the dot, the spin-triplet state, $S = 1$, would have too much electrostatic energy to play any role.) In the second case, the electrostatic potentials on both dots are close, so there is one electron per dot, and in the absence of tunnelling between the dots there will be four degenerate states: a singlet, and a triplet, too many for a qubit. This difficulty was resolved by applying a 100 mT magnetic field, which fixed the preferred direction of the spin projection and produced a Zeeman split between the $|S = 1, m = -1\rangle$ triplet state ($E_Z = -g^*\mu_B B \sim -2.5\ \mu\text{eV}$) and the $|S = 1, m = 1\rangle$ triplet state ($E_Z = +g^*\mu_B B$) on the one hand, and the $|S = 1, m = 0\rangle$ triplet component and the singlet state $|S = 0\rangle$ (for both of which $E_Z = 0$) on the other.[7] It is the latter two states that were used as qubit states. They were manipulated by tuning the tunnelling between the two dots through the gate potential, with J (of order 1 μeV) playing the role of tunable bias in the effective one-qubit Hamiltonian written in the basis $\{|S = 0\rangle, |S = 1, m = 0\rangle\}$,

$$H_{\text{eff,spin}} = \begin{pmatrix} J & B_{\text{nucl}} \\ B_{\text{nucl}} & 0 \end{pmatrix}. \tag{3.83}$$

The readout of the state is performed by spin-charge conversion on the right dot, that is, by again trying to push both electrons on it, and checking through the QPC whether the operation was successful. If yes, then we projected out of the superposition of $|S = 0\rangle$ and $|S = 1, m = 0\rangle$ and measured its spin-singlet component.

---

[7] For GaAs the electron g-factor $g* = -0.44$.

The role of tunnelling, which allows transitions between $|S = 0\rangle$ and $|S = 1, m = 0\rangle$, is played by the random, time-dependent (on a scale of tens of microseconds) hyperfine magnetic fields B_{nucl} describing the hyperfine interaction of the electron spin with the nuclei. They are also a potentially dangerous source of decoherence (see § 5.3.3). Nevertheless, the coherent operations could be performed by tuning J, with, e.g., quantum beats surviving for about 10 ns, and the spin-echo technique allowing researchers to see coherence for as long as 1.2 μs.

3.3 Loops, interferometers and hybrid structures

3.3.1 2DEG loops: Aharonov–Bohm effect

Quantum dots and quantum point contact are the most relevant 2DEG devices from the quantum engineering point of view, providing a way of creating, maintaining, manipulating and measuring quantum superpositions of states. But split-gate techniques allow us to realize other quantum coherent 2DEG devices as well, which may prove useful for future technology, show beautiful physics and help us introduce other useful pieces of formalism.

First, we will consider 2DEG conducting loops. Their realization requires such advanced experimental tricks as the formation of hanging gates, which form the inner part of a loop, but a theoretical understanding of the effect is quite straightforward.

The Aharonov–Bohm effect was predicted when Aharonov and Bohm (1959) considered the question of whether the vector potential **A** of a magnetic field **B**,

$$\mathbf{B} = \nabla \times \mathbf{A} \tag{3.84}$$

has a direct physical meaning. In their *Gedankenexperiment* (for a detailed review and discussion see Olariu and Popescu, 1985) Aharonov and Bohm proposed to send an electronic wave at an infinitely long, impenetrable solenoid, containing magnetic flux Φ. The magnetic field is absent outside the solenoid and the electron is absent inside it. This precludes the direct influence of the magnetic field on the electron's propagation via the Lorentz force. Nevertheless, the behaviour of the electron may still depend on the vector potential, which does *not* disappear outside the solenoid. If for simplicity we consider the limit of an infinitesimally thin solenoid along the z-axis, so that

$$\mathbf{B} = \hat{\mathbf{e}}_z \Phi \delta(x) \delta(y), \tag{3.85}$$

then a convenient choice for **A** is given by

$$\mathbf{A}(\mathbf{r}) = \hat{\mathbf{e}}_y \Phi \theta(x) \delta(y). \tag{3.86}$$

This expression, when substituted in (3.84), obviously yields (3.85). There exist other, equivalent expressions (*gauges*), differing by a gradient of some function and therefore yielding the same magnetic field,[8] but Eq. (3.86) is the most convenient for our purposes.

The most straightforward way of seeing how the field can act where it is not present, and why this only happens in quantum mechanics, is to use the Feynman path integral approach (Feynman and Hibbs, 1965), according to which the probability amplitude for a particle to travel from point $A = (\mathbf{r}_A, t_A)$ to point $B = (\mathbf{r}_B, t_B)$ is given by

$$K(\mathbf{r}_B, t_B | \mathbf{r}_A, t_A) = \int_{\mathbf{r}_A}^{\mathbf{r}_B} \mathcal{D}\mathbf{r} \exp\left[\frac{i}{\hbar} \int_{t_A}^{t_B} dt\, L(\mathbf{r}, \dot{\mathbf{r}}, t)\right]. \tag{3.87}$$

Here $\mathcal{D}\mathbf{r}$ means that the summation is taken over all classical paths connecting the two points.[9]

The integral in the exponent is the classical action, S_{cl}, and $L(\mathbf{r}, \dot{\mathbf{r}}, t)$ is the classical Lagrange function, related to the classical Hamilton function $H(\mathbf{P}, \mathbf{r})$ via (Goldstein, 1980)

$$L(\mathbf{r}, \dot{\mathbf{r}}, t) = \mathbf{P} \cdot \dot{\mathbf{r}} - H(\mathbf{P}, \mathbf{r}). \tag{3.88}$$

Here $\mathbf{P}$ is the canonical momentum, which is replaced with $\frac{\hbar}{i}\nabla$, when the Hamilton function is quantized to become the quantum mechanical Hamiltonian (Messiah, 2003). In one dimension, this would give for the probability amplitude an expression reminiscent of a WKB-type solution for the wave function, $\sim \exp\frac{i}{\hbar}\left[\int P\, dx - \int E\, dt\right]$ (since the classical Hamilton function *is* energy).

The interaction with a magnetic field is introduced through

$$H = \frac{1}{2m}\left(\mathbf{P} - \frac{q}{c}\mathbf{A}\right)^2 + V(\mathbf{r}), \tag{3.89}$$

where q is the electric charge of the particle. The first term in this expression is the kinetic energy, and the second is static potential. The canonical momentum $\mathbf{P} = \mathbf{p} + \frac{q}{c}\mathbf{A} = m\dot{\mathbf{r}} + \frac{q}{c}\mathbf{A}$, where $\mathbf{p}$ is the momentum in the absence of the field.

---

[8] An axially symmetric expression for the vector potential producing (3.85) can be written immediately. If we assume that $\mathbf{A}(\mathbf{r}) = A(r)\hat{\mathbf{e}}_\phi$, where r is the distance from the origin in the xy-plane, and $\hat{\mathbf{e}}_\phi$ is the azimuthal unit vector, then the circulation of the vector $\mathbf{A}$ around a circle of radius r centred at the origin is $\oint \mathbf{A} \cdot d\mathbf{s} = 2\pi r A(r) = \int dx\, dy\, [\nabla \times \mathbf{A}]_z$. On the other hand, by the virtue of Stokes' theorem, $\oint \mathbf{A} \cdot d\mathbf{s} = \int dx\, dy\, B_z = \Phi$. Therefore $\mathbf{A}(\mathbf{r}) = (\Phi/2\pi r)\hat{\mathbf{e}}_\phi$.

[9] The details of the corresponding mathematical procedure do not concern us here and are given in Feynman and Hibbs (1965) and in any number of textbooks, e.g., Zagoskin (1998), §1.2.2, Ryder (1996), Chapter 5 and Kleinert (2006). If $K(\mathbf{r}_B, t_B | \mathbf{r}_A, t_A)$ is known, one can express the wave function of the particle at the moment t_B through its wave function at $t_A < t_B$ as a conventional integral, $\Psi(\mathbf{r}_B, t_B) = \int d\mathbf{r}_A\, K(\mathbf{r}_B, t_B | \mathbf{r}_A, t_A)\Psi(\mathbf{r}_A, t_A)$, and the standard quantum mechanical formalism readily follows.

Then from (3.88) we find that

$$L(\mathbf{r}, \dot{\mathbf{r}}, t; \mathbf{A}) = L(\mathbf{r}, \dot{\mathbf{r}}, t; \mathbf{A} = 0) + \frac{q}{c}\mathbf{A} \cdot \mathbf{r}. \tag{3.90}$$

The vector potential only adds to the action, $S_{\mathrm{cl}} = \int L \, dt$, a term $\frac{q}{c}\int_{t_A}^{t_B} dt \, \dot{\mathbf{r}} \cdot \mathbf{A} = \frac{q}{c}\int_{t_A}^{t_B} d\mathbf{r} \cdot \mathbf{A}$. Therefore, the exponent in (3.87) can be written as $\left[\frac{i}{\hbar}\int_{t_A}^{t_B} dt \, L(\mathbf{r}, \dot{\mathbf{r}}, t; \mathbf{A} = 0) \right] \times \left[\frac{iq}{c\hbar}\int_{t_A}^{t_B} d\mathbf{r} \cdot \mathbf{A} \right]$.

Let us now consider the transition between A and B in the presence of the solenoid. The only difference it makes to (3.87) is through the term $\left[\frac{iq}{c\hbar}\int_{t_A}^{t_B} d\mathbf{r} \cdot \mathbf{A} \right]$ in the exponent. In the gauge (3.86) we chose, any trajectory that crosses the half-plane ($x > 0$, $y = 0$) N_+ times in the positive-y direction and N_- times in the opposite direction, will acquire a phase factor $\exp\left[\frac{iq}{\hbar c}\Phi(N_+ - N_-) \right]$. In classical mechanics there is only one trajectory to take into account: the one which provides the extremum to the action S_{cl} (Goldstein, 1980). This is precisely the trajectory of classical mechanics. The presence of the vector potential of the solenoid (3.86) can only add a constant to the action, and, therefore, will not affect the shape of the extremal trajectory. In classical mechanics indeed, where there is no field, there is no effect.

The situation changes in the quantum case. If, e.g., we take into account two trajectories between the points A and B, one with $N_+ - N_- = 0$ and another with $N_+ - N_- = 1$ (e.g., one passing the solenoid on the right, and another on the left), and assuming for simplicity that they contribute equally to the probability amplitude, we find for the transition probability

$$P(\mathbf{r}_B, t_B | \mathbf{r}_A, t_A) = |K(\mathbf{r}_B, t_B | \mathbf{r}_A, t_A)|^2 \propto \left| \exp\left[\frac{iq}{\hbar c}\Phi \right] + 1 \right|^2, \tag{3.91}$$

which clearly depends on the flux Φ. The non-hand-waving calculations (Aharonov and Bohm, 1959; Olariu and Popescu, 1985) show that indeed the magnetic flux in the solenoid would shift the interference pattern of an electron wave impinging on it. In solid-state devices, the effect was observed as a periodic dependence of conductance of metal loops on the magnetic field (see, e.g., Imry, 2002, §3.2), as well as in 2DEG devices. The period of oscillations is $2\pi\hbar c/q$. For an electron, $q = e$, and the period is $2\Phi_0 = 2hc/2e$, twice the superconducting flux quantum of § 2.1.1.

The theoretical treatment of the problem is best based on Landauer formalism, reducing the transport problem to one of coherent scattering (Büttiker, 1986). In 2DEG loops, we can in addition take advantage of the small number of conducting channels, and consider the system as one-dimensional (Fig. 3.15). The scattering

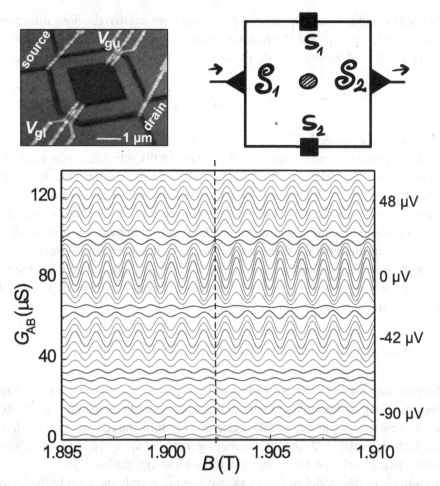

Fig. 3.15. Aharonov–Bohm effects in a 2DEG interferometer. (Top) 2DEG loop device (van der Wiel et al., 2003) and a 1D model: 2×2-matrices S_1, S_2 describe scattering in the branches, 3×3-matrices $\mathcal{S}_1$, $\mathcal{S}_2$ the scattering in the junctions. The scattering in the branches is tuned by the gate voltages. (Bottom) Differential conductance between the source and drain electrodes as a function of the magnetic field and gate voltage. Periodic dependence on the magnetic and electric fields reflects the magnetic and electrostatic Aharonov–Bohm effects (top left and bottom reprinted with permission from van der Wiel et al., 2003. © American Physical Society).

in the loop is described by the two-by-two-matrices

$$S_j = \begin{pmatrix} r_j & \bar{t}_j \\ t_j & \bar{r}_j \end{pmatrix}. \tag{3.92}$$

The effect of the vector potential is easily taken care of if we use a gauge like (3.86), with the discontinuity being crossed only by the bottom branch of the loop. Then we only need to multiply the transmission amplitude from left to right along this branch by the phase factor $\exp[i\pi \Phi/\Phi_0]$, and the transmission amplitude from right to left by $\exp[-i\pi \Phi/\Phi_0]$. In other words, in the presence of the flux the scattering matrix

$$S_2 \to S_2(\Phi) = \begin{pmatrix} r_2 & \bar{t}_2 e^{-i\pi \Phi/\Phi_0} \\ t_2 e^{i\pi \Phi/\Phi_0} & \bar{r}_2 \end{pmatrix}. \tag{3.93}$$

So far we have discussed only the *magnetic* Aharonov–Bohm effect. There exists also the *electrostatic* one (Aharonov and Bohm, 1959), which is easier to discuss directly using the model of Fig. 3.15. Suppose we send an electron (a localized wave packet) through the system from the left to the right, and when it passes through the loop, switch on an electrostatic potential between the top and bottom branches, for a duration t_0. Since the electron cannot be accelerated by the electric field (the branches being horizontal), there can be no field-produced effect. Nevertheless, there will be a phase difference, $\phi = \Delta E t_0/\hbar = eVt_0/\hbar$, induced between the top- and bottom-propagating components of the wave packet, and this will produce the oscillations in the transmission probability,

$$P(\mathbf{r}_B, t_B | \mathbf{r}_A, t_A) = |K(\mathbf{r}_B, t_B | \mathbf{r}_A, t_A)|^2 \propto \left| \exp\left[\frac{ieVt_0}{\hbar}\right] + 1 \right|^2. \tag{3.94}$$

(Aharonov and Bohm (1959) instead considered passing the electron wave through a couple of long and narrow cylindrical electrodes, inside which the electric field does not penetrate.) This can be taken care of by putting the corresponding relative phase factors, $\exp(\pm i\phi/2)$, into the scattering matrices S_1, S_2. Here t_0 will be some characteristic travel time for an electron to pass through the loop.

Returning to the model of Fig. 3.15, we need to couple the loop to the source and drain leads. This is done by using orthogonal 3×3-matrices (Shapiro, 1983)

$$S = \begin{pmatrix} 0 & -1/\sqrt{2} & -1/\sqrt{2} \\ -1/\sqrt{2} & 1/2 & -1/\sqrt{2} \\ -1/\sqrt{2} & -1/2 & 1/2 \end{pmatrix}, \tag{3.95}$$

where the diagonal terms give the reflection amplitudes, and the off-diagonal ones the transmission amplitudes. We see that there is, e.g., zero probability of reflection from the junction back into the left (right) lead, and equal probability of passing on the right or left side of the loop. This simplification is not crucial, since additional scatterers can be always added to the model. Solving for the transmission amplitude through the ring, one then obtains

$$t_{\text{ring}} = 2\frac{t_1 t_2(\bar{t}_1 + \bar{t}_2) + t_1(r_2 - 1)(1 - \bar{r}_2) + t_2(r_1 - 1)(1 - \bar{r}_1)}{(t_1 + t_2)(\bar{t}_1 + \bar{t}_2) - (2 - r_1 - r_2)(2 - \bar{r}_1 - \bar{r}_2)}, \tag{3.96}$$

and, of course,

$$G_{\text{ring}} = \frac{e^2}{\pi \hbar} |t_{\text{ring}}|^2. \tag{3.97}$$

The results of one of the experiments on the Aharonov–Bohm effect in 2DEG interferometers are shown in Fig. 3.15. The branches of the loop, formed in GaAs/AlGaAs 2DEG by etching, were pinched by gates, creating strongly scattering barriers. The periodic dependence on the magnetic field clearly demonstrates the magnetic Aharonov–Bohm effect. In addition, van der Wiel et al. (2003) also made measurements when a finite voltage V was applied between the source and drain electrodes. This had the effect of producing an additional electric Aharonov–Bohm phase, $eVt_0/\hbar$, where the electron travel time across the loop was estimated as $t_0 = 45$ ps. The periodic dependence of the nonlinear conductance on the voltage is also shown in Fig. 3.15.

3.3.2 *Hybrid 2DEG-superconducting structures. Landauer–Lambert formalism*

One more interesting possibility is presented by the fact that the 2DEG layers can be brought to touch with superconductors (e.g., used as normal barriers in superconducting Josephson structures). One such device is shown in Fig. 3.16. A special treatment of the area under the niobium contact allows the formation of a transparent Nb-2DEG interface. Therefore, the current flow through the interface is almost unhindered and is conducted through Andreev reflections (§ 2.6).[10] Electrostatic shaping of the normal area will thus provide a direct control of the Josephson critical current in the system.

One unexpected result of the experiment (Heida et al., 1998) was that the Josephson critical current dependence on the magnetic flux through the contact area was not Φ_0 (as expected from theory and confirmed by all the standard experiments – see, e.g., Tinkham, 2004, § 6.4) – but $\Phi_0/2$. The explanation lies in the large electron scattering length in the normal part of the junction. This means that electron transport through it is ballistic, and the finite width of the 2DEG channel must be taken into account.

The case of Josephson current flow through a short and narrow QPC was considered by Furusaki (1999). Using more advanced techniques than we care to introduce here (like Matsubara Green's functions for superconductors), he obtained the following expression for the Josephson current in a short, but smooth, symmetric

---

[10] This is an excellent demonstration of the fact that as long as the dephasing processes are weak enough, quantum coherence survives in a normal conductor. In this particular case, the surviving quantum correlations are the ones between electrons and Andreev-reflected holes, and vice versa, originating in the superconducting bank. As C.W.J. Beenakker said to the author, "Normal metal is neutral with respect to superconducting correlations."

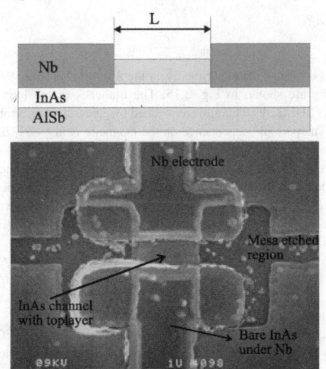

Fig. 3.16. Hybrid superconductor-2DEG device: an SNS Josephson junction. The Nb superconducting electrodes are in contact with 2DEG formed in an InAs/AlSb heterostructure, with electron density $n = 2 \times 10^{12}$ cm^{-2} (reprinted with permission from Heida et al., 1998, © 1998 American Physical Society).

quantum point contact (S-QPC-S contact):

$$I_J(\varphi) = -N\frac{e\Delta_0}{\hbar}\sin\frac{\varphi}{2}\tanh\left[\frac{\Delta_0}{2k_BT}\cos\frac{\varphi}{2}\right]. \qquad (3.98)$$

Here N is the number of open channels in the QPC (so that its Landauer conductance in the normal state would have been $N\frac{e^2}{\pi\hbar}$), and Δ_0 is the magnitude of the superconducting gap far inside the superconducting banks. As the number of open channels is changed by the gate voltage, so should the Josephson critical current. This quantization effect was indeed observed by Takayanagi et al. (1995a,b).

Let us now consider the situation directly relevant for the device shown in Fig. 3.16. Here the size of the 2DEG region greatly exceeded the electron wavelength, which allows us to use quasiclassical considerations. As we have seen in § 2.6, the Josephson current in a clean 1D SNS junction is carried by Andreev levels and has a characteristic sawtooth dependence on the superconducting phase difference. What happens in a 2D case? In a quasiclassical picture, an electron emerging from the superconducting bank into the 2DEG travels along a straight line, and – possibly after several reflections from the potential barriers, which restrict the 2DEG laterally – arrives at the other superconductor. There it undergoes an Andreev reflection and is reflected back as a hole, which will retrace the initial electron's trajectory and end up in the first superconductor. Here it is again Andreev-converted into an electron, and thus an Andreev level is formed. We can think of it as an *Andreev tube*, with a characteristic diameter of λ_F. This quasi-1D "tube" carries the Josephson current precisely as if it were a 1D SNS junction, and we only need to sum the contributions from all possible electron–hole trajectories to find the total Josephson current.[11] The effect of the normal magnetic field **H** through the 2DEG region can be taken into account by adding to the superconducting phase difference between the superconductors the additional phase gain along the Andreev tube:

$$\varphi \to \varphi + \frac{2\pi}{\Phi_0} \int_{\tau_1}^{0} d\tau \, \mathbf{A}(\mathbf{r} - v_F \tau \mathbf{n}) \cdot \mathbf{n}. \tag{3.99}$$

Here **n** is a unit vector along the tube, and the vector potential must be taken in the *Landau gauge*,

$$\mathbf{A} = H y \hat{\mathbf{e}}_z. \tag{3.100}$$

(Of course, we could choose a different gauge, but then we would have to make the phase of superconducting banks position-dependent; this will not change the result, but why make calculations any harder than necessary?) The current across the 2DEG-S boundary (Fig. 3.16) is then, at zero temperature, given by

$$I(\varphi, H) = \frac{e v_F}{W \lambda_F L} \int \int_{-W/2}^{W/2} \frac{dy_1 \, dy_2}{\left[1 + ((y_1 - y_2)/L)^2\right]^{3/2}}$$
$$\frac{2}{\pi} \sum_{k=1}^{\infty} (-1)^{k+1} \frac{\sin k \left[(\pi \Phi / W \Phi_0)(y_1 + y_2) + \varphi\right]}{k}. \tag{3.101}$$

In this expression, for the sake of simplicity, only the trajectories, which directly link the two superconductors, are taken into account. The sum on the right-hand side is the sawtooth we encountered before, Eq. (2.165). The critical current is the maximum value of this expression for φ between zero and 2π, and in the limit of either infinitely wide or infinitely narrow contact it can be calculated explicitly

---

[11] This cavalier approach can actually be justified within the Green's functions formalism (Barzykin and Zagoskin, 1999). It provides a convenient way of calculating Josephson currents in normal layers and can take into account finite temperatures and weak elastic scattering by impurities.

(Barzykin and Zagoskin, 1999):

$$I_c\left(\frac{\Phi}{\Phi_0}\right) = \frac{2W}{\lambda_F}\frac{ev_F}{L}\frac{\left(1-\left\{\frac{\Phi}{\Phi_0}\right\}\right)\left\{\frac{\Phi}{\Phi_0}\right\}}{\left|\frac{\Phi}{\Phi_0}\right|}, \quad L \ll W; \qquad (3.102)$$

$$I_c\left(\frac{\Phi}{\Phi_0}\right) = \frac{W}{\lambda_F}\frac{ev_F}{L}\frac{\left(1-\left\{\frac{\Phi}{2\Phi_0}\right\}\right)^2\left\{\frac{\Phi}{2\Phi_0}\right\}^2}{\left|\frac{\Phi}{2\Phi_0}\right|^2}, \quad L \gg W. \qquad (3.103)$$

Here $\{x\}$ is the fractional part of x; the first expression is Φ_0-periodic and reproduces the known results for a wide SNS junction, while the second one is $2\Phi_0$-periodic, in agreement with the measurements of Heida et al. (1998).

After having dealt with the superconducting current through a hybrid structure, it is logical to ask whether and how the presence of a superconducting reservoir affects *normal* conductance in the system. The answer is yes, and in quite an interesting way. Following Lambert (1991) (see also Zagoskin, 1998, Appendix 1), consider normal transport between the source and drain reservoirs, and assume that the coherent part of the system is also connected to a superconducting reservoir (Fig. 3.17). Compared to the case of normal conductors (§3.1.3), the situation is immediately more complicated, since we must consider electrons and holes separately, and account for the possibility of Andreev reflection. Therefore, instead of simple transition and reflection coefficients, $T + R = 1$, we need a 4×4-matrix of

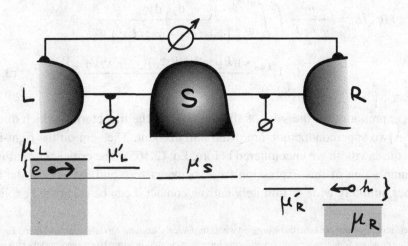

Fig. 3.17. Landauer conductance in a hybrid normal-superconducting (NSN) system.

probabilities:

$$P = \begin{pmatrix} \mathbf{R} & \bar{\mathbf{T}} \\ \mathbf{T} & \bar{\mathbf{R}} \end{pmatrix};$$

(3.104)

$$\mathbf{R} = \begin{pmatrix} R_{e \to e} & R_{h \to e} \\ R_{e \to h} & R_{h \to h} \end{pmatrix}; \quad \mathbf{T} = \begin{pmatrix} T_{e \to e} & T_{h \to e} \\ T_{e \to h} & T_{h \to h} \end{pmatrix}.$$

(3.105)

Here the blocks $\mathbf{T}, \mathbf{R}$ describe the transmission/reflection of electrons and holes incident from the left, and $\bar{\mathbf{T}}, \bar{\mathbf{R}}$, from the right reservoir. Unitarity requires that every column and row of $\mathbf{P}$ adds up to unity, for example

$$R_{e \to e} + R_{h \to e} + \bar{T}_{e \to e} + \bar{T}_{h \to e} = 1; \quad R_{e \to e} + R_{e \to h} + T_{e \to e} + T_{e \to h} = 1. \quad (3.106)$$

The first equation means that an electron can arrive at the left reservoir in only four ways: being normally reflected back by the system; Andreev-reflected; normally transmitted from the right reservoir; or Andreev-transmitted. The second equation similarly counts the ways an electron can leave the left reservoir – to be normally or Andreev-reflected to the left, or normally or Andreev-transmitted to the right.

We will calculate the conductance between the normal reservoirs in the limit of a small voltage drop between them. This allows us to use for the density of states in the 1D lead a constant value, $1/\pi \hbar v_F$, and write, e.g., for the current in the left lead:

$$I = e v_F \frac{1}{\pi \hbar v_F} \left[(\mu_L - \mu_S)(1 - R_{e \to e} + R_{e \to h}) + (\mu_S - \mu_R)(\bar{T}_{h \to h} - \bar{T}_{h \to e}) \right].$$

(3.107)

In the presence of a superconductor with the chemical potential μ_S it is convenient to consider the quasiparticles with energies below and above μ_S separately. Then we can say that the left reservoir injects into the system *electrons* with energies between μ_S and μ_R, while the right one injects *holes* with energies between μ_L and μ_S, and this is reflected in Eq. (3.107). The total current flowing into the system from the left reservoir is carried by the electrons within the band $[\mu_S, \mu_L]$ and equals

$$I_L = \frac{e}{\pi \hbar} (\mu_L - \mu_S)(1 - R_{e \to e} + R_{e \to h} - T_{e \to e} + T_{e \to h})$$

$$= \frac{2e}{\pi \hbar} (\mu_L - \mu_S)(R_{e \to h} + T_{e \to h}),$$

(3.108)

where we have used the unitarity condition (3.106). Similarly, the current from the right reservoir is carried by the holes in the band $[\mu_R, \mu_S]$:

$$I_R = -\frac{2e}{\pi \hbar} (\mu_S - \mu_R)(\bar{R}_{h \to e} + \bar{T}_{h \to e}).$$

(3.109)

The value of μ_S must be determined from the net current flowing in or out of the superconductor. If this current is zero, which is usually the case, that is, if $I_L + I_R = 0$, then μ_S is readily excluded from Eq. (3.107), and the two-point conductance is found as

$$
\begin{aligned}
G &= \frac{eI}{\mu_L - \mu_R} \\
&= \frac{e^2}{\pi\hbar} \frac{(\bar{R}_{h\to e} + \bar{T}_{h\to h})(R_{e\to h} + T_{e\to h}) + (\bar{R}_{h\to e} + \bar{T}_{h\to e})(R_{e\to h} + T_{e\to e})}{\bar{R}_{h\to e} + \bar{T}_{h\to e} + R_{e\to h} + T_{e\to h}}.
\end{aligned}
$$

(3.110)

If there is particle–hole symmetry, that is, normal and Andreev reflection/transmission are the same for electrons and holes ($R_{e\to e} = R_{h\to h} = R_N$, $R_{e\to h} = R_{h\to e} = R_A$, etc.), then (3.110) simplifies to

$$
G = \frac{e^2}{\pi\hbar} \frac{(\bar{R}_A + \bar{T}_N)(R_A + T_A) + (\bar{R}_A + \bar{T}_A)(R_A + T_N)}{\bar{R}_A + \bar{T}_A + R_A + T_A}.
$$

(3.111)

If, in addition, the system is spatially symmetric, that is, there is no difference between barred and non-barred coefficients, the conductance becomes

$$
G = \frac{e^2}{\pi\hbar}(T_N + R_A).
$$

(3.112)

Note that Andreev processes open an additional conductance channel through electron–hole conversion.

For the sake of completeness, let us provide the expression for the four-point conductance,

$$
G^{(4pt)} = \frac{eI}{\mu'_L - \mu'_R},
$$

where the effective chemical potentials of the leads, μ'_L, μ'_R, are determined in the same way as in Eq. (3.39). For particle–hole and spatial symmetry one finds

$$
G^{(4pt)} = \frac{e^2}{\pi\hbar} \frac{T_N + R_A}{T_A + R_N},
$$

(3.113)

which in the absence of a superconductor gives the correct limit $G^{(4pt)} = \frac{e^2}{\pi\hbar}\frac{T}{R}$.

Concluding remarks

We have described the main types of quantum coherent device based on a two-dimensional electron gas, such as are currently available, their theory and useful elements of theoretical formalism of more general applicability. We have seen that their design is flexible and scalable, and that the parameters and the very shape of the structure can be changed during the experiment. This makes this class of devices good, both as quantum block elements (e.g., lateral quantum dots) and as quantum detectors. In the latter role the 2DEG devices enable the electric measurement of charge and spin degrees of freedom, including the single-shot measurement regime. Transparent superconductor-2DEG boundaries, achievable in (InAs/AlSb)/Nb structures, provide an additional possibility of a direct interfacing between 2DEG- and superconductor-based quantum coherent devices. Transport in 2DEG and 2DEG-superconductor structures is naturally described in the framework of Landauer formalism, which reduces the problem to one of coherent scattering and allows us to avoid the introduction of the density matrix and nonunitary evolution altogether.

4

Superconducting multiqubit devices

Conquer with skill, not with numbers.

A. V. Suvorov, from his military maxims,
ca. 1795, translated from Russian by the author

4.1 Physical implementations of qubit coupling

4.1.1 Coupling by linear passive elements. Capacitive coupling

To have good basic elements is not enough – it is necessary to be able to connect them in a controllable way, without losing quantum coherence. Any simple effective coupling Hamiltonian (like in Eqs. (3.81, 3.82)) must be somehow implemented "in metal". Here superconducting circuits provide a wide variety of coupling schemes to choose from (see, e.g., Wendin and Shumeiko, 2005). We will begin with the simplest case, when the interaction between the qubits is realized using linear elements (conventional capacitances and inductances), the coupling circuit stays in its ground state and adiabatically follows the evolution of the qubits (Averin and Bruder, 2003) – that is, it remains "passive". For this to happen, the excitation energy of the coupler, $\hbar\omega_{res}$, must be much higher than the interlevel spacing in the qubits (where ω_{res} is the resonance frequency of the coupler). In other words, the evolution of the coupler is much faster than that of the qubits, and the coupler can indeed adjust to changes in the state of the latter. In the case of a purely capacitive or purely inductive coupling this condition is automatically satisfied, as then $\omega_{res} \to \infty$.

Consider, for example, two phase qubits coupled capacitively (Fig. 4.1). The Hamiltonian of the system can be reduced to the form (Blais et al., 2003)

$$H = \frac{c^2 \Pi_1^2}{2\widetilde{C}_1} - E_J^{(j)} \cos 2\pi \frac{\Phi_1}{\Phi_0} - \frac{I_b^{(1)} \Phi_1}{c} + \frac{c^2 \Pi_2^2}{2\widetilde{C}_2}$$

$$- E_J^{(2)} \cos 2\pi \frac{\Phi_2}{\Phi_0} - \frac{I_b^{(2)} \Phi_2}{c} + \frac{c^2 \Pi_1 \Pi_2}{\widetilde{C}_c}, \qquad (4.1)$$

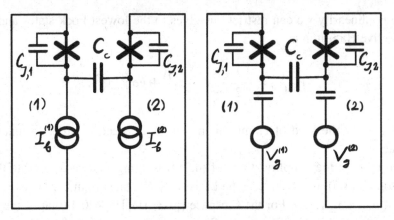

Fig. 4.1. Direct capacitive coupling of phase (left) and charge (right) superconducting qubits.

where the effective capacitances are

$$\tilde{C}_1 = C_{J,1} + (C_{J,2}^{-1} + C_c^{-1})^{-1},$$

$$\tilde{C}_2 = C_{J,2} + (C_{J,1}^{-1} + C_c^{-1})^{-1}, \tag{4.2}$$

$$\tilde{C}_c = C_{J,1} C_{J,2} (C_{J,1}^{-1} + C_{J,2}^{-1} + C_c^{-1}). \tag{4.3}$$

The phase qubits are nonlinear oscillators. It is, therefore, convenient to express the node flux and charge operators through the appropriate Bose operators, as in Eq. (2.106):

$$\hat{\Phi}_j = \frac{a_j + a_j^\dagger}{2} \Lambda_j, \quad \hat{\Pi}_j = \hbar \frac{a_j - a_j^\dagger}{i\Lambda_j}, \quad \Lambda = \sqrt{\frac{2\hbar}{\tilde{C}_j \omega_j}}, \quad (j = 1, 2), \tag{4.4}$$

where $\omega_j = \omega_{0,j}(1 - (I_{b,j}/I_{c,j})^2)^{1/2}$, the "unsoftened" Josephson plasma frequency of the jth phase qubit $\omega_{0,j} = [E_{J,j} E_{C,j}(1 - (I_{b,j}/I_{c,j})^2)]^{1/2}/\hbar$, and the charging energy $E_{C,j} = 2e^2/\tilde{C}_j$. Assuming that the qubits have identical parameters (so that $\tilde{C}_1 = \tilde{C}_2 = \tilde{C}$) and substituting (4.4) in (4.1), we see that the Hamiltonian becomes

$$H = \sum_{j=1}^{2} \hbar \omega_j \left(a_j^\dagger a_j + \frac{1}{2} \right) + (\ldots) - 2g \left(a_1 - a_1^\dagger \right) \left(a_2 - a_2^\dagger \right), \tag{4.5}$$

where the ellipsis stands for the higher-order terms, and the coupling coefficient is

$$g = \frac{\tilde{C}}{2\tilde{C}_c} \hbar \sqrt{\omega_1 \omega_2}. \tag{4.6}$$

Due to nonlinearity we can restrict ourselves to the lowest Fock states and write the effective Hamiltonian

$$H_{int}^{(eff)} = -\frac{1}{2}\sum_{j=1}^{2}\hbar\omega_j\sigma_z^{(j)} + g\sigma_y^{(1)}\sigma_y^{(2)}. \tag{4.7}$$

This is exactly the kind of Hamiltonian one would need in order to link qubits together.

Even without the simplifications of (4.7) it is straightforward to write the full Hamiltonian (4.5) as a matrix in the basis of Fock states (number states) $\{|n_1\rangle \otimes |m_2\rangle\}$, $(m, n = 0, 1, \ldots)$. For the subspace $\{|0\rangle \otimes |1\rangle, |1\rangle \otimes |0\rangle\}$ (that is, the states, where the two qubits simultaneously flip), we find

$$H_2 = \begin{pmatrix} E_0^{(1)} + E_1^{(2)} & g \\ g & E_1^{(1)} + E_0^{(2)} \end{pmatrix}, \tag{4.8}$$

which is equivalent to the $\sigma_y\sigma_y$-term in (4.7). In resonance, that is, when the inter-level spacings are made equal, $E_1^{(1)} - E_0^{(1)} = E_1^{(2)} - E_0^{(2)}$, by tuning the bias currents, the diagonal terms in (4.8) can be dropped (they only produce an overall phase factor), and we obtain the eigenstates

$$|\Psi_\pm\rangle = \frac{|0\rangle \otimes |1\rangle \pm |1\rangle \otimes |0\rangle}{\sqrt{2}}, \tag{4.9}$$

with the energy split g. These states are entangled, that is, if the whole system is in a pure state $|\Psi_\pm\rangle$ and the state of, e.g., qubit 1 is measured in the basis $\{|0\rangle, |1\rangle\}$ (see § 1.2.3), the system will no be longer in a pure state, but in a mixed state, with the density matrix

$$\rho = \frac{1}{2}|0\rangle\langle 0| \otimes |1\rangle\langle 1| + \frac{1}{2}|1\rangle\langle 1| \otimes |0\rangle\langle 0|. \tag{4.10}$$

If we initialize the system in a pure diabatic state, e.g., $|0\rangle \otimes |1\rangle$, the probability of finding qubit 1 in state $|0\rangle$ (or $|1\rangle$) will oscillate with the frequency $\omega = 2g/\hbar$. These are precisely the quantum beats discussed in § 1.4.2, only this time not between the states of a single qubit, but between the states of two coupled qubits.[1] Experimental confirmation of this effect by McDermott et al. (2005) is shown in Fig. 4.2.

Another direct signature of the formation of entangled states (4.9) of two qubits can be obtained from observation of the anticrossing between the diabatic levels,

---

[1] In Blais et al. (2003) their period is called T_{Rabi}. We will see in § 4.2.3 in what sense this kind of quantum beats can be called "vacuum Rabi oscillations".

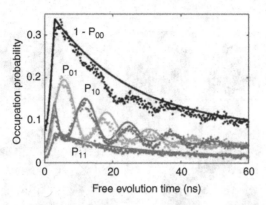

Fig. 4.2. Quantum beats between capacitively coupled phase qubits (from McDermott et al., 2005; reprinted with permission from AAAS). The two qubits were brought in resonance (with $(E_1^{(1,2)} - E_0^{(1,2)})/h = 8.65$ GHz) and initialized in the state $|1\rangle \otimes |0\rangle$. Repeated measurements allowed determination of the probabilities of finding the system in states $|11\rangle$, $|01\rangle$, $|10\rangle$, and $|00\rangle$ after interval t_{free} of free evolution. The quantum beats are clearly seen in the behaviour of P_{01} and P_{10}. The curve for P_{00} reflects the relaxation of qubits to the ground state with the single-qubit relaxation time $T_1 = 25$ ns. The solid lines are the results of numerical simulations, taking into account the finite state preparation time of 5 ns (rising sections on the left), finite measurement fidelity of 70%, and microwave cross-coupling (influence of the microwave pulse applied to qubit 1 to put it in state $|1\rangle$, on qubit 2, with a finite probability of exciting it).

as in the tunnelling splitting in a single qubit, Fig. 2.7. This can be done, e.g., using spectroscopic measurements, with the advantage of not needing time-domain operations with their precise timing. Notably, in the experiment of Berkley et al. (2003) (Fig 4.2), quantum entanglement was established between the qubits separated by quite a macroscopic distance of about 1 mm.

Returning to Eq. (4.8), we see that in resonance this block of the total Hamiltonian acts as $\exp[-ig\sigma_x t/\hbar]$ in the subspace $\{|0\rangle \otimes |1\rangle, |1\rangle \otimes |0\rangle\}$, while producing just some phase factors for the other states. When the two qubits are taken out of resonance, the coupling g becomes negligible, as the eigenstates of (4.8) approach the diabatic states (cf. Eq. (1.133)).

Direct capacitive coupling of two charge qubits, of a charge and phase qubit, of quantronium qubits, etc. is not really different from the case just described. For example, for two charge qubits (Figs. 4.1, 4.3) the coupling Hamiltonian is

$$H_{\text{int}} = \frac{e^2}{2C_x}\sigma_x^{(1)}\sigma_x^{(2)}. \tag{4.11}$$

After strongly biasing one of the qubits (i.e., fixing it in a given charge state), one can manipulate the other qubit, and the qubit–qubit interaction (4.11) is only effective

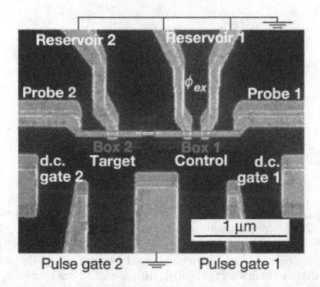

Fig. 4.3. Capacitively coupled charge qubits fabricated of AlO_x (reprinted by permission from *Nature*, Yamamoto et al. (2003), © 2003 Macmillan Publishers Ltd). The Josephson energy of charge qubit 1 can be tuned by the magnetic flux ϕ_{ex}. When one of the qubits is biased away from the co-degeneracy point, it produces almost no influence on the evolution of the other qubit. The ground and excited states of each qubit are the linear combination of the corresponding charge states, $|g(e)\rangle = (|0\rangle \pm |1\rangle)/\sqrt{2}$.

when both qubits are near their co-degeneracy point $n_{g1} = n_{g2} = 0.5$. Using this coupling, Yamamoto et al. (2003) realized a two-qubit quantum CNOT gate for charge qubits (see the Appendix).

In principle, a purely capacitive coupling is impossible, since the connecting elements will have finite geometric inductances (roughly 1 nH per 1 μm of length; more if the kinetic inductance of superconductors is taken into account). Taking $L \sim 1$ nH and a characteristic capacitance $C \sim 1$ fF, we obtain the excitation frequency of the capacitive link ~ 100 GHz. As long as the interlevel separation in qubits and other characteristic energies are well below this level, the capacitive coupling can indeed be considered as passive.

4.1.2 Passive inductive coupling of flux qubits

This coupling seems quite straightforward: the external flux through a flux qubit loop changes the energies of the states with clockwise/anticlockwise directions of persistent current in the opposite direction, thus affecting the σ_z-term in the qubit Hamiltonian (1.85). If the amplitude of the circulating current in the jth qubit is

I_j, and their mutual inductance is M, the coupling term, written as before in the "physical" (diabatic) basis, is

$$H_{\text{int}} = 2MI_1 I_2 \sigma_z^{(1)} \sigma_z^{(2)} \equiv J\sigma_z^{(1)} \sigma_z^{(2)}. \tag{4.12}$$

From a simple consideration of magnetic force lines, this will be "antiferromagnetic" coupling. A drawback of zz-coupling (4.12) compared to xx- or yy-coupling, Eqs (4.7, 4.11), is that now it is impossible to tune out one of the qubits by biasing it away from the co-degeneracy point. An example of such coupling is shown in Fig. 4.4. We postpone the discussion of another important feature of this particular design (a pick-up coil used for both readout and insulation of the qubits from external noise) until Sections 5.3 and 5.5.

When discussing the inductive coupling of persistent current flux qubits, we run into a contradiction. On the one hand, the mutual inductance of the qubits must be nonzero. On the other hand, it cannot exceed the self-inductance of the qubits, which was, nevertheless, neglected in our consideration of this device, see Eqs (2.110, 2.112, 2.114). Self-inductance enters the Lagrangian as $1/L$, and the proper taking of the limit $L \to 0$ is tricky. Fortunately, the inconsistency does not lead to serious problems.[2]

In order to increase the coupling strength, the persistent current qubits can be fabricated with a shared branch (so-called "galvanic coupling" (Majer et al., 2005)), thereby adding to the magnetic (geometric) mutual inductance of the qubit loops the kinetic inductance of the branch (see, e.g., Schmidt, 2002, §10).

4.1.3 Coupling by nonlinear passive elements. Tunable coupling

A significant increase in the inductive coupling strength for flux qubits can be achieved by making two flux qubits share a Josephson junction. This will add to the Hamiltonian a coupling term of order E_J, sensitive to the states of both qubits. Physically, this kind of coupling makes use of the Josephson inductance (2.27).[3]

Since we assume that the coupler is passive, we do not need to go through all the motions of § 2.3.3 in order to find the strength of this coupling in the leading approximation. It is enough to analyse the potential energy of the system of two qubits sharing a junction, Fig. 4.5a,

$$U = U_a(\phi_{a1}, \phi_a, \widetilde{\phi}_a + \varphi) + U_b(\phi_{b1}, \phi_{b2}, \widetilde{\phi}_b + \varphi) - E_c \cos\varphi \tag{4.13}$$

---

[2] A consistent perturbative theory of single and inductively coupled persistent current qubits for small, but finite, self-inductances was developed by Maassen van den Brink (2005). It turns out that the coupling (4.12) remains valid, with renormalized mutual inductance; self-inductances and effective Josephson couplings within each qubit will also be renormalized. The corrections are generally small and below the current experimental uncertainty of the devices' parameters.

[3] This approach was used in the first device with four superconducting qubits (Grajcar et al., 2006).

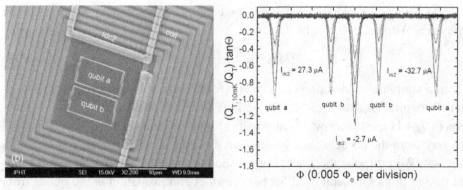

Fig. 4.4. Inductive coupling between flux qubits (reprinted with permission from Izmalkov et al., 2004b, © 2004 American Physical Society). (Left) A micrograph of two aluminium persistent current qubits placed inside a niobium pick-up coil. The qubit parameters were as follows: self-inductance $L_{a,b} \approx 39$ pH, critical currents of the two larger junctions $I_c \approx 400$ nA, Coulomb energies $E_C \approx 6.4$ GHz, reduction parameter for the smaller junction $\alpha \sim 0.8 - 0.9$, estimated mutual inductance $M_{ab} \approx 2.7$ pH; the persistent currents $I_p = 320$ nA in either qubit, and the tunnelling splitting $\Delta_{a,b} \approx 1$ GHz. The Nb coil with $L_T \approx 130$ nH, connected to an external capacitor $C_T \approx 470$ pF, forms a high-quality LC circuit (tank circuit) with resonant frequency $\omega_T/2\pi = 20.139$ MHz and quality factor $Q_T = \omega_T \tau_T = 1680$ at 10 mK ($\tau_T = R_T C_T$ is the relaxation time). The LC circuit plays the additional role of a filter, protecting the qubits from the external noise (see Section 5.3). The bias magnetic field in the qubits was created by the dc component I_{dc1} of the current in the pick-up coil and by the dc current I_{dc2} in the additional loop (Π-shaped wire in the micrograph). The measured interqubit coupling constant (Eq. 4.12) is $J/h = 420$ MHz and corresponds to "antiferromagnetic" coupling. (Right) The quantum inductance of a qubit depends on its state and influences the effective inductance of the LC circuit. Therefore, the response of the tank circuit to the external drive at near-resonant frequency can be used to measure the state of the qubits (impedance measurement technique (IMT); see § 5.5.3). Here the normalized tangent of the phase angle between the current and voltage in the tank is plotted versus the external magnetic flux bias for different temperatures (the nominal temperature, from the lower to the upper curve, is 10, 50, 90, 160 mK, respectively). The external flux Φ is changed by varying I_{dc1}, while the current I_{dc2} allows us to produce a relative bias, i.e., a difference between the magnetic fluxes penetrating the qubits a and b. tan Θ is proportional to the minus inverse effective inductance of the system, which includes a qubit state-sensitive quantum inductance term (Eqs (2.142, 2.143)): tan $\Theta \sim -\partial^2_{\Phi\Phi} E(\Phi)$, where $E(\Phi) = \langle H \rangle$ is the external flux-dependent energy. Therefore, tan Θ will have a sharp dip near the degeneracy point (i.e., the anticrossing between the ground and excited states). The sharpness of the dip depends on the tunnelling matrix element, Δ, and the temperature. The three sets of curves correspond to different relative biases between the qubits. The central set, when both of them are simultaneously at the degeneracy point, shows the formation of the entangled ground state of the two-qubit system: otherwise the dip would have been just a sum of single-qubit dips.

(here it is more convenient to use the phase variables $\phi_j = 2\pi \Phi_j / \Phi_0$). We neglect the self-inductances and mutual inductances of the qubits, which allows us to eliminate phases ϕ_{a3} and ϕ_{b3} and only include the Josephson coupling term. We also explicitly use the convenient fact that, according to the rules of formalism of § 2.3.3, the phase φ enters expressions for qubit potential energies $U_{a,b}$ in the same way as the corresponding external fluxes, $\tilde{\phi}_{a,b} = 2\pi \tilde{\Phi}_{a,b} / \Phi_0$.

Assuming that the coupling Josephson energy $E_c \gg E$, where $E \sim U_{a,b}$ is the Josephson energy of the larger junctions in flux qubits, we will now find the minimum of (4.13) with respect to the phase φ by expanding it in powers of (E/E_c). Writing the minimum condition as

$$\sin \varphi + \frac{1}{E_c} \frac{\partial}{\partial \varphi} U_a(\phi_{a1}, \phi_a, \tilde{\phi}_a + \varphi) + \frac{1}{E_c} \frac{\partial}{\partial \varphi} U_b(\phi_{b1}, \phi_b, \tilde{\phi}_b + \varphi) = 0, \quad (4.14)$$

we see that $\varphi_{\min} = (E/E_c)\varphi^{(1)} + \ldots$. Therefore, when calculating the next order term, we can take the derivatives of qubit potentials at $\varphi = 0$, and the internal qubit phases correspond to the minima of single-qubit potentials. These derivatives yield the amplitudes $I_{pa,b}$ of the persistent current in the qubits, and

$$\varphi^{(1)} = \pm \frac{I_{pa} \Phi_0}{2\pi c E_c} \pm \frac{I_{pb} \Phi_0}{2\pi c E_c}. \quad (4.15)$$

The signs depend on whether the qubit is in its "left" or "right" physical state. Substituting this in (4.13), we obtain the effective coupling from the term

$$-E_c \cos(\varphi^{(1)} + \ldots) = -E_c + \frac{\Phi_0^2}{8\pi^2 c^2 E_c} \left(I_{pa} \pm I_{pb}\right)^2 + \ldots. \quad (4.16)$$

It is clear that the coupling will be "antiferromagnetic", since the energy is lowered if the persistent currents in qubit loops flow in opposite directions. Therefore, the effective coupling term in the Hamiltonian will be

$$H_{\text{eff}} = J \sigma_z^{(a)} \sigma_z^{(b)}; \quad J = \frac{\Phi_0^2}{4\pi^2 c^2 E_c} I_{pa} I_{pb} + \ldots. \quad (4.17)$$

A logical extension of this idea is to insert, instead of a single junction, a coupler loop (Fig. 4.5a), the Josephson energy of which can be tuned by an external magnetic flux $\tilde{\phi}_c$ through the coupler loop. (An additional junction with the Josephson energy $\alpha_c E$ is necessary to allow this tuning, if the coupler loop is small; otherwise, due to the flux quantization in the coupler, the coupler flux will directly influence the qubits.)

Neglecting all mutual and self-inductances and following the same steps as before gives the effective coupling of Eq. (4.17), only this time the coupling amplitude depends on the coupler flux $\widetilde{\phi}_c$ (van der Ploeg et al., 2007):

$$J(\widetilde{\phi}_c) = \frac{\hbar}{2e} \frac{I'(\widetilde{\phi}_c)}{I_c^2 - I^2(\widetilde{\phi}_c)} I_{pa} I_{pb} + \dots. \tag{4.18}$$

Here $I_c = (2e/\hbar c)E$ is the critical Josephson current in the coupler junctions shared with the qubits, and the current circulating in the coupler is $I(\widetilde{\phi}_c) = I_c \sin(\bar{\varphi})$. The phase difference $\bar{\varphi}$, to the same accuracy, is the same across the shared junctions and is found from the condition

$$\sin \bar{\varphi} = \alpha_c \sin(\widetilde{\phi}_c - 2\bar{\varphi}), \tag{4.19}$$

(i.e., current conservation and flux quantization). Obviously, the coupling coefficient (4.18) can even change sign, thus switching between ferro- and antiferromagnetic coupling (and, of course, allowing us to completely decouple the qubits from each other). This conclusion is not really dependent on the approximations we made; still, the experimental results (Fig. 4.5) show that (4.18) is a pretty good approximation.

The tuning of passive qubit–qubit coupling, up to the switching of its sign, is not limited to the coupling through a shared Josephson junction. The coupling scheme equivalent to the one of Fig. 4.5a was first proposed for purely inductive coupling via an intermediate rf SQUID (Maassen van den Brink et al., 2005) and was demonstrated in the classical regime by Zakosarenko et al. (2007), thus confirming that the nonlinear Josephson inductance, responsible for the tuning, is not a quantum phenomenon. (That is, it does not depend on the superconducting phase and particle number behaving as quantum variables: the bulk superconductivity itself *is* a (many-body) quantum phenomenon.) In the quantum regime it has been realized by Allman et al. (2010) (Fig. 4.6).

On the other hand, we have seen in Section 2.5 that in the quantum regime both inductance *and* capacitance are sensitive to the quantum state of the system. This property is utilized by the "variable electrostatic transformer" (Averin and Bruder, 2003) shown in Fig. 4.7. The Hamiltonian of the system is

$$H = \sum_{i=1,2} \left[-E_{Ji} \cos \phi_i + E_{Ci} \left(\hat{N}_i - n_i^* \right)^2 \right] - E_J \cos \varphi + E_C \left[\hat{N} - \hat{n}^*(\hat{N}_1, \hat{N}_2) \right]^2, \tag{4.20}$$

where $\hat{N}, \hat{N}_i$ are the Cooper pair number operators. The charging energies of the charge qubits and the coupler ("transformer junction") are determined by the

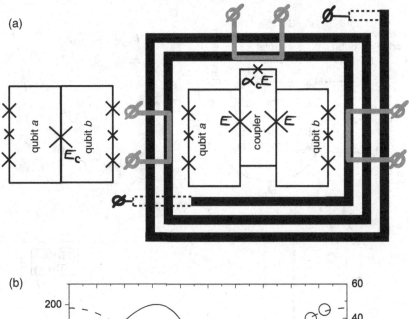

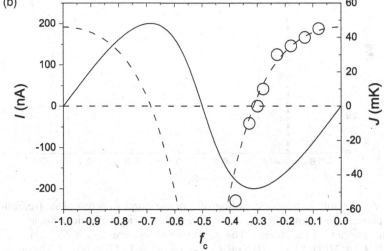

Fig. 4.5. (a) Josephson coupling between persistent current flux qubits: a simple coupling through a shared Josephson junction with Josephson energy E_c (left) and a tunable coupling through a coupler loop (right). In the experiment, the qubits and coupler were fabricated out of aluminium. The black spiral is a part of the niobium pick-up coil; Π-shaped grey wires are used to bias individual qubits. (b) Theoretical fit and experimental data (reprinted with permission from van der Ploeg et al., 2007, © 2007 American Physical Society). The circulating current in the coupler loop (solid line), and the coupling energy (4.18) (dashed line) as a function of external flux through the coupler loop, $\tilde{\phi}_c = 2\pi f_c$, are calculated for $\alpha_c = 0.2$ and $I_c = 1\,\mu\text{A}$. The experimental values of $J(f_c)$ are marked by circles.

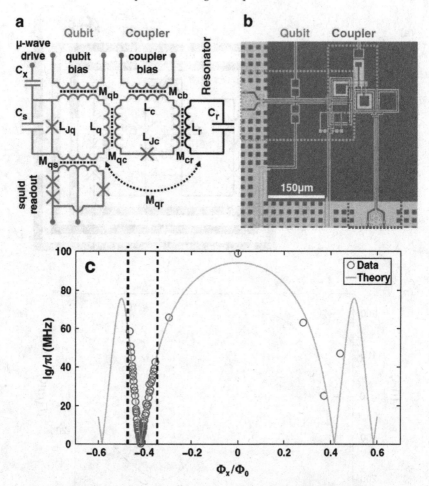

Fig. 4.6. Tunable inductive coupling of a phase qubit to a resonator (reprinted with permission Allman et al., 2010, © 2010 American Physical Society). (a) Circuit diagram of the device. The qubit parameters are $L_{Jq} \approx 550$ pH, $I_{q0} \approx 0.6\,\mu$A, $L_q \approx 1000$ pH, $C_s \approx 0.6$ pF, $C_{Jq} \approx 0.3$ pF, and $M_{qc} \approx 60$ pH. The coupler parameters are $L_{Jc} \approx 370$ pH, $I_{c0} \approx 0.9\,\mu$A, $L_c \approx 200$ pH, and $C_{Jc} \approx 0.3$ pF. The resonator parameters are $L_r \approx 1000$ pH, $C_r \approx 0.4$ pF, and $M_{cr} \approx 60$ pH. (b) Optical micrograph of the circuit. (c) Measured coupling strength as a function of applied coupler flux (circles) along with the theoretical fit.

effective capacitances:

$$E_C = \frac{2e^2}{C_\Sigma - \sum_i C_{mi}^2/C_{\Sigma i}}; \quad E_{Ci} = \frac{2e^2}{C_{\Sigma i}}; \tag{4.21}$$

$$C_{\Sigma i} = C_i + C_{gi} + C_{mi}; \quad C_\Sigma = C + C_{m1} + C_{m2}.$$

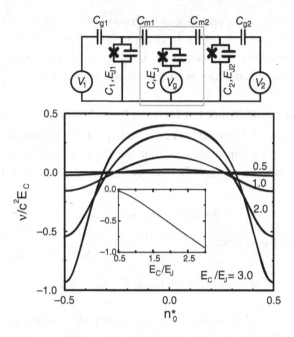

Fig. 4.7. Tunable capacitive coupling (Averin and Bruder, 2003). (Top) Diagram of two charge qubits coupled through a variable electrostatic transformer (marked by the grey rectangle). (Bottom) Coupling energy v of the two charge qubits as a function of the quasicharge induced on the transformed junction. Inset: coupling energy as a function of the transformer junction parameters. (Reprinted with permission from Averin and Bruder, 2003, © 2003 American Physical Society.)

The effective charges (in units of $2e$) induced by the external potentials on the qubits, the gate capacitor and the transformer junction are, respectively

$$n_i^* = \frac{C_{gi} V_i}{2e}, \quad n_g^* = \left[\sum_i C_{mi} \left(1 - \frac{C_{mi}}{C_{\Sigma i}} \right) \right] \frac{V_g}{2e};$$

$$\hat{n}^* = n_g^* - \sum_i \left(\hat{N}_i - n_i^* \right) \frac{C_{mi}}{C_{\Sigma i}}. \tag{4.22}$$

The latter wears a hat since it depends on the Cooper pair number operators. As pointed out by Averin and Bruder (2003), as long as the dynamics of the coupler is much faster than that of the coupled qubits, the former will adiabatically adjust to the latter. In other words, if the energy gap between the ground state and the first excited state of the coupler greatly exceeds the qubit level splitting,

$$\Delta E_{\text{coupl}} \gg \Delta E_{\text{qb}}, \tag{4.23}$$

we can assume that the coupler, once in its ground state, will remain there, by virtue of the adiabatic theorem (§ 1.4.5). In charge qubits, the level splitting is given by their Josephson energies, E_{Ji}, while the level splitting in the transformer is given by either E_J if the transformer is in the charge regime ($E_C \gg E_J$), or $\sqrt{E_J E_C}$ otherwise.

When the charge qubits are not far from their degeneracy points ($n^*_{1,2} \sim 1/2$), their dynamics can be restricted to the two adjacent charge states, and, thus, $\hat{N}_i = (1 + \sigma_i^z)/2$. Assuming for simplicity that the structure of Fig. 4.7 is symmetric ($C_{mi}/C_{\Sigma i} = c$), we can replace the transformer contribution to the Hamiltonian (4.20) by

$$v \sigma_1^z \sigma_2^z + \delta(\sigma_1^z + \sigma_2^z) + \mu, \qquad (4.24)$$

where the coupling coefficient,

$$v = \frac{1}{4}\left[\epsilon_0(n_0^* + c) + \epsilon_0(n_0^* - c) - 2\epsilon_0(n_0^*)\right], \qquad (4.25)$$

is a "discretized second derivative" of the ground state energy of the transformer junction, $\epsilon(n^*)$, with respect to the quasicharge (in units of $2e$), taken at the point $n_0^* = n_g^* + c\sum_i(n_i^* - 1/2)$.[4] It is indeed related to the quantum contribution to the inverse capacitance; cf. Eq. (2.155). The remaining coefficients in (4.24) are $\delta = (1/4)[\epsilon_0(n_0^* + c) - \epsilon_0(n_0^* - c)]$ and the irrelevant constant shift[5] $\mu = \epsilon_0(n_0^* + c) + \epsilon_0(n_0^* - c) + 2\epsilon_0(n_0^*)$.

The function $\epsilon_0(n^*)$ is periodic in the quasicharge $2en^*$ (2.136), and, therefore, the sign of the coupling constant v, Eq. (4.25), can be changed by the gate voltage V_g (Fig. 4.7). This also means, of course, that the qubits can be decoupled altogether. This method can be applied to other types of qubit (e.g., the quantronium, or the phase qubit). When more than two qubits are coupled this way, one can no longer neglect the constant term, μ, in (4.24), since it will have different values for different qubit pairs and introduce a time-dependent phase difference between their quantum states.

4.1.4 Quantum buses

Instead of pairwise couplings, it has been proposed that all qubits are coupled to a shared passive element – a "quantum bus" (Makhlin et al., 1999, 2001); for example, a set of M identical charge qubits can be capacitively coupled to an inductor; the capacitance of the resulting LC circuit is the summary capacitance of the qubit

---

[4] It is actually the second difference of $\tilde{\epsilon}_0(n^*/c) \equiv \epsilon_0(c \cdot (n^*/c))$; cf. Eq. (3.73).
[5] Irrelevant because, though it depends on the gate voltage via n_0^*, it is always the same for both qubits.

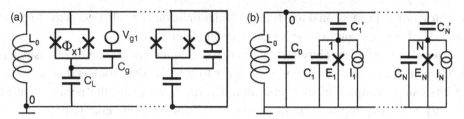

Fig. 4.8. A set of qubits coupled to a common quantum bus (LC circuit): (a) tunable charge qubits (after Makhlin et al., 1999); (b) phase qubits (after Zagoskin et al., 2006).

gate electrodes and Josephson junctions (Fig. 4.8a).[6] The Hamilton function of this system in terms of the node charges and fluxes can be written as

$$\mathcal{H} = \frac{Q^2}{2NC_{qb}} + \frac{\Phi^2}{2L} + \sum_{i=1}^{M}\left\{\frac{(Q_i - C_g V_{gi})^2}{2(C_J + C_g)} - E_J(\widetilde{\Phi}_i)\cos\left[\frac{2\pi}{\Phi_0}\left(\Phi_i - \frac{C_{qb}}{C_J}\Phi\right)\right]\right\},$$

(4.26)

where Φ is the flux in the bus inductor, $\widetilde{\Phi}_i$ the external flux, which tunes the Josephson energy of the ith qubit, and $C_{qb} = (C_J^{-1} + C_g^{-1})^{-1}$. After replacing the charges with the canonical momenta, and the latter with operators, we will obtain from (4.26) the quantum Hamiltonian. Assuming that the excitation energy of the bus, $\hbar\omega_{LC} = \hbar c/\sqrt{MLC_{qb}}$, exceeds the characteristic energies of the qubits, and restricting the bus to its ground state, and the qubits to two charge states, the effective Hamiltonian becomes (Makhlin et al., 1999, 2001)

$$H = -\frac{1}{2}\sum_{i=1}^{M}\left(\epsilon_i\sigma_z^i + E_J(\widetilde{\Phi}_i)\sigma_x^i\right) - \sum_{i<j}\frac{E_J(\widetilde{\Phi}_i)E_J(\widetilde{\Phi}_j)}{E_L}\sigma_y^i\sigma_y^j + \text{const}, \quad (4.27)$$

where the single charge qubit biases, ϵ_i, are tuned by the gate voltages (cf. Eq. (2.129)), and

$$E_L = \left(\frac{C_J}{C_{qb}}\right)^2\frac{\Phi_0^2}{\pi^2 L}.$$

(4.28)

The coupling term in (4.27) is, physically, the magnetic energy of the ac current flowing in the LC circuit (that is, through the qubits, gate capacitors and the ground). This is the ground state current, since by assumption the excitation energy $\hbar\omega_{LC}$ greatly exceeds the other energies in the system. This requirement limits the maximal number M of qubits which can be coupled this way: the total capacitance of

---

[6] Including an additional capacitor would be rather counterproductive, since we will want the resonance frequency of the oscillator mode to be as high as possible.

the structure is proportional to M, and its total inductance is also expected to grow with M, if only because of the longer coupling leads.

The advantage of the above scheme is that each qubit can be completely isolated from the rest by tuning its Josephson energy to zero. Its drawback is that only two qubits can be coupled at the same time. On the other hand, the requirement of exactly zero coupling may be too restrictive. If the coupling energy between the qubits i and j, E_J^2/E_L, is much smaller than the single qubit energies E_J, E_C, then there will be just a small perturbation (which can be counteracted, if necessary, by single-qubit operations) – that is, unless the two qubits are in resonance *with each other* (Makhlin et al., 1999).

Let us consider the latter possibility for a set of phase qubits coupled to the LC circuit, Fig. 4.8b (Zagoskin et al., 2006). The Lagrangian of the system is

$$\mathcal{L} = \frac{C_{\text{bus}}\dot{\Phi}_{\text{bus}}^2}{2c^2} + \sum_{j=1}^{M}\left(\frac{C_j\dot{\Phi}_j^2}{2c^2} + \frac{C_j'(\dot{\Phi}_j - \dot{\Phi}_{\text{bus}})^2}{2c^2}\right) - \frac{\Phi_{\text{bus}}^2}{2L}$$

$$-\sum_{j=1}^{M}\left(-\frac{I_{cj}\Phi_j}{c} - E_{Jj}\cos 2\pi\frac{\Phi_j}{\Phi_0}\right) \equiv \frac{1}{2c^2}\sum_{j,k=\text{bus},1,2...}^{M} C_{jk}\dot{\Phi}_j\dot{\Phi}_k - U(\Phi_j). \tag{4.29}$$

The quantum Hamiltonian following from (4.29) is

$$H = \sum_{j,k=\text{bus},1,2...}^{M} c^2\frac{\mathcal{C}_{jk}^{-1}\hat{\Pi}_j\hat{\Pi}_k}{2} + \sum_{j=\text{bus},1,2...}^{M} \frac{\omega_j^2\hat{\Phi}_j^2}{2c^2\,\mathcal{C}_{jj}^{-1}} + \cdots \tag{4.30}$$

Here $\mathcal{C}^{-1}$ is the inverse capacitance matrix, ω_j is the bias-dependent plasma frequency of the jth phase qubit, and the ellipsis denotes the nonlinear terms. Using the second quantization representation, Eq. (2.106) (or (4.4)), with $\Lambda_j = [2\hbar\mathcal{C}_{jj}^{-1}/\omega_j]^{1/2}$, we reduce (4.30) to the convenient form

$$H = \sum_{j=\text{bus},1,2...}^{M} \hbar\omega_j\left(a_j^\dagger a_j + \frac{1}{2}\right) - \sum_{k>j} g_{jk}\left(a_j - a_j^\dagger\right)\left(a_k - a_k^\dagger\right) + \cdots, \tag{4.31}$$

with the coupling (cf. Eq. (4.6))

$$g_{jk} = \frac{\hbar(\omega_j\omega_k)^{1/2}\mathcal{C}_{jk}^{-1}}{2\left(\mathcal{C}_{jj}^{-1}\mathcal{C}_{kk}^{-1}\right)^{1/2}}. \tag{4.32}$$

So far we did not make any assumptions, except the relative smallness of the nonlinear terms (which still can be restored to Eq. (4.31) to an arbitrary degree of accuracy). If now we use the condition $\omega_{\text{bus}} \gg \omega_j$, project the Hamiltonian (4.31) on the ground state of the bus, and also use the nonlinearity of the phase qubits to restrict them to their first swo states, $|0\rangle_j$ and $|1\rangle_j$, we obtain the Hamiltonian

$$H = -\frac{1}{2}\sum_{j=1}^{M}\hbar\omega_j\sigma_z^j + \sum_{k>j=1}^{M} g_{jk}\sigma_y^j\sigma_y^k, \tag{4.33}$$

which, not surprisingly, is almost the same as the ones we have obtained before (Eqs (4.7, 4.27)). Of course, here we cannot suppress the coupling by zeroing g_{jk} as we could in Eq. (4.27) for tunable charge qubits, since this would mean putting $\omega_j = 0$, that is, eliminating the very qubit states we need to use. But we easily enough find that in the interaction representation with respect to the single-qubit part of the Hamiltonian, $H_0 = -\sum_j(\hbar\omega_j\sigma_z^j/2)$, the σ_y-operators become

$$\sigma_y^j \to \sigma_y^j(t) = e^{\frac{i}{\hbar}H_0 t}\sigma_y^j e^{-\frac{i}{\hbar}H_0 t} = \sigma_y^j(0)\cos\omega_j t - \sigma_x^j(0)\sin\omega_j t. \tag{4.34}$$

If the coupling strength g_{jk} is small, its effects will be negligible (as those of a small perturbation oscillating with a comparatively high frequency $|\omega_j - \omega_k|$), unless the two qubits in question are in resonance, $\omega_j = \omega_k$. (This is, of course, valid in the RWA.) In resonance, the effective interaction term is

$$\tilde{H}_{\text{int,eff}}^{jk} = g_{jk}\frac{\sigma_x^j\sigma_x^k + \sigma_y^j\sigma_y^k}{2}. \tag{4.35}$$

From the point of view of quantum computing, this is a perfectly good coupling allowing all the universal two-qubit manipulations, since the operator $(\sigma_x^j\sigma_x^k + \sigma_y^j\sigma_y^k)/2$ acts exactly as the σ_x-operator on the subspace of quantum states spanned by the states $|0\rangle_j \otimes |1\rangle_k$ and $|1\rangle_j \otimes |0\rangle_k$, and is zero outside it.

Any number of qubit pairs, connected to the same bus, can be coupled *simultaneously* this way. "Any" is, of course, limited by the requirement that the frequencies of different pairs did not overlap, $\Delta\omega \gg \Gamma$, where Γ is the effective qubit linewidth. The total number M of qubits on the same bus is limited, as before, by the "passivity" condition, $\omega_{\text{bus}} \gg \omega_j$.

4.1.5 Active coupling

As soon as the excitation energy of the coupling circuit stops being much larger than any other energy in the problem (e.g., the Josephson and charging energies of the individual qubits), we are no longer allowed to neglect its dynamics. Consider, e.g., the coupling of a charge or phase qubit to a tunable LC circuit (the role of which can be played by a single large Josephson junction): see Blais et al. (2003).

We can save ourselves some time by following the same steps as those that led us to Eqs (4.31, 4.32) in Section 4.1.5. Now, instead of projecting everything on the ground state of the bus, we will consider the dynamics of the bus explicitly. The whole idea of using the active bus is to bring it sequentially in resonance with the qubits, thus first swapping the excitation between the jth qubit and the bus, and then between the bus and the kth qubit. This differs from performing two-qubit operations in, e.g., a chain of coupled qubits in two important points. First, the qubits share the bus, and, therefore, the operation above involves only one intermediate swap (with the bus) for any two qubits. Second, the bus does not need to be a two-level system. As a matter of fact, there is certain advantage to it being an almost linear oscillator (LC circuit), but we will first consider the case from Blais et al. (2003), where the bus is a (nonlinear) current-biased Josephson junction, and where the bus and the phase qubit have only three metastable levels each: $|0\rangle_{\text{bus(qb)}}$, $|1\rangle_{\text{bus(qb)}}$ and $|2\rangle_{\text{bus(qb)}}$. The decay rates from the topmost states $|2\rangle$ into the continuum, $\Gamma_{2,\text{bus(qb)}}$, are large, and the leakage to these states from the subspace spanned by $\{|0\rangle_{\text{bus}}, |1\rangle_{\text{bus}}\} \otimes \{|0\rangle_{\text{qb}}, |1\rangle_{\text{qb}}\}$ will lead to the nonunitarity of the system's evolution (the effects of direct decay from the lower states can be neglected by comparison). One should also take into account relaxation and dephasing in the qubit and bus, characterized by the times $T_{1,\text{bus(qb)}}$ and $T_{2,\text{bus(qb)}}$.

We have already discussed this situation in the context of the capacitive coupling between two phase qubits (see Fig. 4.2 and Eq. 4.1). We need only change the indices, $1 \to$ qubit, $2 \to$ bus, to see that, e.g., the Hamiltonian (4.8) allows us to perform a swap between the state of the qubit and that of the bus by bringing them in resonance for a specific length of time. If several qubits (in arbitrary quantum states) are connected to the bus in the latter's ground state, a universal two-qubit operation (a so-called $\sqrt{\text{SWAP}}$: see the Appendix) can be performed between any two of them by the consecutive qubit-bus state swaps, which will in the end leave the bus in its initial state (Blais et al., 2003).

So far we have not been concerned with the question of how well justi-fied is it disregard to the leakage from the "qubit" subspace, $\{|0\rangle_{\text{qb}}, |1\rangle_{\text{qb}}\} \otimes \{|0\rangle_{\text{bus}}, |1\rangle_{\text{bus}}\}$. This leakage is produced by the coupling term in the Hamiltonian (4.1), $c^2 \Pi_{1(\text{qb})} \Pi_{2(\text{bus})}/\widetilde{C}_c$, and is restricted by the nonlinearity of the Josephson circuits. The numerical estimates made for the realistic choice of circuit parameters are presented in Fig. 4.9[7] and show that in this particular case the effects of leakage are negligible. This, of course, does not give us an indulgence to neglect them in the general case.

---

[7] The decoherence time estimates shown there were calculated using the decoherence models, which will be discussed in the next chapter.

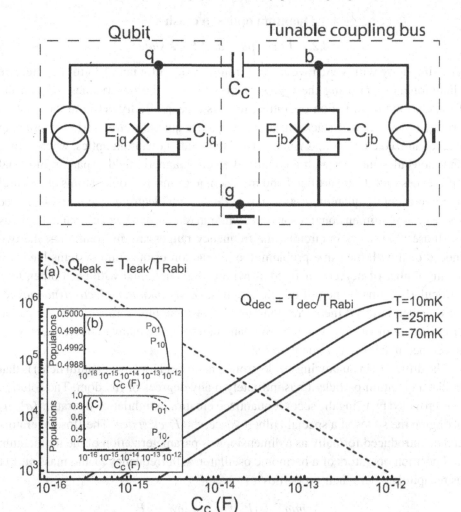

Fig. 4.9. (Reprinted with permission from Blais et al., 2003, © 2003 American Physical Society). (Top) A phase qubit coupled to an active tunable bus. (Bottom) Numerical estimates of the quality of the system, when the qubit and the bus have identical parameters ($C_j = 6\,\mathrm{pF}$, $I_c = 21\,\mu\mathrm{A}$). (a) Decoherence time T_{dec} (full lines) and leakage time T_{leak} (dashed line) in the units of the vacuum Rabi oscillation period T_{Rabi} are plotted vs. the coupling capacitance C_c. (For $C_c = 25\,\mathrm{fF}$, calculated $T_{\mathrm{Rabi}} \approx 40\,\mathrm{ns}$.) The qubits and bus are in resonance ($I_{\mathrm{bias,qb}} = I_{\mathrm{bias,bus}} = 20.8$ $\mu\mathrm{A}$, and each well contains three metastable levels. The leakage time $T_{\mathrm{leak}}^{-1} = \Gamma_2(1 - P_{01} - P_{10})$, where Γ_2 is the calculated decay rate of the metastable state $|2\rangle$, which is close to the top of the potential barrier. By P_{jk} we denote the population of the state $|j\rangle_{\mathrm{qb}} \otimes |k\rangle_{\mathrm{bus}}$. The decoherence time, $T_{\mathrm{dec}}^{-1} = T_1^{-1} + T_2^{-1}$, is calculated using the oscillator bath model (§ 5.3.2) with the impedance $\hat{Z}(\omega) \approx 560\,\mathrm{k}\Omega$, which for the chosen parameters yields in resonance $T_1 \sim T_2 \sim 1\,\mathrm{ms}$ at 25 mK. (b) Populations P_{01}, P_{10} in resonance. (c) Same when off resonance: $I_{\mathrm{bias,qb}} = 20.8$ $\mu\mathrm{A}$, $I_{\mathrm{bias,bus}} = 20.43\,\mu\mathrm{A}$ (solid lines); $I_{\mathrm{bias,bus}} = 20.74\,\mu\mathrm{A}$ (dotted lines).

4.2 Quantum optics: a crash course

4.2.1 Fock space and Fock states

The simplicity with which we moved from the Hamiltonian (4.1) to the effective Hamiltonian (4.7) using the representation (4.4) of the observables through the Bose creation/annihilation operators, reflects a deep affinity between the class of problems we meet when tackling qubits coupled to quantum circuits (i.e., interacting with quantized currents and voltages) and the field of quantum optics, where atoms and molecules interact with a quantized electromagnetic field. Apart from trivial differences (qubits are artificial and mesoscopic or macroscopic, atoms are natural and microscopic; quantum optics mostly deals with electromagnetic fields in free space or three-dimensional resonators, while in our case the electromagnetic fields are those of currents in circuits; the frequency ranges are different; etc.) the two indeed deal with the same problem: the interaction of quantum systems having a small number of degrees of freedom with a Bose field and with each other. One particularly affine branch of quantum optics, *cavity quantum electrodynamics* (c-QED), even induced the corresponding part of quantum engineering to be called *circuit* quantum electrodynamics. We can, therefore, use many tools from the well-developed machinery of quantum optics.

The difficulty in quantizing an electromagnetic field (or any quantum field) is that unlike a quantum particle it has inifinitely many degrees of freedom. The solution was provided by using the second-quantized creation/annihilation operators $(a^\dagger, a)$ acting on the states in a special Hilbert space, the *Fock space*. The Bose operators can be introduced formally as a dimensionless parameterization of the momentum and position operators of a harmonic oscillator with frequency ω and mass m. Let us recapitulate here their properties[8]:

$$a = \frac{m\omega\hat{X} + i\hat{P}}{\sqrt{2m\hbar\omega}}; \quad a^\dagger = \frac{m\omega\hat{X} - i\hat{P}}{\sqrt{2m\hbar\omega}}. \tag{4.36}$$

Their commutation relation is

$$[a, a^\dagger] = 1, \tag{4.37}$$

and the Hamiltonian of the oscillator becomes

$$H = \frac{\hat{P}^2}{2m} + \frac{m\omega^2\hat{X}^2}{2} \equiv \hbar\omega\left(a^\dagger a + \frac{1}{2}\right). \tag{4.38}$$

The *Fock states* are the energy eigenstates:

$$H|n\rangle = \hbar\omega\left(n + \frac{1}{2}\right)|n\rangle. \tag{4.39}$$

---

[8] See, e.g., Landau and Lifshitz (2003), § 64, Orszag (1999), Chapter 3.

They are also the eigenstates of the particle (excitation) number operator,

$$\hat{N}|n\rangle \equiv a^\dagger a |n\rangle = n|n\rangle, \tag{4.40}$$

and are, therefore, also called *number states*. The excitations of the electromagnetic field are the photons. The matrix elements of the creation–annihilation operators between the Fock states are

$$\langle n|a|n+1\rangle = \sqrt{n+1}, \quad \langle n+1|a^\dagger|n\rangle = \sqrt{n+1} \tag{4.41}$$

and zero otherwise. Any Fock state can be obtained from the *vacuum state* $|0\rangle$ by the repeated application of the creation operator:

$$|n\rangle = \frac{(a^\dagger)^n}{\sqrt{n!}}|0\rangle. \tag{4.42}$$

The prefactor is needed to maintain the normalization condition,

$$\langle n|m\rangle = \delta_{mn}. \tag{4.43}$$

The vacuum state is annihilated by the annihilation operator,

$$a|0\rangle = 0. \tag{4.44}$$

For a single oscillator the Fock space is the Hilbert space spanned by the states $|0\rangle, |1\rangle, \ldots |n\rangle, \ldots$ with arbitrarily large *excitation numbers* n. For a Bose field – like an electromagnetic field – each of the infinitely many classical modes of the field (e.g., its spatial Fourier components) is represented by a quantized oscillator mode; the resulting Fock space is the product of these one-mode Fock spaces, and any state in this space can be written as an expansion

$$|\Psi\rangle = \sum C_{n1,n2,\ldots n_q,\ldots}|n_1, n_2, \ldots, n_q, \ldots\rangle \equiv \sum C_{n1,n2,\ldots n_q,\ldots}|n_1\rangle$$
$$\otimes |n_2\rangle \cdots \otimes |n_q\rangle \ldots. \tag{4.45}$$

Note that though the numbers n_j and their sum can be arbitrarily high, they are not *actually* infinite. This subtlety is important for resolving the field quantization problem (see, e.g., Umezawa et al., 1982), but not essential for what we will be doing here. In practice, when one uses the second quantization approach for numerical calculations, the Fock space anyway has to be truncated to some manageable size. For example, the calculations that led to Fig. 4.9 involved just one mode with $n \leq 20$.[9]

[9] A very useful add-on to MATLAB (Quantum Optics Toolbox; Tan, 2002) is freely available on the web. It automates the task of obtaining the master equations, with the Lindblad terms, directly from the second-quantized form of the Hamiltonian or Liouvillian, incorporates several numerical and visualization routines, and contains many helpful examples.

4.2.2 Jaynes–Cummings model

Let us now consider the situation when a qubit interacts with a field mode. For example, if a superconducting charge qubit with the Hamiltonian (2.129) includes a tunable junction, the latter's Josephson energy depends on the external magnetic flux through the junction loop as $E_J(\widetilde{\Phi}) = E_{J,\max} \cos \pi \frac{\Phi_{dc}+\widetilde{\Phi}}{\Phi_0}$ (see Eq. (2.32)), where Φ_{dc} is a static bias flux, and $\widetilde{\Phi}$ is the time-dependent flux created by the quantum field mode – i.e., by the current in the quantum circuit connected to the qubit. Then in the linear approximation the field–qubit interaction can be written as

$$H_{int} = \hbar g(a + a^\dagger)\sigma_x. \qquad (4.46)$$

Adding this to the terms for the qubit and oscillator energy, we obtain the textbook (two-level) atom-field Hamiltonian from quantum optics (e.g., cf. Orszag, 1999, Eq. (8.23)),

$$H = \left[-\frac{\hbar\Omega}{2}\sigma_z\right]_{\text{qubit}} + \left[\frac{\hbar\omega}{2}\left(a^\dagger a + \frac{1}{2}\right)\right]_{\text{field}} + \hbar g(a + a^\dagger)\sigma_x. \qquad (4.47)$$

Earlier (Eq. (1.64)) we introduced the matrices σ_+ and σ_-,

$$\sigma_+ = \frac{\sigma_x + i\sigma_y}{2} = \begin{pmatrix} 0 & 1 \\ 0 & 0 \end{pmatrix}; \quad \sigma_- = \frac{\sigma_x - i\sigma_y}{2} = \begin{pmatrix} 0 & 0 \\ 1 & 0 \end{pmatrix}. \qquad (4.48)$$

These matrix operators flip the qubit:

$$\sigma_-\begin{pmatrix} 1 \\ 0 \end{pmatrix} = \begin{pmatrix} 0 \\ 1 \end{pmatrix}; \quad \sigma_+\begin{pmatrix} 0 \\ 1 \end{pmatrix} = \begin{pmatrix} 1 \\ 0 \end{pmatrix}. \qquad (4.49)$$

Obviously, $\sigma_x = \sigma_+ + \sigma_-$. In terms of these operators the interaction term (4.46) describes four possible processes: (A) a photon is absorbed and the qubit is excited $(a\sigma_-)$[10]; (B) a photon is emitted and the qubit is excited $(a^\dagger\sigma_-)$; (C) a photon is absorbed and the qubit is relaxed to its ground state $(a\sigma_+)$; and, finally, (D) a photon is emitted and the qubit is relaxed $(a^\dagger\sigma_+)$.

Let us switch to the interaction representation, with the unperturbed Hamiltonian $H_0 = H_{\text{qubit}} + H_{\text{field}}$. The Bose operators will become $a \to a(t) = a\exp[-i\omega t]$; $a^\dagger \to a^\dagger(t) = a^\dagger \exp[i\omega t]$. For the $\sigma_\pm$-operators we calculate

$$\sigma_\pm(t) = e^{-i\Omega t\sigma_z/2}\sigma_\pm e^{i\Omega t\sigma_z/2} \qquad (4.50)$$

---

[10] Note that throughout we use the convention that the qubit Hamiltonian is written as $-\hbar\Omega\sigma_z/2 + \ldots$; therefore, the "topmost" state, $\begin{pmatrix} 1 \\ 0 \end{pmatrix}$, is actually the ground state, with energy $-\hbar\Omega/2$. The convenience of such a choice becomes obvious as soon as we, e.g., start explicitly writing operator matrices for multilevel systems.

using

$$e^{iq\sigma_z} = \cos q + i\sigma_z \sin q \tag{4.51}$$

(which is immediately verified by the Taylor expansion, since $\sigma_z^2 = \hat{1}$; cf. (1.96))
and

$$\sigma_z \sigma_{x,y} \sigma_z = -\sigma_{x,y}. \tag{4.52}$$

Then

$$\sigma_\pm(t) = \cos \Omega t \sigma_\pm \mp i \sin \Omega t \sigma_\pm = e^{\mp i\Omega t}\sigma_\pm, \tag{4.53}$$

and the Hamiltonian (4.47) in the interaction representation becomes

$$\tilde{H} = e^{i\frac{H_0}{\hbar}t} H e^{-i\frac{H_0}{\hbar}t} - H_0 = \hbar g(a(t) + a^\dagger(t))(\sigma_+(t) + \sigma_-(t))$$
$$= \hbar g(a\sigma_+ e^{-i(\omega+\Omega)t} + a^\dagger\sigma_- e^{i(\omega+\Omega)t} + a\sigma_- e^{-i(\omega-\Omega)t} + a^\dagger\sigma_+ e^{i(\omega-\Omega)t}). \tag{4.54}$$

Eq. (4.54) follows from the general expression for a time-dependent unitary transformation of a Hamiltonian: if we want to keep the Schrödinger equation intact, the transformation $|\psi\rangle \to |\tilde{\psi}\rangle = U(t)|\psi\rangle$ should be accompanied by

$$H \to \tilde{H} = U(t)HU^\dagger(t) - i\hbar U(t)\frac{\partial}{\partial t}U^\dagger(t); \tag{4.55}$$

compare with Eq. (1.102), which describes the evolution of a Bloch vector in a rotating frame.

Returning to Eq. (4.54), we see that its second line contains two "fast" terms with frequency $\omega + \Omega$ and two "slow" ones, with $\omega - \Omega$. Near resonance between the field mode and the qubit, when $|\omega - \Omega| \ll \omega, g$, we can invoke the rotating wave approximation and drop the fast terms. This leaves us with the *Jaynes–Cummings Hamiltonian*,

$$\tilde{H}_{JC} = \hbar g(a\sigma_- e^{-i(\omega-\Omega)t} + a^\dagger\sigma_+ e^{i(\omega-\Omega)t}). \tag{4.56}$$

This Hamiltonian conserves energy in each elementary event, since only the events (A) (excitation of the qubit by the absorbed photon) and (D) (emission of a photon by the relaxing qubit) are included. We can make the same approximation in the laboratory frame as well:

$$H_{JC} = \left[-\frac{\hbar\Omega}{2}\sigma_z\right]_{\text{qubit}} + \left[\hbar\omega\left(a^\dagger a + \frac{1}{2}\right)\right]_{\text{field}} + \hbar g(a\sigma_- + a^\dagger\sigma_+). \tag{4.57}$$

4.2.3 Quantum Rabi oscillations. Vacuum Rabi oscillations

The eigenstates of the Hamiltonian (4.57) can be obtained from the eigenstates of the unperturbed Hamiltonian H_0, $|s,n\rangle \equiv |s\rangle_{\text{qubit}} \otimes |n\rangle_{\text{field}}$. Here $s = 0(1)$ denotes the qubit in the ground (excited) state, and n is the number of photons. Due to the structure of the Jaynes–Cummings Hamiltonian, only transitions between the states $|0, n + 1\rangle$ and $|1, n\rangle$ are allowed.[11] Moreover, near resonance these states are almost degenerate: a qubit's downflip with energy change $-\hbar\Omega$ produces an extra photon with energy $\hbar\omega$, and vice versa.

Isolating one block of the Jaynes–Cummings Hamiltonian, which acts on the subspace $\{|0, n + 1\rangle, |1, n\rangle\}$, we find

$$H_n = \begin{pmatrix} -\hbar\Omega/2 + (n + 1/2 + 1)\hbar\omega & \hbar g\sqrt{n+1} \\ \hbar g\sqrt{n+1} & \hbar\Omega/2 + (n + 1/2)\hbar\omega \end{pmatrix}$$

$$= \begin{pmatrix} \hbar\delta/2 + (n + 1)\hbar\omega & \hbar g\sqrt{n+1} \\ \hbar g\sqrt{n+1} & -\hbar\delta/2 + (n + 1)\hbar\omega \end{pmatrix}, \quad (4.58)$$

where the detuning $\delta = \omega - \Omega$ (cf. Eq. (1.104)).

This matrix is readily diagonalized, with eigenvalues

$$E_{n,\pm} = \hbar\omega(n + 1) \pm \left[\left(\frac{\hbar\delta}{2} \right)^2 + \hbar^2 g^2(n + 1) \right]^{1/2} \quad (4.59)$$

and eigenstates

$$|n+\rangle = \cos\theta_n |1, n\rangle + \sin\theta_n |0, n + 1\rangle;$$

$$|n-\rangle = -\sin\theta_n |1, n\rangle + \cos\theta_n |0, n + 1\rangle; \quad (4.60)$$

$$\tan 2\theta_n = \frac{2g\sqrt{n+1}}{\delta}.$$

These *dressed states* are the solutions (in the Jaynes–Cummings model) to the problem of a qubit interacting with the field mode. In other words, they describe the Rabi oscillations, when the electromagnetic field that causes them is itself quantized. Indeed, we see from Eqs (4.59) and (4.60) that in the presence of the field with frequency ω close to the interlevel spacing, our two-level system can no longer stay in a given state ($|0\rangle$ or $|1\rangle$), since the states are linear combinations of the dressed states, $|n+\rangle, |n-\rangle$. (The splitting of the previously degenerate states $|0, n + 1\rangle$ and $|1, n\rangle$ into doublets of dressed states, $|n\pm\rangle$, is called *dynamical Stark splitting*.) The probability of finding the qubit, say, in its ground state, $P_0(t)$, will oscillate with the frequency

$$\omega_R = \frac{E_{n,+} - E_{n,-}}{\hbar} = \left[\delta^2 + 4g^2(n + 1) \right]^{1/2}; \quad \omega_R|_{\omega=\Omega} = 2g\sqrt{n+1}. \quad (4.61)$$

---

[11] Formally the situation is the same as with the charge qubit near the degeneracy point, when only the transitions between the states with n and $(n + 1)$ Cooper pairs are possible; see Eq. (2.124).

Comparing this with Eq. (1.110) and recalling that the field amplitude is proportional to the square root of the number of photons,[12] we see that this is indeed the quantum Rabi frequency. The simplicity with which we have arrived at this result, compared to the work required for Eq. (1.110) is, in a sense, an illusion: the key rotating wave approximation is already included in the structure of the Jaynes–Cummings Hamiltonian.

Eq. (1.110) is only valid as long as the field amplitude is at least comparable to the detuning; of course, we would not expect any Rabi oscillations in the absence of the field. On the contrary, in the quantum case the field amplitude is never zero: zero-point oscillations will produce *vacuum Rabi oscillations* with the frequency

$$\omega_R^{\text{vac}} = \left[\delta^2 + 4g^2\right]^{1/2}. \tag{4.62}$$

Quantum Rabi oscillations can now be seen to be a special case of quantum beats, the beats between the states of the combined "qubit+field" system.

Vacuum Rabi mode splitting and Rabi oscillations have been observed in several experiments with superconducting qubits, e.g., a flux qubit inductively coupled to an LC circuit (Johansson et al., (2006c)), a charge/flux qubit capacitively/galvanically coupled to a 1D transmission line resonator (Wallraff et al., 2004; Abdumalikov et al., 2008, see next section), and a phase qubit coupled to a resonator (Hofheinz et al., 2008, 2009). The results from Johansson et al. (2006c) are shown in Fig. 4.10.

4.2.4 Dispersive regime. Schrieffer–Wolff transformation

The above results do not apply farther away from the resonance, while the super-conducting qubits or the LC circuits they are coupled to can be tuned in a rather wide frequency range (which is one of their advantages). In order to deal with the *dispersive regime*, where detuning exceeds the coupling energy:

$$|\omega - \Omega| \gg g, \tag{4.63}$$

we need a different approach. Let us canonically transform the field-qubit Hamiltonian (4.47),

$$H \rightarrow \tilde{H} = e^S H e^{-S} \equiv e^S (H_0 + H_{\text{int}}) e^{-S}, \tag{4.64}$$

and choose S in such a way as to eliminate the terms linear in the coupling constant, g, which parameterizes the interaction term (4.46). This transformation was introduced by Schrieffer and Wolff (1966) for a different problem.

---

[12] Since the field energy, $n\hbar\omega$, is quadratic in the field amplitude in the classical limit.

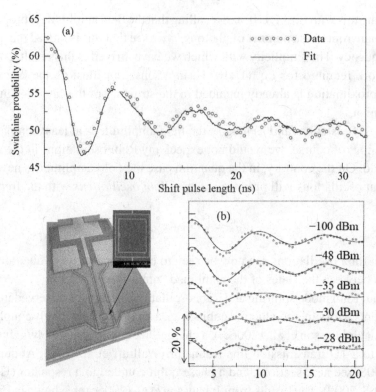

Fig. 4.10. Quantum Rabi oscillations in a persistent current flux qubit coupled to an LC circuit (reprinted with permission from Johansson et al., 2006c, © 2006 American Physical Society). The persistent current flux qubit (in the centre of the STM micrograph) was enclosed by the dc SQUID loop, used for the readout. The conductors running in parallel to the dc SQUID wire form a part of the LC circuit, to which the qubit is inductively coupled. (The large square electrodes on the top of the picture, together with the large bottom plate, under the insulating substrate, are the LC circuit's capacitors.) The circuit parameters are $L = 140$ pH, $C = 10$ pF, and qubit-oscillator mutual inductance $M = 5.7$ pH. The qubit parameters at the temperature of measurements ($T = 20$ mK) are $\Delta/h = 2.1$ GHz (tunnelling splitting) and $I_p = 350$ nA. The qubit and oscillator relaxation and dephasing rates and the qubit-oscillator coupling constant were extracted from fitting the experimental data and are $\Gamma_{qb} = 0.2$ MHz, $\Gamma_{qb,\phi} = 0.1$ GHz, $\Gamma_{osc} = 0.02$ GHz, $\Gamma_{osc,\phi} = 0.3$ GHz, $g/2\pi = 0.22$ GHz. (a) Vacuum Rabi oscillations. The LC circuit was allowed to relax to its ground state, the qubit was initialized in the excited state and shifted to the degeneracy point between the states $|1\rangle_{qubit} \otimes |0\rangle_{field}$ and $|0\rangle_{qubit} \otimes |1\rangle_{field}$. The rise time of the shift pulse, $\tau_{rise} = 0.8$ ns, was much shorter than the expected Rabi oscillation period, but longer than the qubit and oscillator characteristic times ($1/\Delta$, LC/c^2), which allowed exciting the system to be avoided during the process. The probability of the qubit switching to the ground state is plotted as a function of the pulse duration and is in quantitative agreement with Eq. (4.62). (b) Quantum Rabi oscillations in the presence of finite occupation of the oscillator mode. The fits show a good agreement with the theoretical $\sqrt{n+1}$-dependence of the Rabi frequency, Eq. (4.61).

After expanding (4.64) and choosing S to ensure that $[S, H_0] = -H_{int}$, we see that the term H_{int} cancels out, and the qubit-field interaction only appears in the second and higher orders. We can solve the equation

$$[S, H_0] = -H_{int} \qquad (4.65)$$

explicitly for S in the form

$$S = (u\sigma_x + v\sigma_y)a + (\widetilde{u}\sigma_x + \widetilde{v}\sigma_y)a^\dagger, \qquad (4.66)$$

using the commutation relations $[a^\dagger a, a] = -a$, $[a^\dagger a, a^\dagger] = a^\dagger$, $[\sigma_z, \sigma_x] = 2i\sigma_y$ and $[\sigma_z, \sigma_y] = -2i\sigma_x$. We find

$$u = \frac{g\omega}{\Omega^2 - \omega^2}; \quad v = -iu\frac{\Omega}{\omega} = -i\frac{g\Omega}{\Omega^2 - \omega^2};$$

$$\widetilde{u} = -u; \quad \widetilde{v} = i\widetilde{u}\frac{\Omega}{\omega} = v;$$

$$S = \frac{g\omega}{\Omega^2 - \omega^2}\left[\left(\sigma_x - i\frac{\Omega}{\omega}\sigma_y\right)a - \left(\sigma_x + i\frac{\Omega}{\omega}\sigma_y\right)a^\dagger\right]. \qquad (4.67)$$

If we are far enough from resonance, but not too far, i.e., when $\omega, \Omega \gg |\Omega - \omega| \gg g$, we can further simplify (4.67) to

$$S \approx \frac{g\omega}{2\omega(\Omega - \omega)}\left[\left(\sigma_x - i\sigma_y\right)a - \left(\sigma_x + i\sigma_y\right)a^\dagger\right] = \frac{g}{\Omega - \omega}\left[\sigma_- a - \sigma_+ a^\dagger\right]. \qquad (4.68)$$

(Obviously Eq. (4.68) relies on RWA, while Eq. (4.67) does not.) From (4.67) and (4.68) we see that the effect of the Schrieffer–Wolff transformation should be to replace the expansion parameter g with $g^2/|\delta|$ (the detuning $\delta = \omega - \Omega$), which will indeed be much smaller away from the resonance, where $|\delta| \gg g$.

Expanding the exponents in (4.64) and keeping only the lowest (quadratic) term in g, we find

$$\widetilde{H} = H_0 + [S, H_{int}] + \frac{1}{2}\{S^2, H_0\} + \cdots \approx H_0 + \frac{\hbar g^2}{\delta}\sigma_z\left(a^\dagger a + \frac{1}{2}\right). \qquad (4.69)$$

Now it is clear that away from resonance, the interaction of the field mode with the qubit does not lead to Rabi oscillations, but to a mutual frequency shift: depending on the state of the qubit, the mode frequency is shifted by $\pm g^2/|\delta|$, while the interlevel spacing in the qubit now contains a correction dependent on the number of photons in the mode, $(\hbar g^2/\delta)(n + 1/2)$. Since the resonant frequency of a high-quality oscillator can be measured to a high accuracy, this provides a way of measuring the state of the qubit (Blais et al., 2004; Gambetta et al., 2006).[13]

---

[13] Another important effect of the oscillator on the qubit is that, depending on the tuning, it can either suppress or increase the qubit's decoherence rate; see § 5.4.4.

4.3 Circuit quantum electrodynamics

4.3.1 Circuit implementation of cavity QED

The situation when an "atom" – i.e., a quantum object with a discrete energy spectrum – interacts with a single mode (or a few modes) of an electromagnetic field is typical for *cavity quantum electrodynamics (cavity QED)* (see, e.g., Walther et al., 2006), a branch of quantum optics that specifically investigates atoms in resonators (cavities). The frequency range depends on the relevant atomic transition and can actually be in the optical or microwave range (the latter for Rydberg atoms, i.e., atoms in highly excited states (Scully and Zubairy, 1997, Chapters 5, 13)). A qubit, or a set of qubits, coupled to a high-quality LC circuit presents the same physical picture, but with several positive twists (see Table 4.1). The "cavity" is one-dimensional, which leads to a smaller effective volume of the field mode and a relatively stronger field, than in 3D cavities of cavity QED. The qubit-mode coupling is stronger or comparable in absolute numbers, and significantly stronger relatively. (This compensates for the comparatively low quality of the superconducting circuits and – so far – shorter decoherence times of mesoscopic solid-state qubits compared to natural atoms.) The "atoms" can be designed and fabricated, and tuned over a wide range of parameters. Last but not least, while in cavity QED one has to deal with a stream of atoms flying through the resonator, which limits the interaction time by the transit time and forces the atom to pass the regions with varying field intensity, in circuit QED the "atom" sits tight and its position can be chosen to maximize its coupling to a particular field mode. Several qubits placed in the same resonator can be coherently coupled and entangled through their interaction with the field modes (Blais et al., 2007), and so on.

In a series of experiments circuit QED was implemented, using a transmission line resonator as a cavity (or a bus), Fig. 4.11. This system is not the lumped-elements circuit we met in Section 2.3, but the necessary changes to the mathematical formalism are minimal. We use the standard lumped-elements approximation of a transmission line (Fig. 4.11c), quite reasonable given the large (centimetre) wavelengths involved. Assuming a uniform, straight transmission line of length l, with the reduced inductance, $\widetilde{L} = L/l$, and capacitance, $\widetilde{C} = C/l$, each section in Fig. 4.11c has capacitance $\Delta C = \widetilde{C}\Delta x$ and inductance $\Delta L = \widetilde{L}\Delta x$. A superconducting qubit, be it charge or flux, has a scale of $10\,\mu\text{m}$ and can be safely considered as a point-like object. Taking the limit $\Delta x \to 0$, we immediately obtain the continuous version of Eq. (2.84):

$$\mathcal{L}(\Phi(x,t), \dot{\Phi}(x,t)) = \int_{-l/2}^{l/2} dx \left[\frac{\widetilde{C}(\dot{\Phi}(x,t))^2}{2c^2} - \frac{(\nabla\Phi(x,t))^2}{2\widetilde{L}} \right]. \qquad (4.70)$$

Table 4.1. Comparison between 3D cavity QED (a,b) and typical 1D circuit QED (cQED) parameters for charge (c), flux (d) and transmon (e) qubits coupled to a transmission line resonator (Fig. 4.11a,b). Source: (a) Hood et al. (2000); (b) Raimond et al. (2001); (c) Blais et al. (2004); (d) Abdumalikov et al. (2008); (e) Fink et al. (2008); (f) Abdumalikov et al. (2010). The data (a, b, c) were compiled by Blais et al. (2004), Table 1.

Parameter		3D optical[a]	3D microwave[b]	1D cQED[c]	1D cQED[d]	1D cQED[e]
Resonance frequency	$\omega/2\pi$	350 THz	51 GHz	10 GHz	9.907 GHz	6.94 GHz
Vacuum Rabi frequency	g/π	220 MHz	47 kHz	100 MHz	240 MHz	308 MHs
Coupling strength	g/ω	3×10^{-7}	1×10^{-7}	5×10^{-3}	0.012	0.022
Transition dipole	d/ea_0	~1	1×10^3	2×10^4		
Resonator lifetime	$1/\kappa$	10 ns	1 ms	≥ 160 ns	110 ns	180 ns
Resonator quality	Q	3×10^7	3×10^8	$\geq 1 \times 10^4$	7×10^3 ns	180 ns
"Atom" lifetime	$1/\gamma$	61 ns	30 ms	2 μs	14 ns[f]	1.57μs*)
"Atom" transit time	t_{tran}	$\geq 50\mu$s	100 μs	∞	∞	∞
Number of vacuum Rabi flops	$n_R = \frac{2g}{\kappa+\gamma}$	~10	~5	~10^2	~10^2	~10^2

*) Limited by the transmon relaxation; transmon dephasing time ≈ 3 μs, see §5.4.3.

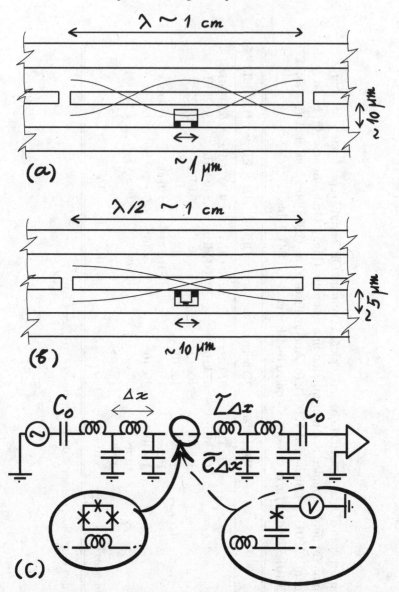

Fig. 4.11. Circuit QED with superconducting qubits: a schematic layout. Thin solid lines indicate amplitudes of voltage oscillations of the lowest relevant mode on the centre trace of the 1D transmission line resonator (full-wave or half-wave coplanar waveguide). (a) Charge qubit (after Blais et al., 2004) capacitively coupled to the central trace at the voltage node. (b) Flux qubit (after Abdumalikov et al., 2008) galvanically (i.e., through the kinetic inductance) coupled to the central trace at the voltage antinode (i.e., the current node). (c) Lumped-elements circuit of a 1D transmission line resonator with flux or charge qubit coupled to it.

When expanding $\Phi(x)$ in a Fourier series on the interval $[-l/2, l/2]$, we must take into account that there is no electric charge leakage through the ends of the resonator. Since the current in an elementary section of the transmission line is given by $c(\Phi(x + \Delta x) - \Phi(x))/\Delta L$, the corresponding boundary conditions are

$$\nabla \Phi|_{x=\pm l/2} = 0, \tag{4.71}$$

and, therefore,

$$\Phi(x, t) = \frac{\phi_0(t)}{2} + \sum_{k=1}^{\infty}\left[\phi_k^{(c)}(t)\cos\frac{2\pi k x}{l} + \phi_k^{(s)}(t)\sin\frac{(2k-1)\pi x}{l}\right]. \tag{4.72}$$

Each of the field modes $\phi_k^{(c,s)}(t)$ is an oscillator mode with frequency $\omega_k^{(c)} = 2\pi k c/l\sqrt{\widetilde{L}\widetilde{C}} = 2\pi k c/\sqrt{LC}$ or $\omega_k^{(s)} = (2k-1)\pi c/l\sqrt{\widetilde{L}\widetilde{C}} = (2k-1)\pi c/\sqrt{LC}$ and can be second-quantized in the usual way. Recalling that $\dot{\Phi}(x, t)/c = V(x, t)$, the local voltage between the line and the ground, we find for the quantized mode $\phi_1^{(c)}$ (Blais et al., 2004)[14]

$$\hat{V}(x, t) = \sqrt{\frac{\hbar\omega}{C}}\cos\frac{2\pi x}{l}\left(\frac{a^\dagger(t) - a(t)}{i}\right), \tag{4.73}$$

where $\omega = \omega_1^{(c)}$, $a^\dagger$ and a are, respectively, the frequency and Bose creation/annihilation operators of the corresponding field mode. The choice in Eq. (4.73) of a particular mode (the lowest even mode, with the voltage antinode in the middle of the resonator) was dictated by the experimental use of a capacitively coupled charge qubit (Wallraff et al., 2004, 2005; Schuster et al., 2005, 2007), Fig. 4.11a. For a centrally placed flux qubit, galvanically coupled to the resonator (Fig. 4.11b), the relevant mode would be the lowest odd mode, $\phi_1^{(s)}$ (Abdumalikov et al., 2008; Bourassa et al., 2009). In general, one has to substitute the field amplitude of the appropriate mode at the location of the qubit (the size of which is negligible compared to the field wavelength), see, e.g., Fig. 4.12.

The Hamiltonian of the charge qubit coupled to the field mode can now be written in the basis of the qubit's diabatic states as

$$H_{\text{qb+int}} = -\frac{E_C}{2}\left(1 - 2n_g^{*(\text{dc})}\right)\sigma_z - \frac{E_J}{2}\sigma_x - e\frac{C_g}{C_\Sigma}\sqrt{\frac{\hbar\omega}{C}}\left(\frac{a^\dagger - a}{i}\right)(1 - 2n_g^* - \sigma_z). \tag{4.74}$$

---

[14] Blais et al. (2004) use a different choice of variables (charges as coordinates and currents as velocities). As noted in Section 2.3, this only leads to the potential and kinetic energies being relabelled, the Lagrangian changing sign, and – in keeping with our earlier convention for the quantization of Φ – to the appearance of $(a^\dagger - a)/i$ instead of $(a + a^\dagger)$ in Eq. (4.73) and below. Of course, the classical and quantum equations of motion, all the observables, etc., do not depend on this choice.

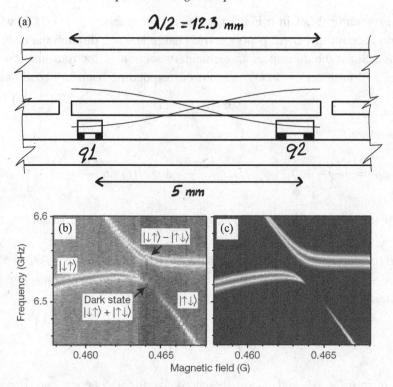

Fig. 4.12. (a) Two charge qubits coupled through a transmission line resonator. In the actual setup of Majer et al. (2007) the distance between the qubits was approximately 5 mm because of the zigzag design of the central part of the half-wavelength resonator. The cavity resonant frequency was $\omega/2\pi = 5.19$ GHz, with the linewidth $\kappa/2\pi = 33$ MHz. The charge qubits, employed were *transmons* (tunable charge qubits, with *large* E_J/E_C ratio (see Section 5.3)) with $E_{C1}/h = 1.7$ GHz, $E_{J1}^{\max}/h = 14.9$ GHz, and $E_{C2}/h = 1.77$ GHz, $E_{J2}^{\max}/h = 18.9$ GHz. Areas of the dc SQUID loop in the qubits were made different (ratio $\approx 5/8$) to simplify differential tuning of their Josephson energies. The coupling coefficients for the indicated field mode were $g/2\pi \approx 105$ MHz for both qubits. (b,c) Spectroscopy of two-qubit states (reprinted by permission from *Nature*, Majer et al., 2007, © 2007 Macmillan Publishers Ltd). (b) The measured cavity transmission coefficient shows an anticrossing between the symmetric and antisymmetric eigenstates of the two-qubit system, with the minimum distance $2g/h = 26$ MHz. The spectroscopic signal is antisymmetric, which results in the symmetric state being dark near the degeneracy point. (c) Theoretical fit to the experiment, using the Markovian master equation for the density matrix and taking into account higher harmonics in the resonator.

Here $2en_g^* = -C_g V_g$ is the effective charge induced on the island by the gate voltage; $C_\Sigma = C_J + C_g$ is the total capacitance of the qubit; and we have explicitly accounted for the qubit interaction with the quantized field. By changing the basis to the eigenbasis of the first two terms of (4.74), denoting Pauli matrices in this basis by σ_z', σ_x', and adding the Hamiltonian of the field, we find (Blais et al., 2004)

$$H = \hbar\omega\left(a^\dagger a + \frac{1}{2}\right) - \frac{\hbar\Omega}{2}\sigma_z' - e\frac{C_g}{C_\Sigma}\sqrt{\frac{\hbar\omega}{C}}\left(\frac{a^\dagger - a}{i}\right)(1 - 2n_g^* + \sigma_z'\cos\Theta - \sigma_x'\sin\Theta),$$

(4.75)

where the mixing angle Θ and the qubit energy splitting Ω are

$$\Theta = \arctan\left[\frac{E_J}{E_C(1 - 2n^{*(\mathrm{dc})})}\right]; \quad \Omega = \frac{[E_J^2 + E_C^2(1 - 2n^{*(\mathrm{dc})})^2]^{1/2}}{\hbar}.$$

(4.76)

By tuning the gate voltage we can make Θ a multiple of π and eliminate the σ_x'-term. Then the Hamiltonian (4.75) is easily reduced to the familiar Jaynes–Cummings form (and we will lose the dashes on the Pauli matrices, to unclutter the formulas). In experiments, such systems have demonstrated all the theoretically expected effects, such as, e.g., vacuum Rabi splitting (Wallraff et al., 2004), formation of photon number states (Schuster et al., 2007) or $\sqrt{n+1}$-dependence of the Rabi frequency (Fink et al., 2008). Of course, flux qubits behave in the same way (Abdumalikov et al., 2008).

4.3.2 Qubits coupled through a resonator

If now several qubits are connected to the same resonator, Fig. 4.12, we can couple them either actively, through the real excitations of the resonator, or passively, through its virtual excitations. In the latter case the qubits should be tuned in resonance with each other, but away from the resonator's frequency, as the situation considered before in a slightly different form (see Eqs (4.29) to (4.35)). The effective Hamiltonian for the resonator and two such qubits can be obtained either as there, by a direct projection to the vacuum state of the resonator, or better by using the Schrieffer–Wolff transformation, Eqs. (4.64, 4.68): then we can allow whatever state the resonator is in and obtain

$$H_{jk}^{\mathrm{eff}} = \left[\hbar\omega\left(a^\dagger a + \frac{1}{2}\right)\right]_{\mathrm{field}} + \left[-\frac{\hbar\Omega}{2}\left(\sigma_z^{(j)} + \sigma_z^{(k)}\right)\right]_{\mathrm{qubits}}$$
$$+ \frac{\hbar g^2}{\delta}\left(a^\dagger a + \frac{1}{2}\right)\left(\sigma_z^{(j)} + \sigma_z^{(k)}\right) - \frac{\hbar g^2}{\delta}\left(\sigma_+^{(j)}\sigma_-^{(k)} + \sigma_-^{(j)}\sigma_+^{(k)}\right). \quad (4.77)$$

The terms in the square brackets correspond to the unperturbed resonator mode and qubits; the next term is the frequency shift of the resonator and qubits due to their interaction, which depends on the quantum state of the whole system, precisely as in Eq. (4.69). The last term, reminiscent of Eq. (4.35), is the effective coupling between the qubits j and k. In the frame rotating with resonant frequency Ω for the qubits (that is, in the interaction representation, where we take as H_0 *only* the "qubits" term in (4.77)), the Hamiltonian (4.77) produces the evolution in the 4×4-subspace of states of qubits j and k (Blais et al., 2004):

$$U(t) = e^{-it\frac{g^2}{\delta}(a^\dagger a+\frac{1}{2})(\sigma_z^{(j)}+\sigma_z^{(k)})} \begin{pmatrix} 1 & 0 & 0 & 0 \\ 0 & \cos\frac{g^2 t}{\delta} & i\sin\frac{g^2 t}{\delta} & 0 \\ 0 & i\sin\frac{g^2 t}{\delta} & \cos\frac{g^2 t}{\delta} & 0 \\ 0 & 0 & 0 & 1 \end{pmatrix}. \tag{4.78}$$

Obviously, this evolution does not change the state of the resonator mode, while it allows swapping of the states of qubits, with the time required for such a $\sqrt{\text{iSWAP}}$ operation (see the Appendix)

$$t = \frac{\pi|\delta|}{4g^2} \sim 50\,\text{ns} \tag{4.79}$$

for characteristic circuit parameters (Blais et al., 2004).

4.4 *Phase space formalism of quantum optics

4.4.1 *Coherent states

The mathematical apparatus of second quantization is quite sufficient for the quantum description of the electromagnetic field in general and electric circuits in particular, and their interaction with atoms or qubits. Nevertheless, it does not necessarily give the clearest physical picture, or the most convenient mathematical description. The equivalent, but in many cases more advantageous, approach based on *coherent states* was first developed in quantum optics (Glauber, 1963), and for a very good reason. Photons, being bosons, are waves in the classical limit, and a formalism better suited to the description of such an essential property of a wave as its phase, all the way across the quantum-classical boundary, was required. A detailed description of these *phase space* methods with many applications can be found, e.g., in Scully and Zubairy (1997), Orszag (1999), Gardiner and Zoller (2004), Perelomov (1986), Schleich (2001). Here we just outline the basics.

Consider again a one-dimensional harmonic oscillator, or a single mode of an electromagnetic field, with frequency ω:

$$H = \hbar\frac{\omega}{2}\left(aa^\dagger + a^\dagger a\right). \tag{4.80}$$

If this oscillator represents, e.g., an LC circuit, then we can express the circuit variables (charge and flux) through $a, a^\dagger$ using Eq. (2.106), with $\Lambda = [2\hbar/C\omega]^{1/2}$

and $\omega = c[LC]^{-1/2}$. (We could, of course, use voltage and current instead.) Recall that the Fock states, $|n\rangle$, are the eigenstates of the excitation number operator,

$$\hat{N}|n\rangle = a^{\dagger}a|n\rangle = n|n\rangle,$$

while $\langle n|a|n\rangle = \langle n|a^{\dagger}|n\rangle = 0$. Therefore, the observable values of flux and charge operators (and current and voltage as well) will be zero in any Fock state:

$$\langle n|\hat{\Phi}|n\rangle = 0; \quad \langle n|\hat{Q}|n\rangle = \langle n|c\hat{\Pi}|n\rangle = 0. \tag{4.81}$$

Only average *squares* will be nonzero. On the other hand, in any *classical* state an LC circuit will have well-defined flux and charge (or current and voltage). To describe them, one would require infinitely many Fock states, like in the BCS wave function (2.5), and the transition from quantum to the quasiclassical and classical cases becomes rather cumbersome.

Instead let us introduce a *coherent state* as an eigenstate of the annihilation operator,

$$a|\alpha\rangle = \alpha|\alpha\rangle; \quad \langle \alpha|\alpha\rangle = 1, \tag{4.82}$$

where α is a complex number.

Let us expand $|\alpha\rangle$ over the Fock states and substitute this expansion in (4.82):

$$|\alpha\rangle = \sum_{n=0}^{\infty} c_n|n\rangle; \quad a|\alpha\rangle = \sum_{n=0}^{\infty} c_n a|n\rangle = \sum_{n=0}^{\infty} c_{n+1}\sqrt{n+1}|n\rangle. \tag{4.83}$$

This gives a recurrence relation for the coefficients,

$$c_{n+1} = \frac{\alpha}{\sqrt{n+1}}c_n = \cdots = \frac{\alpha^{n+1}}{\sqrt{(n+1)!}}c_0. \tag{4.84}$$

The normalization condition fixes c_0 (up to a phase factor), and finally

$$|\alpha\rangle = e^{-\frac{1}{2}|\alpha|^2}\sum_{n=0}^{\infty}\frac{\alpha^n}{\sqrt{n!}}|n\rangle; \quad \langle n|\alpha\rangle = e^{-\frac{1}{2}|\alpha|^2}\frac{\alpha^n}{\sqrt{n!}}. \tag{4.85}$$

From this expression it is clear that $|\alpha\rangle$ is *not* an eigenstate of the creation operator $a^{\dagger}$, and that the creation operator cannot have eigenstates at all, because its action on an expansion like Eq. (4.85) would lead to a series lacking the term proportional to $|0\rangle$, the vacuum state. This is the same obstacle that did not allow us to introduce the phase operator in § 2.2.1. We can, of course, define the action of $a^{\dagger}$ *from the right* on the conjugate state, $\langle \alpha|$,

$$\langle \alpha|a^{\dagger} = \langle \alpha|\alpha^*. \tag{4.86}$$

From (4.85), the scalar product of two coherent states is

$$\langle\beta|\alpha\rangle = e^{\beta^*\alpha-\frac{1}{2}|\beta|^2-\frac{1}{2}|\alpha|^2} = e^{-\frac{1}{2}|\alpha-\beta|^2}e^{i\operatorname{Im}(\beta^*\alpha)} \neq \delta(\beta-\alpha), \tag{4.87}$$

so they are not orthonormal. On the other hand, they constitute a complete (more exactly, *overcomplete*) set of states, over which any quantum state of the oscillator can be expanded. Indeed, let us invert (4.85) to express the number state $|n\rangle$ through the coherent states. We should expect an expression like $|n\rangle = ``\sum_\beta"|\beta\rangle\langle\beta|n\rangle$, where the "sum" over the coherent states is actually an integral, $C\int d\beta\, d\beta^* \equiv C\int dx\, dy$, with $\beta = x+iy, \beta^* = x-iy$, and C some normalization constant. Using (4.85), we find

$$|n\rangle \stackrel{?}{=} C\int d\beta d\beta^*|\beta\rangle\langle\beta|n\rangle = C\sum_m\left[\int d\beta d\beta^*\frac{e^{-|\beta|^2}}{\sqrt{m!n!}}\beta^m(\beta^*)^n\right]|m\rangle. \tag{4.88}$$

In polar coordinates $\beta = x+iy = re^{i\theta}$,

$$\int d\beta d\beta^*(\cdots) \equiv \int dx dy(\cdots) = \int_0^\infty dr r\int_0^{2\pi} d\theta(\cdots), \tag{4.89}$$

and

$$C\sum_m\left[\int_0^\infty dr r\int_0^{2\pi} d\theta\frac{e^{-r^2}}{\sqrt{m!n!}}r^{m+n}e^{i(n-m)\theta}\right]|m\rangle = C\pi|n\rangle = |n\rangle \tag{4.90}$$

if $C = 1/\pi$. We have successfully inverted Eq. (4.85):

$$|n\rangle = \frac{1}{\pi}\int d\beta d\beta^*|\beta\rangle\langle\beta|n\rangle = \frac{1}{\pi}\int d\beta d\beta^*|\beta\rangle\left[e^{-\frac{1}{2}|\beta|^2}\frac{(\beta^*)^n}{\sqrt{n!}}\right]. \tag{4.91}$$

Since the Fock states constitute a basis of the Fock space, over which any physically meaningful state can be expanded, and each of them in its turn can be expanded over the coherent states, we have established the completeness of the set of coherent states.

The *closure relation* for the coherent states, which expresses their completeness, can be written as

$$\hat{1} = \frac{1}{\pi}\int d\alpha d\alpha^*|\alpha\rangle\langle\alpha|, \tag{4.92}$$

where $\hat{1}$ is the identity operator. Its use enables an automatic transformation of an arbitrary quantum state or operator to the basis of coherent states:

$$|\Psi\rangle = \hat{1}|\Psi\rangle = \frac{1}{\pi}\int d\alpha d\alpha^*|\alpha\rangle\langle\alpha|\Psi\rangle; \tag{4.93}$$

$$A = \hat{1}A\hat{1} = \frac{1}{\pi^2}\int d\alpha d\alpha^*\int d\beta d\beta^*|\alpha\rangle\langle\alpha|A|\beta\rangle\langle\beta|. \tag{4.94}$$

The coefficients $\alpha|\Psi\rangle$ and matrix elements $\langle\alpha|A|\beta\rangle$ are directly calculated using Eq. (4.85).

The *over*completeness of the coherent state basis is revealed, e.g., in the fact that the inversion of Eq. (4.85) given by the formula (4.91) is an overkill. Instead of integrating over the whole of the complex plane, it is enough to integrate over a circle of an *arbitrary* radius $|\beta| \equiv r$:

$$|n\rangle = \frac{\sqrt{n!}r^{-n}e^{\frac{1}{2}r^2}}{2\pi} \int_0^{2\pi} d\theta |re^{i\theta}\rangle e^{-in\theta}. \tag{4.95}$$

This equation is immediately verified by substituting for the coherent state $|re^{i\theta}\rangle$ its expansion (4.85) and taking the integral. Another result of the overcompleteness is that, even though the expansion (4.94) includes the expressions $\langle\alpha|A|\beta\rangle$ for all complex numbers α, β, the full information about the operator A (all its matrix elements $\langle m|A|n\rangle$ in the basis of Fock states) can be extracted from its diagonal matrix element, $\langle\alpha|A|\alpha\rangle$ (see, e.g., Gardiner and Zoller, 2004, §4.3g).

4.4.2 *Physical significance of coherent states*

Let us introduce the *quadrature operators*,

$$\hat{X} = \frac{a+a^\dagger}{2}, \quad \hat{Y} = \frac{a-a^\dagger}{2i}. \tag{4.96}$$

From comparison with Eq. (4.4) we see that for an LC circuit $\hat{X}$ is proportional to the flux (i.e., current), and $\hat{Y}$ to its canonically conjugate momentum, i.e., charge. For a mechanical oscillator we would have position and momentum instead. The expectation values of $\hat{X}, \hat{Y}$ in a coherent state $|\alpha\rangle$ are (using the definition (4.82) and Eq. (4.86))

$$\langle X\rangle_\alpha = \text{Re}\,\alpha \equiv x; \quad \langle Y\rangle_\alpha = \text{Im}\,\alpha \equiv y. \tag{4.97}$$

Therefore, a single coherent state, unlike Fock states, describes the state of a system with nonzero average coordinate and conjugate momentum. In other terms, a coherent state $|\alpha\rangle \equiv |re^{i\theta}\rangle$ can be characterized by its amplitude, $r = \sqrt{|\alpha|^2} = \sqrt{x^2 + y^2}$ and phase, $\theta : \tan\theta = y/x$, almost like in a classical oscillator.

The solution to the Schrödinger equation of a harmonic oscillator can be written as Eq. (1.2)

$$|\psi(t)\rangle = e^{-\frac{i}{\hbar}Ht}|\psi(0)\rangle$$

with the Hamiltonian (4.80). If the initial state was a coherent state $|\alpha\rangle$, then we can write

$$a|\psi(t)\rangle = ae^{-\frac{i}{\hbar}Ht}|\psi(0)\rangle = e^{-\frac{i}{\hbar}Ht}\left[e^{\frac{i}{\hbar}Ht}ae^{-\frac{i}{\hbar}Ht}\right]|\alpha\rangle$$

$$= e^{-\frac{i}{\hbar}Ht}\left[e^{-i\omega t}a\right]|\alpha\rangle = \alpha e^{-i\omega t}|\psi(t)\rangle. \tag{4.98}$$

Therefore, we can identify $|\psi(t)\rangle$ with the time-dependent coherent state $|\alpha e^{-i\omega t}\rangle$. Not only coherent states are states of a harmonic oscillator with (almost) definite amplitude and phase; such a state will correctly evolve, describing a circle of given amplitude with the frequency of the oscillator.

Of course, where a classical oscillator would be represented by a point in the complex α plane, a quantum one could not, due to the Heisenberg uncertainty relations. The commutator

$$[\hat{X}, \hat{Y}] = \frac{i}{2}, \tag{4.99}$$

and, therefore,

$$\Delta X \Delta Y \equiv \sqrt{\langle(\hat{X} - \langle\hat{X}\rangle)^2\rangle}\sqrt{\langle(\hat{Y} - \langle\hat{Y}\rangle)^2\rangle} \geq \frac{1}{2}\left|\left\langle\frac{1}{i}[\hat{X}, \hat{Y}]\right\rangle\right| = \frac{1}{4}. \tag{4.100}$$

After calculating the quadrature uncertainties in a coherent state $|\alpha\rangle$,

$$\Delta X_\alpha = \left\langle\alpha\left|\left(\hat{X} - \langle X\rangle_\alpha\right)^2\right|\alpha\right\rangle^{1/2} = \frac{1}{2}, \quad \Delta Y_\alpha = \left\langle\alpha\left|\left(\hat{Y} - \langle Y\rangle_\alpha\right)^2\right|\alpha\right\rangle^{1/2} = \frac{1}{2},$$

$$\tag{4.101}$$

we see from (4.100) that a coherent state is a *minimum uncertainty* state:

$$\Delta X_\alpha \Delta Y_\alpha = \frac{1}{4}. \tag{4.102}$$

Together with Eq. (4.97) this explains the significance of coherent states: they describe the state of an oscillator with given expectation values of amplitude and phase, with the minimal uncertainty allowed by quantum mechanics. In the complex plane α, this state can be represented by a dot of diameter $1/2$ around the point α, which in the classical limit contracts to a point, as the commutator (4.99) disappears.[15] The time evolution of a quantum oscillator is described by the rotation of this dot around the origin with frequency ω.

---

[15] Rewriting Eqs (4.96)–(4.102) in terms of the dimensional flux-charge variables $\hat{\Phi} \propto \hat{X}$, $\hat{\Pi} \propto \hat{Y}$ (2.106) instead of dimensionless quadratures would restore the $\hbar$-factor in the right-hand side of the commutator, thus making the quantum-classical transition more obvious.

A coherent state with $\alpha = 0$ can be denoted by $|0\rangle$ without confusion, because it is indeed the vacuum state, as can be seen from the definition (4.82) or the explicit expansion (4.85). Any coherent state $|\alpha\rangle$ can be thought of as a *displaced vacuum state*,

$$|\alpha\rangle = D(\alpha)|0\rangle, \tag{4.103}$$

where the *displacement operator*

$$D(\alpha) = e^{\alpha a^\dagger - \alpha^* a} \tag{4.104}$$

is to be understood in the usual way, as a formal Taylor expansion. Using the Baker–Hausdorff formula (3.59), we can reduce it to

$$D(\alpha) = e^{\alpha a^\dagger} e^{-\alpha^* a} e^{\alpha a^\dagger - \alpha^* a - \frac{1}{2}[\alpha a^\dagger, -\alpha^* a]} = e^{-\frac{1}{2}|\alpha|^2} e^{\alpha a^\dagger} e^{-\alpha^* a}. \tag{4.105}$$

Since $a|0\rangle = 0$, then $\exp[-\alpha^* a]|0\rangle = |0\rangle$. Therefore, substituting (4.105) in (4.103) and expanding $\exp[\alpha a^\dagger]$, we see that Eq. (4.103) is identical with the formula Eq. (4.85) for the coherent state $|\alpha\rangle$.

Using the Baker–Hausdorff formula, we can also directly check that

$$e^{\alpha a^\dagger - \alpha^* a} e^{\beta a^\dagger - \beta^* a} = e^{(\alpha+\beta)a^\dagger - (\alpha^*+\beta^*)a} e^{\frac{1}{2}[\alpha a^\dagger - \alpha^* a, \beta a^\dagger - \beta^* a]}$$

$$= e^{(\alpha+\beta)a^\dagger - (\alpha^*+\beta^*)a} e^{\frac{1}{2}(\alpha\beta^* - \alpha^*\beta)}, \tag{4.106}$$

and, therefore,

$$D(\alpha+\beta) = e^{-i\operatorname{Im}\alpha\beta^*} D(\alpha)D(\beta). \tag{4.107}$$

4.4.3 *Wigner function for the oscillator states*

We could introduce this function using the definition (1.22) and the expression of the creation/annihilation operators through the operators of position and momentum, similar to Eq. (2.106). But an easier and more straightforward way is through the *characteristic function*, defined as the expectation value of the displacement operator,

$$\chi_W(\eta, \eta^*) = \langle D(\eta) \rangle = \left\langle e^{\eta a^\dagger - \eta^* a} \right\rangle = \operatorname{tr}\left[\rho e^{\eta a^\dagger - \eta^* a}\right]. \tag{4.108}$$

Then the Wigner function is its Fourier transform,

$$W(\alpha, \alpha^*) = \frac{1}{\pi^2} \int d\eta d\eta^* \, e^{-\eta \alpha^* + \eta^* \alpha} \chi_W(\eta, \eta^*). \tag{4.109}$$

We can check that $W(\alpha, \alpha^*)$ has the property (1.27), i.e.,

$$\int d\alpha d\alpha^* \, W(\alpha, \alpha^*) \alpha^m (\alpha^*)^n = \langle \{a^m (a^\dagger)^n\}_{\text{sym}} \rangle. \tag{4.110}$$

Indeed, integrating (4.109) by parts, we obtain the relation

$$\int d\alpha d\alpha^* \, W(\alpha, \alpha^*) \alpha^m (\alpha^*)^n = \left[\frac{\partial^m}{\partial(-\eta^*)^m} \frac{\partial^n}{\partial \eta^n} \chi_W(\eta, \eta^*) \right]_{\eta, \eta^*=0} . \qquad (4.111)$$

Noticing that

$$(\eta a^\dagger - \eta^* a)^{n+m} = \sum_{m,n} \{ a^m (a^\dagger)^n \}_{\mathrm{sym}} (-\eta^*)^m \eta^n, \qquad (4.112)$$

and, therefore,

$$\chi_W(\eta, \eta^*) = \sum_{m,n} \frac{(-\eta^*)^m \eta^n}{m! n!} \left\langle \{ a^m (a^\dagger)^n \}_{\mathrm{sym}} \right\rangle, \qquad (4.113)$$

we come to Eq. (4.110). Using (4.110), we can, e.g., find the average energy of a quantum state,

$$\langle E \rangle = \frac{\hbar\omega}{2} \langle aa^\dagger + a^\dagger a \rangle = \hbar\omega \int d\alpha d\alpha^* \, |\alpha|^2 W(\alpha, \alpha^*), \qquad (4.114)$$

and the average number of quanta in it,

$$\langle n \rangle = \frac{1}{2} \langle aa^\dagger + a^\dagger a \rangle - \frac{1}{2} = \int d\alpha d\alpha^* \, |\alpha|^2 W(\alpha, \alpha^*) - \frac{1}{2}. \qquad (4.115)$$

The Wigner function allows a convenient visualization of oscillator states.[16] For example, a coherent state $|\beta\rangle$ has the density matrix $\rho = |\beta\rangle\langle\beta|$. Its characteristic function is

$$\chi_W(\eta, \eta^*) = \mathrm{tr}\,(|\beta\rangle\langle\beta| D(\eta)) = \mathrm{tr}\,(\langle\beta| D(\eta) D(\beta) |0\rangle)$$

$$= e^{i \mathrm{Im}\eta\beta^*} \langle\beta|\eta + \beta\rangle = e^{-\frac{1}{2}|\eta|^2 - \eta^*\beta + \eta\beta^*}. \qquad (4.116)$$

We have used here Eqs (4.107) and (4.87). Substituting (4.116) in (4.109), we find a Gaussian function,

$$W(\alpha, \alpha^*) = \frac{1}{\pi^2} \int d\eta d\eta^* \, e^{-\eta(\alpha^*-\beta^*)+\eta^*(\alpha-\beta)-\frac{1}{2}|\eta|^2} = \frac{2}{\pi} e^{-2|\alpha-\beta|^2}, \qquad (4.117)$$

with the dispersion determined by the quantum uncertainty relation (4.102) (see Fig. 4.13a).

---

[16] Besides the Wigner function, there are other representations of the density matrix in the basis of the coherent states (P- and Q-distribution functions), which may be more convenient for certain problems (see, e.g., Orszag, 1999, Chapter 7; Schleich, 2001; Gardiner and Zoller, 2004). The advantage of the Wigner function is that it exists as a sufficiently smooth function (with no worse singularities than a δ-function) for any density matrix, and that every such Wigner function corresponds to some density matrix.

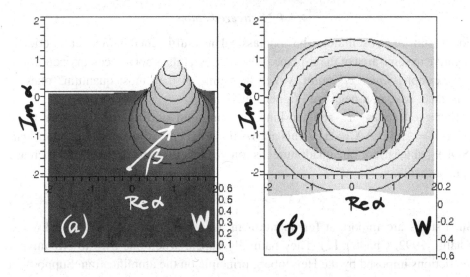

Fig. 4.13. Wigner function of a coherent state with $\beta = 1 + i$ (a), and Fock state with $n = 3$ (b).

By a direct calculation it can be shown (Gardiner and Zoller, 2004, §4.4.2) that the Wigner function of a thermal state, with

$$\rho = (1 - \exp[-\hbar\omega/k_B T]) \sum_n \exp[-n\hbar\omega/k_B T]|n\rangle\langle n|, \tag{4.118}$$

is a Gaussian,

$$W(\alpha, \alpha^*) = \frac{2}{\pi} \tanh \frac{\hbar\omega}{2k_B T} e^{-2|\alpha|^2 \tanh \frac{\hbar\omega}{2k_B T}}, \tag{4.119}$$

while for a Fock state, $\rho = |n\rangle\langle n|$,

$$W(\alpha, \alpha^*) = \frac{2}{\pi}(-1)^n e^{-2|\alpha|^2} L_n(4|\alpha|^2). \tag{4.120}$$

Here $L_n(x)$ is the nth Laguerre polynomial (Gradshteyn and Ryzhik, 2000, §8.970):

$$L_0 = 1, \ L_1 = -x + 1, \ L_2 = \frac{1}{2}(x^2 - 4x + 2), \ \dots L_n = \sum_{m=0}^{n}(-1)^m \frac{n!}{(n-m)!(m!)^2} x^m. \tag{4.121}$$

Note that in the latter case the Wigner function can take negative values (Fig. 4.13b), which underlines the essentially quantum character of Fock states. The vacuum Wigner function, in agreement with (4.117), (4.119), or (4.120), is given simply by

$$W(\alpha, \alpha^*) = \frac{2}{\pi} e^{-2|\alpha|^2}. \tag{4.122}$$

4.4.4 *Squeezed states*

For the state of an oscillator to be nonclassical, it is sufficient if its Wigner function is negative in some area of the α-plane. Nevertheless, this is not a necessary condition. For example, *squeezed states*, which are some of the "most quantum" ones in existence, have non-negative Wigner functions.

Squeezed states are another kind of minimum uncertainty state: for them $\Delta X \Delta Y = \frac{1}{4}$, but $\Delta X \neq \Delta Y$. This allows them to beat the *standard quantum limit* (SQL) – at least, for one quadrature, as long as this is compensated by an increase in the other, e.g.,

$$\Delta X = \frac{e^{-r}}{2}, \quad \Delta Y = \frac{e^r}{2}. \tag{4.123}$$

Such states are important for quantum measurements (see, e.g., Braginsky and Khalili, 1992, Chapter 12). They naturally appear when one tries to circumvent restrictions imposed by the Heisenberg principle on the amplification. Suppose an ideal amplifier amplifies both the current and voltage of a harmonic signal in a quantum LC-circuit. In terms of the quadrature operators $\hat{X}, \hat{Y}$, this means that $\hat{X}$ is multiplied by g_X and $\hat{Y}$, by g_Y. It can be shown that if the only source of noise in the amplifier is due to quantum fluctuations (*quantum-limited amplifier*), generally, noise must be added to both quadratures; the exception is the case, when $g_X = 1/g_Y$. Then the Heisenberg uncertainty relation is satisfied without any added noise (Caves, 1982), which is obvious from Eq. (4.123). Such *phase-sensitive* amplifiers have been realized; one recent example is a parametric microwave amplifier based on a tunable Josephson metamaterial (Castellanos-Beltran et al., 2008).

Mathematically, a squeezed state can be obtained from a *squeezed vacuum*,

$$|0, \xi\rangle = S(\xi)|0\rangle, \tag{4.124}$$

where the unitary *squeeze operator*

$$S(\xi) = e^{\frac{1}{2}\xi^* a^2 - \frac{1}{2}\xi(a^\dagger)^2}, \tag{4.125}$$

and $\xi = r \exp[i\theta]$ is a complex *squeeze parameter*. Since $[a^2, (a^\dagger)^2]$ is an operator and does not commute with a^2 and $(a^\dagger)^2$, the Baker–Hausdorff formula will not help us here. Instead let us modify the creation/annihilation operators and introduce

$$A = S(\xi)aS^\dagger(\xi); \quad A^\dagger = S(\xi)a^\dagger S^\dagger(\xi). \tag{4.126}$$

From this definition it is obvious that

$$A|0, \xi\rangle = S(\xi)a|0\rangle = 0; \quad \langle 0, \xi|A^\dagger = \langle 0|a^\dagger S^\dagger(\xi) = 0; \quad [A, A^\dagger] = 1, \tag{4.127}$$

that is, the new operators are bosonic creation/annihilation operators with respect to the squeezed vacuum state, $|0, \xi\rangle$.

Note that

$$\frac{1}{2}[a^2, a^\dagger] = a; \quad \frac{1}{2}[(a^\dagger)^2, a] = -a^\dagger. \tag{4.128}$$

Then we can use the operator formula (e.g., Orszag, 1999, Appendix A, Eq. (A.1))

$$e^{\epsilon A} B e^{-\epsilon A} = B + \epsilon[A, B] + \frac{\epsilon^2}{2!}[A, [A, B]] + \cdots + \frac{\epsilon^n}{n!}\underbrace{[A, [A, \ldots [A, B]]}_{n \text{ times}} + \cdots, \tag{4.129}$$

where ϵ is a number. Applying it to A and $A^\dagger$ of Eq. (4.126), we see that due to (4.128) the expansions will contain only (alternating) terms proportional to a and $a^\dagger$. In other words, Eq. (4.126) defines a linear transformation:

$$A = u^* a + v a^\dagger; \quad A^\dagger = u a^\dagger + v^* a. \tag{4.130}$$

Instead of collecting the terms in the expansion, we can now directly use the unitarity condition, which means, in particular, that the commutation relations must hold, $[A, A^\dagger] = [a, a^\dagger]$. Therefore, $|u|^2 - |v|^2 = 1$, and we can write (since the overall phase factor is irrelevant)[17]

$$u = \cosh v, \quad v = e^{i\theta} \sinh v. \tag{4.131}$$

Expanding this to the second order in v and comparing with the corresponding terms in the expansion of (4.126) in ξ, we establish the relation

$$v = |\xi| = r; \quad \theta = \arg \xi, \tag{4.132}$$

where the θ is the same as in the definition for the squeeze parameters, χ_i. Inverting the transformation (4.130), we now find

$$a = A \cosh r - A^\dagger \sinh r e^{i\theta}; \quad a^\dagger = A^\dagger \cosh r - A \sinh r e^{-i\theta}; \tag{4.133}$$

$$\hat{X} = \frac{1}{2}\left[\left(\cosh r - \sinh r e^{-i\theta}\right) A + \left(\cosh r - \sinh r e^{i\theta}\right) A^\dagger\right];$$

$$\hat{Y} = \frac{1}{2i}\left[\left(\cosh r + \sinh r e^{-i\theta}\right) A - \left(\cosh r + \sinh r e^{i\theta}\right) A^\dagger\right]. \tag{4.134}$$

---

[17] Eq. (4.130) is an example of a *Bogoliubov transformation*. Such transformations, where old creation/annihilation operators become linear combinations of new creation *and* annihilation operators, play a very important role in the theory of superconductivity. In particular, they are responsible for the appearance of bogolons, which we briefly met when talking about Andreev reflection, and the BCS wave function (2.5). The coherence factors (2.6) are the counterparts of u, v in Eq. (4.130), with the difference that they must preserve the *anti*commutation relations of the fermionic operators.

The expectation values and variances of the quadratures in a squeezed vacuum state are now readily found:

$$\langle X \rangle = \langle Y \rangle = 0; \tag{4.135}$$

$$\langle \Delta X^2 \rangle = \frac{1}{4}(\cosh 2r - \sinh 2r \cos\theta); \tag{4.136}$$

$$\langle \Delta Y^2 \rangle = \frac{1}{4}(\cosh 2r + \sinh 2r \cos\theta). \tag{4.137}$$

The uncertainty relation is satisfied

$$\Delta X \Delta Y = \frac{1}{4}\sqrt{\cosh^2 2r - \sinh^2 2r \cos\theta} \geq \frac{1}{4}, \tag{4.138}$$

and the minimum is reached when $\theta = 0$ or π, and, respectively, $\Delta X = e^{-r}/2$, $\Delta Y = e^{r}/2$ or vice versa.

The character of a squeezed vacuum state is clear if we look at its Wigner function (Fig 4.14), given by the expression

$$W(\alpha, \alpha^*) = \frac{2}{\pi} \exp\left\{-2\left[e^{2r}\left(\text{Re}[\alpha e^{-i\theta/2}]\right)^2 + e^{-2r}\left(\text{Im}[\alpha e^{-i\theta/2}]\right)^2\right]\right\}. \tag{4.139}$$

The circular symmetry is gone: one quadrature is indeed squeezed, the other stretched, and the whole "Gaussian cigar" (Schleich, 2001, §4.3.1) is rotated by the angle $\theta/2$. We see that a squeezed vacuum is a minimum uncertainty state for any θ: all the angular dependence in Eq. (4.138) comes from the fact that we calculated variances in the wrong frame of reference, the coordinate axes of which are not parallel to the axes of symmetry of the cigar.

The energy of a squeezed vacuum is by no means the half-quantum of zero-point oscillations: it is clear already from the definition of $|0, \xi\rangle$ that it includes Fock states with arbitrarily high occupation numbers. From Eq. (4.114) we see that

$$\langle E \rangle_{\text{sq.vac.}} = \frac{\hbar\omega}{2}\cosh 2r \geq \frac{\hbar\omega}{2}. \tag{4.140}$$

In the limit of infinite squeezing the squeezed vacuum will be a state with a definite phase and indefinite number of quanta. The state with a definite number of quanta is, of course, a Fock state $|n\rangle$, which has a radially symmetric Wigner function, and thus an indefinite phase. The transition between the two can give us an idea of the transition between the BCS ground state of a bulk superconductor to the ground state of a finite one, with a fixed number of particles.

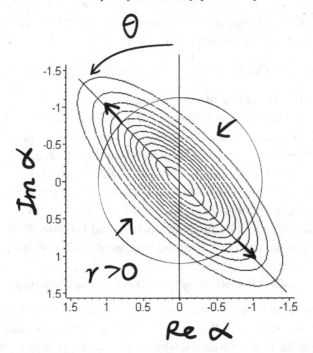

Fig. 4.14. Wigner function of a squeezed vacuum state ("Gaussian cigar", Eq. (4.139)) for $r = 0.5, \theta = \frac{\pi}{2}$. For negative r the squeezed and extended quadratures change places. The circle indicates the shape of the unsqueezed vacuum.

A general squeezed state can be obtained from a squeezed vacuum in the same way as a coherent state from the usual vacuum, by displacing it:

$$|\alpha, \xi\rangle = D(\alpha)|0, \xi\rangle = D(\alpha)S(\xi)|0\rangle. \qquad (4.141)$$

The Wigner function of the resulting state will be given by (4.139), but centered at a given complex point α.

Using (4.129) one can prove that

$$S^\dagger(\xi)D(\alpha)S(\xi) = D(\beta), \qquad (4.142)$$

where

$$\beta = \alpha \cosh r + \alpha^* \sinh r e^{i\theta}. \qquad (4.143)$$

From here it follows that

$$A|\alpha, \xi\rangle = AD(\alpha)S(\xi)|0\rangle = AS(\xi)D(\beta)|0\rangle = S(\xi)a|\beta\rangle =$$
$$\beta S(\xi)a|\beta\rangle = \cdots = \beta|\alpha, \xi\rangle = (\alpha \cosh r + \alpha^* \sinh r e^{i\theta})|\alpha, \xi\rangle. \qquad (4.144)$$

With the help of (4.144) and the conjugate formula for $A^\dagger$, we can calculate, e.g., the expectation values for the quadratures (4.134):

$$\langle X \rangle = \langle \alpha, \xi | \hat{X} | \alpha, \xi \rangle = \text{Re } \alpha; \quad \langle Y \rangle = \langle \alpha, \xi | \hat{Y} | \alpha, \xi \rangle = \text{Im } \alpha, \tag{4.145}$$

which indeed gives the centre of the displaced Gaussian cigar.

A general squeezed state can be thought of as an intermediate between the eigenstates of conjugate variables, which can be envisaged as the limiting cases of Gaussian cigars, squeezed to zero in one direction and stretched to infinity in an orthogonal one (but in such a way as to keep the product $\Delta X \Delta Y$ constant). Strictly speaking, this does not apply to, e.g., the position and momentum of a free particle, since the corresponding eigenfunctions, $\propto \exp[\pm ipx/\hbar]$, are not normalizable (i.e., they can only be normalized to a delta-function (Landau and Lifshitz, 2003, § 5)), but we can consider an arbitrarily weak harmonic potential to see how the analogy works.

4.4.5 *Equations of motion for Wigner function

One big advantage of using the phase space methods, and in particular, the Wigner function, is that the master equation for the density matrix can be reduced to a partial differential equation of few variables, which can be easier to solve or provide a better insight into the physics of the problem. For example, a quasiclassical mode of an LC circuit, which interacts with a qubit, and the measurable phase and amplitude of current in this circuit, can be characterized by a single complex number α.

The translation is done using the *operator correspondences* between the action of creation/annihilation operators on the density matrix and the differentiations of the Wigner function. They are based eventually on Eqs (4.82) and (4.86) and are as follows (see Gardiner and Zoller, 2004, §4.5):

$$
\begin{aligned}
a\rho \quad &\longleftrightarrow \quad \left(\alpha + \tfrac{1}{2} \tfrac{\partial}{\partial \alpha^*} \right) W(\alpha, \alpha^*) \\
\rho a \quad &\longleftrightarrow \quad \left(\alpha - \tfrac{1}{2} \tfrac{\partial}{\partial \alpha^*} \right) W(\alpha, \alpha^*) \\
a^\dagger \rho \quad &\longleftrightarrow \quad \left(\alpha^* - \tfrac{1}{2} \tfrac{\partial}{\partial \alpha} \right) W(\alpha, \alpha^*) \\
\rho a^\dagger \quad &\longleftrightarrow \quad \left(\alpha^* + \tfrac{1}{2} \tfrac{\partial}{\partial \alpha} \right) W(\alpha, \alpha^*).
\end{aligned}
\tag{4.146}
$$

Here, as usual,

$$
\frac{\partial}{\partial \alpha} = \frac{\partial x}{\partial \alpha} \frac{\partial}{\partial x} + \frac{\partial y}{\partial \alpha} \frac{\partial}{\partial y} = \frac{1}{2} \left[\frac{\partial}{\partial x} - i \frac{\partial}{\partial y} \right],
$$

$$
\frac{\partial}{\partial \alpha^*} = \frac{1}{2} \left[\frac{\partial}{\partial x} + i \frac{\partial}{\partial y} \right],
\tag{4.147}
$$

$$
\alpha = x + iy, \; \alpha^* = x - iy.
$$

Consider, for example, the master equation for the density matrix of an oscillator in contact with a thermal reservoir:

$$\frac{d}{dt}\rho = \frac{\omega}{2i}[aa^\dagger + a^\dagger a, \rho] - \frac{\kappa}{2}(\overline{n(\omega)} + 1)\left[a^\dagger a\rho + \rho a^\dagger a - 2a\rho a^\dagger\right]$$
$$- \frac{\kappa}{2}\overline{n(\omega)}\left[aa^\dagger\rho + \rho aa^\dagger - 2a^\dagger\rho a\right]. \tag{4.148}$$

The last two terms are the Lindblad terms (1.60), with $B_+ \to a^\dagger$, $B_- \to a$ and the thermal average of the bosonic bath excitations

$$\overline{n(\omega)} = \frac{1}{e^{\hbar\omega/k_B T} - 1}. \tag{4.149}$$

The operator correspondences (4.146) translate Eq. (4.148) into

$$\frac{\partial}{\partial t}W(\alpha, \alpha^*) = \frac{\omega}{i}\left[\alpha^*\frac{\partial}{\partial\alpha^*} - \alpha\frac{\partial}{\partial\alpha}\right]W(\alpha, \alpha^*)$$
$$+ \frac{\kappa}{2}\left[\frac{\partial}{\partial\alpha}(\alpha W(\alpha, \alpha^*)) + \frac{\partial}{\partial\alpha^*}(\alpha^* W(\alpha, \alpha^*))\right]$$
$$+ \frac{\kappa}{2}(2\overline{n(\omega)} + 1)\frac{\partial^2}{\partial\alpha\partial\alpha^*}W(\alpha, \alpha^*), \tag{4.150}$$

which is essentially a two-dimensional Fokker–Planck equation (Gardiner and Zoller, 2004, § 5.3) of the kind usually obtained for classical probability distributions.[18] It is convenient to rewrite (4.150) in the variables $x = \text{Re}\,\alpha$, $y = \text{Im}\,\alpha$:

$$\frac{\partial}{\partial t}W(x, y) = \omega\left[x\frac{\partial}{\partial y} - y\frac{\partial}{\partial x}\right]W(x, y) + \frac{\kappa}{2}\left[x\frac{\partial}{\partial x} + y\frac{\partial}{\partial y}\right]W(x, y)$$
$$+ \frac{\kappa}{8}(2\overline{n(\omega)} + 1)\left(\frac{\partial^2}{\partial x^2} + \frac{\partial^2}{\partial y^2}\right)W(x, y). \tag{4.151}$$

The first line here describes a rotation of the Wigner function around the origin with angular velocity ω, as one would expect from, e.g., Eq. (4.98). The second line is responsible for its eventual relaxation. One can check directly that the stationary solution of (4.151) is given by the thermal Wigner function (4.119) with the temperature of the reservoir.

---

[18] If the system includes some other quantum objects (e.g., qubits) coupled to one or several oscillator modes, its density matrix can be transformed to the form $W_{mm',nn',\dots}(\alpha, \alpha^*; \beta, \beta^*; \dots)$, where the matrix indices take care of the non oscillator degrees of freedom. The master equation for it will reduce to a set of differential equations like (4.150).

4.4.6 *Parametric generation of squeezed states*

Phase space formalism finds wide applications in quantum optics (see especially Schleich, 2001; Gardiner and Zoller, 2004) and, increasingly, in circuit QED and other related problems (e.g., Gambetta et al., 2006). Consider, for example, a linear oscillator with variable frequency. If initially it was in a coherent state, and then we change its frequency, what state will it be in? Consider a mechanical oscillator. Expressing the position and momentum operators of Eq. (4.36) as

$$\hat{X} = \sqrt{\frac{\hbar}{2m\omega}}(a + a^\dagger), \ \hat{P} = \frac{\sqrt{2m\hbar\omega}}{2i}(a - a^\dagger), \tag{4.152}$$

we see that, as the oscillator frequency grows, the quantum uncertainty of the position in, e.g., a vacuum state, decreases, and of the momentum increases. Physically this is self-evident: in a stiffer oscillator the position is better fixed. If an oscillator is in a coherent state, and then its frequency is instantaneously changed, the quantum state $|0\rangle$ of the oscillator will not change, but now it will have to be expanded over the state vectors belonging to a *new* Fock space, corresponding to the changed frequency ω_f. From the point of view of this space, the state $|0\rangle$ is no longer a vacuum: if $\omega_f > \omega$, the uncertainty of position in this state is now larger, and the uncertainty of momentum, less, than in the new vacuum state, $|0\rangle_f$. If $\omega_f < \omega$, it will be the other way around. As a matter of fact, a sudden frequency change transforms a coherent state into a squeezed state, with the squeezing proportional to $|\omega_f/\omega|$ – a *parametric squeezing* – while a slow, adiabatic change, does not (Graham, 1987; Agarwal and Kumar, 1991; Janszky and Adam, 1992; Kiss et al., 1994). A repeated series of fast and slow changes of the oscillator frequency could, therefore, allow a high degree of squeezing even if the frequency change is not large (Abdalla and Colegrave, 1993; Averbukh et al., 1994). This is an especially interesting possibility when we have at our disposal such coherently tunable circuits as, e.g., a superconducting resonator coupled to a Josephson inductance.

Let us consider the general case of a linear oscillator with time-dependent frequency, $\omega(t)$. Let us also, slightly inconsistently, denote the creation/annihilation operators belonging to the Fock state of an oscillator with the frequency $\omega(t = 0)$ by a_0, $a_0^\dagger$, and those corresponding to $\omega(t)$, by a_ω, $a_\omega^\dagger$. In order for the Hamiltonian to maintain its standard form,

$$H(t) = \hbar\omega(t)\left(a_\omega^\dagger a_\omega + \frac{1}{2}\right), \tag{4.153}$$

and to keep the commutation relations between a_ω, $a_\omega^\dagger$, we easily find that

$$a_\omega = \frac{(\omega(t) + \omega(0))a_0 - (\omega(t) - \omega(0))a_0^\dagger}{2\sqrt{\omega(t)\omega(0)}}, \tag{4.154}$$

and the Hermitian conjugate expression for $a_\omega^\dagger$. This is an example of the familiar Bogoliubov transformation, like Eq. (4.130). It can also be explicitly written as a unitary transformation (Graham, 1987):

$$a_0 = V^\dagger(t)a_\omega V(t); \quad a_0^\dagger = V^\dagger(t)a_\omega^\dagger V(t) \tag{4.155}$$

where

$$V(t) = e^{-\frac{1}{4}\left(\ln\frac{\omega(0)}{\omega(t)}\right)\left(a_\omega^2 - (a_\omega^\dagger)^2\right)}; \quad V^\dagger(t) = e^{\frac{1}{4}\left(\ln\frac{\omega(0)}{\omega(t)}\right)\left(a_\omega^2 - (a_\omega^\dagger)^2\right)}. \tag{4.156}$$

From here one can see directly that the change of frequency can produce a squeezed state, since this transformation $V(t)$ looks precisely like the squeeze operator S (Eq. (4.125)).

It is unpleasant to work in a continually changing basis, and we will instead transform everything into the initial Fock space at $t = 0$ by applying the transformation $V^{-1}(t) = V^\dagger(t)$. With the Hamiltonian we should be cautious, since the transformation is time-dependent (cf. Eq. (1.102)):

$$H(t) \to \tilde{H}(t) = V(t)H(t)V^\dagger(t) - i\hbar V(t)\frac{\partial}{\partial t}V^\dagger(t) =$$

$$\hbar\omega(t)\left(a_0^\dagger a_0 + \frac{1}{2}\right) + i\hbar\frac{\dot\omega(t)}{\omega(t)}\left(a_0^2 - (a_0^\dagger)^2\right). \tag{4.157}$$

Now we can write the master equation for the Wigner function, with α, α^* always referring to the coherent states in the same Fock space (at $t = 0$). Doing this, we find (Zagoskin et al., 2008)

$$\frac{\partial}{\partial t}W(\alpha, \alpha^*) = 2\omega(t)\mathrm{Im}\left(\alpha^*\frac{\partial}{\partial\alpha^*}\right)W(\alpha, \alpha^*) + \frac{\partial\ln\omega(t)}{\partial t}\mathrm{Re}\left(\alpha\frac{\partial}{\partial\alpha^*}\right)W(\alpha, \alpha^*), \tag{4.158}$$

or

$$\frac{\partial}{\partial t}W(x, y) = \omega(t)\left(x\frac{\partial}{\partial y} - y\frac{\partial}{\partial x}\right)W(x, y) + \frac{1}{2}\frac{\partial\ln\omega(t)}{\partial t}\left(x\frac{\partial}{\partial x} - y\frac{\partial}{\partial y}\right)W(x, y). \tag{4.159}$$

For times much shorter than the decoherence time we can neglect the non-unitary Lindblad terms, which allows us to find simple solutions for the two limiting cases: fast and slow frequency changes. Indeed, Eq. (4.159) is a first-order linear equation and can be solved by the method of characteristics (see, e.g., Courant and Hilbert, 1989, Chapter II, § 1): using the ansatz $W(x, y, t) \equiv W(x(t), y(t))$ we find from (4.159) the *characteristic equations*,

$$\frac{dx}{dt} = \frac{\dot\omega}{2\omega}x - \omega y; \quad \frac{dy}{dt} = \omega x - \frac{\dot\omega}{2\omega}y. \tag{4.160}$$

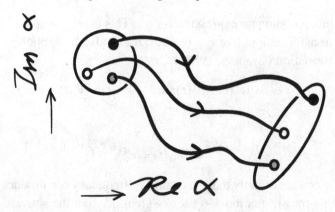

Fig. 4.15. Unitary evolution of the Wigner function in the absence of decoherence. The initial value of $W(\alpha, \alpha^*)$ is dragged along the characteristic.

Solving them, one finds the characteristic curves, or trajectories, $x(t)$, $y(t)$, determined by the set of initial conditions, $x(0)$, $y(0)$. Equivalently, at any given time t for any x, y we can go back along the characteristic curve passing through this point and find its point of origin, x_0, y_0. Therefore the Wigner function $W(t)$ can be presented as $W(x_0(x, y, t), y_0(x, y, t))$: its evolution in time is due to the initial distribution being "dragged" along the characteristic curves (Fig. 4.15). This is a unitary process: the total phase volume occupied by the system obviously will not change during such an evolution. Of course, as soon as the non-unitary processes are included, this picture would break down, as we have duly noted above.

In the slow limit, the time derivative of the oscillator frequency in Eq. (4.160) can be neglected, and

$$\frac{dx}{dt} \approx -\omega(t)y; \quad \frac{dy}{dt} \approx \omega(t)x. \tag{4.161}$$

These equations obviously define circular characteristics, so that under an adiabatic change of the oscillator frequency the Wigner function will simply rotate as a whole with the frequency $\omega(t)$.

In the fast limit, instead the terms with $\dot{\omega}$ dominate, and the equations (4.160) uncouple:

$$\frac{dx}{dt} \approx \frac{\dot{\omega}}{2\omega}x; \quad \frac{dy}{dt} \approx -\frac{\dot{\omega}}{2\omega}y. \tag{4.162}$$

The solutions are $x(t) = x(0)[\omega(t)/\omega(0)]^{1/2}$ and $y(t) = y(0)[\omega(t)/\omega(0)]^{-1/2}$, and it is straightforward to check that, starting from a coherent state, we will indeed arrive at a squeezed state (or we could unsqueeze a squeezed state if so inclined).

5

Noise and decoherence

How is't with me, when every noise appals me?
William Shakespeare, *Macbeth*, Act II, Scene ii

5.1 Quantum noise

5.1.1 Autocorrelation function and spectral density

Any observable quantity is generally subject to fluctuations (due to uncontrollable interactions with the rest of the world, and – in the quantum case – also due to the intrinsic quantum randomness expressed in the Heisenberg uncertainty relations). These fluctuations are intimately related to the eventual loss of quantum coherence by the system, and are, therefore, of primary interest for our subject: it is important to know what influences the decoherence time of the system in question and how it can be extended, even though the final outcome is known beforehand.

The standard and convenient description of fluctuations is given by the auto-correlation function of the quantity in question, which we have already used in Section 1.2 (Eq. (1.41)).[1] Generally,

$$K_A(t+\tau, t) = \langle A(t+\tau)A(t)\rangle \equiv \text{tr}\,[A(t+\tau)A(t)\rho(t)]. \qquad (5.1)$$

For *stationary* fluctuations, which is the only (and sufficiently general) case we will be discussing, the autocorrelator only depends on the time difference,

$$K_A(t+\tau, t) \equiv K_A(\tau), \qquad (5.2)$$

which does not necessarily imply that the density matrix itself is time-independent (though this is usually the case). If the average value of $A(t)$ is zero (or if we subtract

---

[1] The mathematical theory of random processes and random fields, which underpins our discussion of noise, is described with a varying degree of scope, rigour and detail in many textbooks, e.g., Buckingham (1983), Rytov et al. (1987, 1988, 1989), Gardiner (2003), Gardiner and Zoller (2004).

it from $A(t)$ and work with the *autocovariance* instead of the autocorrelator), then $K_A(t+\tau,t)$ will usually become negligible for $|\tau| \gg \tau_c$, where τ_c is the so-called *correlation time*.

Since the observable is an operator, $[A(t+\tau), A(t)] \neq 0$, and $A(t+\tau)A(t)$ can be written as the sum of a Hermitian and an anti-Hermitian operator:

$$A(t+\tau)A(t) = \frac{1}{2}\{A(t+\tau), A(t)\} + i\frac{1}{2i}[A(t+\tau), A(t)]. \tag{5.3}$$

The autocorrelator $K_A(t+\tau, t)$ is, therefore, a complex number, expressed through the symmetrized and antisymmetrized correlators:

$$K_A(t+\tau,t) = \mathrm{Re}K_A + i\mathrm{Im}K_A \equiv K_{A,s}(t+\tau,t) + K_{A,a}(t+\tau,t); \tag{5.4}$$

$$K_{A,s(a)}(t+\tau,t) = \frac{\langle A(t+\tau)A(t) \pm A(t)A(t+\tau)\rangle}{2}.$$

The symmetrized autocorrelator survives in the classical limit and is related to a directly measurable quantity, the classical spectral density of the fluctuations. Consider, for example, a classical electric circuit, where the current or voltage (whichever we choose as the observable $A(t)$) is continually monitored. In the stationary case the instantaneous value will fluctuate around some average value $\langle A \rangle$, which we, as before, assume to be zero to simplify the expressions. After recording $A(t)$ during the time interval of length T we can calculate the Fourier transform,

$$\tilde{A}_T(\omega; t_0) = \int_{t_0}^{t_0+T} dt\, e^{i\omega t} A(t) \equiv \int_{-\infty}^{\infty} dt\, e^{i\omega t} A_T(t; t_0); \tag{5.5}$$

here $A_T(t; t_0) = A(t)\theta(t - t_0)\theta(t_0 + T - t)$ is the "windowed" variable. From Parseval's theorem of Fourier analysis, for any two absolutely integrable functions[2]

$$\int_{-\infty}^{\infty} dt\, f(t)g^*(t) = \int_{-\infty}^{\infty} \frac{d\omega}{2\pi} \tilde{f}(\omega)\tilde{g}^*(\omega). \tag{5.6}$$

Substituting $f(t) = g(t) = A_T(t; t_0)$, we find

$$\int_{-\infty}^{\infty} dt\, A_T^2(t; t_0) = \int_{-\infty}^{\infty} \frac{d\omega}{2\pi} |\tilde{A}_T(\omega; t_0)|^2. \tag{5.7}$$

Our use of the windowed variables (i.e., a finite observation time) guarantees the absolute integrability required by Eq. (5.6). The expression on the left determines the total energy of the fluctuations during the observation time T. (E.g., if a resistor

---

[2] A function is absolutely integrable on an (finite or infinite) interval if the integral of its absolute value over that interval is finite.

R is included in the circuit, the instantaneous fluctuation power dissipated on this resistor is $I^2(t)R$ or $V^2(t)/R$.) The average power is then characterized by

$$\frac{1}{T}\left\langle \int_{-\infty}^{\infty} dt\, A_T^2(t; t_0) \right\rangle = \int_{-\infty}^{\infty} \frac{d\omega}{2\pi} \frac{\left\langle \left| \tilde{A}_T(\omega; t_0) \right|^2 \right\rangle}{T}. \tag{5.8}$$

Taking the limit $T \to \infty$ we will, due to stationarity, lose the dependence on t_0 and arrive at the spectral density of the fluctuations

$$S_A(\omega) = \lim_{T \to \infty} \frac{\left\langle \left| \tilde{A}_T(\omega; t_0) \right|^2 \right\rangle}{T}. \tag{5.9}$$

This is a directly measurable quantity: connecting to the circuit a classical spectrometer (i.e., a classical quadratic detector connected through a band-pass filter), we can measure the power of the signal, proportional to $S_A(\omega_0)\Delta\omega$, where ω_0 is the filter centre frequency, and $\Delta\omega$ is the filter bandwidth. In keeping with our notation, we should take into account both positive and negative filter bands, centred at $\pm|\omega_0|$.

On the other hand, choosing in (5.6) $f(t) = A_T(t + \tau; t_0)$ and $g(t) = A_T(t; t_0)$, we see that

$$\lim_{T \to \infty} \frac{1}{T} \int_{-\infty}^{\infty} dt\, \langle A_T(t + \tau; t_0) A_T(t; t_0) \rangle = \int_{-\infty}^{\infty} \frac{d\omega}{2\pi} e^{-i\omega\tau} \lim_{T \to \infty} \frac{\left\langle \left| \tilde{A}_T(\omega; t_0) \right|^2 \right\rangle}{T}, \tag{5.10}$$

that is, the *Wiener–Khinchin theorem*

$$K_A(\tau) = \int_{-\infty}^{\infty} \frac{d\omega}{2\pi} e^{-i\omega\tau} S_A(\omega), \tag{5.11}$$

which relates the autocorrelation function to the spectral density of the fluctuations.[3] Note that since the classical autocorrelator is real, the spectral density is an even function of frequency, which allows us to consider it only for positive frequencies (as is often done in electrical engineering[4]).

---

[3] See, e.g., Buckingham (1983), § 2.5, Gardiner (2003), § 1.4.2; a rigorous mathematical treatment can be found in Richtmyer (1978, § 4.6).

[4] For the "electrical engineering" definition of spectral density, using the linear frequency $f = \omega/2\pi$ as well, we would have instead

$$K_A(\tau) = \int_0^{\infty} df\, S_A^{(EE)}(f) \cos 2\pi f\tau; \quad S_A^{(EE)}(f) = 4 \int_0^{\infty} d\tau\, K_A(\tau) \cos 2\pi f\tau.$$

If we rewrite Eq. (5.11) as[5]

$$K_A(\tau) = \langle A(\tau) A(0) \rangle = \left\langle \int_{-\infty}^{\infty} \frac{d\omega}{2\pi} \widetilde{A}(\omega) e^{-i\omega\tau} \int_{-\infty}^{\infty} \frac{d\omega'}{2\pi} \widetilde{A}^*(\omega') \right\rangle$$

$$= \int_{-\infty}^{\infty} \frac{d\omega}{2\pi} \int_{-\infty}^{\infty} \frac{d\omega'}{2\pi} \langle \widetilde{A}(\omega) \widetilde{A}^*(\omega') \rangle e^{-i\omega\tau},$$

and compare it with Eq. (5.11), we arrive at another expression for the spectral density, which is sometimes more convenient than (5.9):

$$2\pi \delta(\omega - \omega') S_A(\omega) = \langle \widetilde{A}(\omega) \widetilde{A}^*(\omega') \rangle, \tag{5.12}$$

where the right-hand side is to be understood as $\lim_{T \to \infty} \langle \widetilde{A}_T(\omega) \widetilde{A}_T^*(\omega') \rangle$.

The *white noise*, for which $S_A(\omega)$ is constant, and correspondingly $K_A(\tau) \propto \delta(\tau)$, is a convenient mathematical abstraction of a stationary random process with a short correlation time. For white noise, the mean square of the fluctuations, $\overline{A^2} = K(0) = \int_{-\infty}^{\infty} \frac{d\omega}{2\pi} S(\omega)$, would diverge. Real physical processes will always have some minimal correlation time (respectively, cutoff frequency), but with this caveat white noise is a very useful model.

In the quantum case the spectral density of fluctuations can be introduced directly by inverting Eq. (5.11):

$$S_A(\omega) = S_{A,s}(\omega) + S_{A,a}(\omega) = \int_{-\infty}^{\infty} d\tau e^{i\omega\tau} K_A(\tau); \tag{5.13}$$

$$S_{A,s(a)}(\omega) = \int_{-\infty}^{\infty} d\tau e^{i\omega\tau} K_{A,s(a)}(\tau). \tag{5.14}$$

We have already seen that the symmetric part of the quantum spectral density, $S_{A,s}(\omega) = S_{A,s}(-\omega)$, enters the fluctuation–dissipation theorem, Eq. (1.46), which relates the intensity of equilibrium fluctuations to the response of the system to a small external perturbation. In the classical limit the symmetric part of the correlator K_A obviously reduces to the classical correlator, and therefore S_A tends to the classical spectral density (5.9), and $S_{A,a}/S_{A,s} \to 0$.

From the definition (5.13) one can readily obtain the equilibrium spectral density for, e.g., an LC circuit with resonant frequency ω_0. First, express the observable through the creation/annihilation operators. For example, if A is the flux (or current) in the circuit, then $A \propto (a + a^\dagger) \propto \hat{X}$, the quadrature operator (cf. Eqs (4.4, 4.96)); if A is the voltage (or charge), then $A \propto (a - a^\dagger)/i \propto \hat{Y}$. Let us for simplicity take $A = \hat{X} = (a + a^\dagger)/2$. As usual when talking about equilibrium without going into

---

[5] Strictly speaking, we would have to replace $A(t)$ with the "windowed" variable $A_T(t)$, to make the integrals converge, and then take the limit $T \to \infty$.

specifics, we assume that the system is weakly coupled to the thermal reservoir (bath) at fixed temperature T. For an LC circuit this role may be played by the ambient black-body radiation.

Since for an unperturbed oscillator $a(t) = a\exp[-i\omega_0 t]$, $a^\dagger(t) = a^\dagger \exp[i\omega_0 t]$, and in equilibrium at temperature T the average $\langle a^\dagger a \rangle$ is the Bose distribution, $n_B(\hbar\omega_0/k_B T) = (\exp(\hbar\omega_0/k_B T) - 1)^{-1}$, then the correlation function is immediately found:

$$K_X(\tau) = \langle \hat{X}(\tau)\hat{X}(0) \rangle = [2n_B(\hbar\omega_0/k_B T) + 1]\cos\omega_0\tau - i\sin\omega_0\tau \quad (5.15)$$
$$\equiv K_{X,s}(\tau) + K_{X,a}(\tau).$$

Taking the Fourier transform, we find

$$S_{X,s}(\omega) = [2n_B(\hbar\omega_0/k_B T) + 1]\pi(\delta(\omega - \omega_0) + \delta(\omega + \omega_0)); \quad (5.16)$$
$$S_{X,a}(\omega) = \pi(\delta(\omega - \omega_0) - \delta(\omega + \omega_0)). \quad (5.17)$$

Comparing the total spectral density for positive and negative frequencies,

$$S_X(|\omega|) = 2\pi[n_B(\hbar\omega_0/k_B T) + 1]\delta(\omega - \omega_0);$$
$$S_X(-|\omega|) = 2\pi n_B(\hbar\omega_0/k_B T)\delta(-|\omega| + \omega_0), \quad (5.18)$$

we notice that the former describes the emission of quanta by the system into the thermal reservoir, while the latter describes their absorption. The asymmetry of the quantum spectral density has, therefore, the same origin as the asymmetry of quantum emission and absorption coefficients: the presence of spontaneous emission, represented by the unity in the brackets and first reflected in Einstein's A and B coefficients (see, e.g., Saleh and Teich, 2007, §13.3). At high temperatures, $k_B T \gg \hbar\omega_0$, the unity can be neglected compared to $n_B \gg 1$, the asymmetry is wiped out, and we return to the classical, frequency-symmetric spectral density. At zero temperature $n_B = 0$, and the asymmetry of $S_X(\omega)$ simply means that energy cannot be absorbed from, but can be emitted into, the vacuum state of the electromagnetic field.

From (5.17) we see that

$$S_{X,s}(\omega) = [2n_B(\hbar|\omega|/k_B T) + 1]\sin\omega\, S_{X,a}(\omega) = \coth\frac{\hbar\omega}{2k_B T}\, S_{X,a}(\omega). \quad (5.19)$$

This is a particular case of the general relation (1.44), valid in equilibrium, which we derived when proving the fluctuation–dissipation theorem, Eq. (1.46). We recall that the antisymmetric part of the spectral density (and autocorrelator) are related to the imaginary part of the linear response of the system to an external perturbation (i.e., to damping).

It is a widespread practice in electrical engineering to use (5.16) or its classical counterpart in order to approximate the spectral density of fluctuations at a given frequency, even in a clearly nonequilibrium case:

$$S_{X,s}(\omega)\Delta\omega = \pi \left[2n_B \left(\frac{\hbar\omega}{k_B T^*} \right) + 1 \right] = \pi \coth \frac{\hbar\omega}{2k_B T^*}. \tag{5.20}$$

The "temperature"

$$T^*(\omega) = \frac{\hbar\omega}{2k_B \operatorname{arccoth}(S_{X,s}(\omega)\Delta\omega/\pi)}, \tag{5.21}$$

which follows from such a fitting, is called the *noise temperature* and is just a parameter, which characterizes the noise in a given frequency band. In the classical limit, when $S_{X,s}(\omega)\Delta\omega/\pi \gg 1$, this simplifies to

$$T^*(\omega) \approx \frac{\hbar\omega}{2k_B} \frac{S_{X,s}(\omega)\Delta\omega}{\pi} \gg \frac{\hbar\omega}{k_B}. \tag{5.22}$$

Another kind of effective temperature can be introduced if we want to characterize the positive–negative frequency asymmetry of the noise spectrum. From (5.18) we find that $S_X(|\omega|)/S_X(-|\omega|) = \exp[\hbar|\omega|/k_B T]$, and, therefore, one can introduce (see Clerk et al., 2010)

$$T^{**}(\omega) = \frac{\hbar\omega}{k_B \ln[S_X(\omega)/S_X(-\omega)]}. \tag{5.23}$$

When $S_X(\omega) \approx S_X(-\omega)$ (i.e., $S_{X,s}(\omega) \gg S_{X,a}(\omega)$), this reduces to

$$T^{**}(\omega) \approx \frac{\hbar\omega}{2k_B} \frac{S_{X,s}(\omega)}{S_{X,a}(\omega)}. \tag{5.24}$$

5.1.2 Physical meaning of quantum spectral density of fluctuations

We have seen that the classical spectral density can be directly measured, at least in the idealized case. The same turns out to be true in the case of the quantum spectral density. Of course, one must employ some kind of *quantum* spectrometer, that is, a quantum system, which can be coupled to the noise source in such a way that the occupation probabilities of the spectrometer's energy eigenstates depend on the noise, while the influence of the noise on the eigenstates themselves can be neglected (see, e.g., Schwinger, 1961; Dykman, 1978; Aguado and Kouwenhoven, 2000; Gavish et al., 2000; Schoelkopf et al., 2003). The positive-(negative-)frequency component of the spectral density is then revealed in the stationary relaxation (respectively, excitation) rate of the system, which one can, in principle, measure. We will follow here Schoelkopf et al. (2003) and Clerk et al. (2010).

The Hamiltonian of a quantum spectrometer can be written as

$$H(t) = H_0 - \hat{f}(t)\hat{B}, \tag{5.25}$$

where H_0 is the unperturbed spectrometer Hamiltonian, and the second term represents the coupling of the qubit to the noisy system. Unlike in Eq. (1.36), the generalized force $\hat{f}$ is now an operator, which represents the fluctuations in the noisy system, which influence the qubit. We are not concerned with the dynamics of $\hat{f}(t)$, beyond its correlation function and spectral density, which are assumed to be unperturbed by the interaction with the spectrometer. We therefore do not need to include the corresponding term in (5.25). We also assume that the spectrometer has a discrete energy spectrum, $\{|m\rangle\}$, $m = 0, 1, \ldots$

In the interaction representation with respect to the unperturbed spectrometer Hamiltonian H_0, the spectrometer's wave function in the first order is given by

$$|\psi(t)\rangle = |\psi(0)\rangle + \frac{\mathrm{i}}{\hbar} \int_0^t \mathrm{d}\tau\, \hat{f}(\tau)\hat{B}(\tau)|\psi(0)\rangle, \tag{5.26}$$

where the operator $\hat{B}$ acquires time dependence, $\hat{B}(\tau) = \mathrm{e}^{\mathrm{i}H_0\tau/\hbar}\hat{B}\mathrm{e}^{-\mathrm{i}H_0\tau/\hbar}$. The probability amplitude finding the spectrometer at time t in the energy eigenstate $|m\rangle$ is then

$$\langle m|\psi(t)\rangle = \langle m|\psi(0)\rangle + \frac{\mathrm{i}}{\hbar} \int_0^t \mathrm{d}\tau\, \langle m|\hat{f}(\tau)\hat{B}(\tau)|\psi(0)\rangle. \tag{5.27}$$

Taking $|\psi(0)\rangle = |n\rangle$, $n \neq m$, we find the transition probability

$$
\begin{aligned}
P_{n\to m}(t) &= \left\langle \left. |\langle m|\psi(t)\rangle|^2 \right|_{|\psi(0)\rangle=|n\rangle} \right\rangle_{\text{noise}} \\
&= \frac{1}{\hbar^2} \int_0^t \int_0^t \mathrm{d}\tau'\mathrm{d}\tau\, \langle n|\hat{B}(\tau')|m\rangle\langle m|\hat{B}(\tau)|n\rangle \left\langle \hat{f}(\tau')\hat{f}(\tau) \right\rangle \tag{5.28} \\
&= \frac{1}{\hbar^2} \int_0^t \int_0^t \mathrm{d}\tau\mathrm{d}\tau'\, \mathrm{e}^{-\mathrm{i}(E_m-E_n)(\tau'-\tau)/\hbar} \left| \langle m|\hat{B}|n\rangle \right|^2 K_f(\tau'-\tau).
\end{aligned}
$$

Here we have assumed the stationarity of the fluctuations of $\hat{f}(t)$, Eq. (5.2).[6] Now, changing the variables of integration to $\bar{\tau} = (\tau' + \tau)/2$, $\xi = \tau' - \tau$, we get

$$P_{n \to m}(t) = \frac{1}{\hbar^2} \left| \langle m | \hat{B} | n \rangle \right|^2 \left[\int_0^{t/2} d\bar{\tau} \int_{-2\bar{\tau}}^{2\bar{\tau}} d\xi \, e^{-i(E_m - E_n)\xi/\hbar} K_f(\xi) \right.$$

$$\left. + \int_{t/2}^t d\bar{\tau} \int_{-2(t-\bar{\tau})}^{2(t-\bar{\tau})} d\xi \, e^{-i(E_m - E_n)\xi/\hbar} K_f(\xi) \right]. \tag{5.29}$$

For time intervals t much longer than the noise correlation time, but short enough for the perturbation scheme (5.26) to be still applicable, we can extend the limits of integration over ξ to $\pm\infty$. Then, recalling the definition (5.13), we arrive at the expression for the *transition rate* in the spectrometer,

$$\Gamma_{n \to m} \equiv \frac{d}{dt} P_{n \to m}(t) = \frac{1}{\hbar^2} \left| \langle m | \hat{B} | n \rangle \right|^2 S_f \left(-\frac{E_m - E_n}{\hbar} \right). \tag{5.30}$$

Now it becomes evident that the quantum asymmetry between the negative- and positive-frequency values of the spectral density is reflected in the asymmetry between the excitation and relaxation rates between the same two levels in the spectrometer: if $m > n$, $E_m - E_n = \hbar\Omega > 0$, then

$$\Gamma_\uparrow = \Gamma_{n \to m} = \frac{1}{\hbar^2} \left| \langle m | \hat{B} | n \rangle \right|^2 S_f(-\Omega);$$

$$\Gamma_\downarrow = \Gamma_{m \to n} = \frac{1}{\hbar^2} \left| \langle m | \hat{B} | n \rangle \right|^2 S_f(\Omega). \tag{5.31}$$

This quite agrees with our earlier conclusions from Eq. (5.18). The ratio

$$\frac{\Gamma_\uparrow}{\Gamma_\downarrow} = \frac{S_f(-\Omega)}{S_f(\Omega)} \equiv e^{-\frac{\hbar\Omega}{k_B T^{**}}} \tag{5.32}$$

is independent of the spectrometer and only reflects the quantum asymmetry of the noise spectrum.

Let us apply Eq. (5.30) to two particularly relevant cases, when the role of the quantum spectrometer is played by a two-level system (e.g., a qubit) or a linear oscillator (e.g., an LC circuit). A *qubit spectrometer*, with level splitting $\hbar\Omega$, can be coupled to the noise source via

$$\hat{B} = g\sigma_x, \tag{5.33}$$

[6] We have written in Eq. (5.28) $\langle n | \hat{B}(\tau') | m \rangle \langle m | \hat{B}(\tau) | n \rangle$, and not $\langle m | \hat{B}(\tau) | n \rangle \langle n | \hat{B}(\tau') | m \rangle$ (which would also exchange $\hat{f}(\tau)$ and $\hat{f}(\tau')$), because the initial state of the system is $|n\rangle$, and the evolution represented by the sequence of operators $(\hat{f}(\tau') \hat{B}(\tau'))^\dagger \hat{f}(\tau) \hat{B}(\tau)$ must start and end there. A more coherent answer would require an unjustifiably long excursion into the field of nonequilibrium (Keldysh) Green's functions (see, e.g., Zagoskin, 1998, §3.4).

where g is the appropriate coupling strength, dependent on the character of the qubit and of the noise operator $\hat{f}(t)$. Then the matrix element $\langle 0|\hat{B}|1\rangle = \langle 1|\hat{B}|0\rangle = g$, and

$$\Gamma_\uparrow = \Gamma_{0\to 1} = \frac{g^2}{\hbar^2} S_f(-\Omega); \quad \Gamma_\downarrow = \Gamma_{1\to 0} = \frac{g^2}{\hbar^2} S_f(\Omega). \qquad (5.34)$$

For an oscillator spectrometer with the resonant frequency ω_0 we can have, for example,

$$\hat{B} = g(a + a^\dagger), \qquad (5.35)$$

which within our conventions, Eqs. (4.4, 4.96), would correspond to the noise operator coupled to the flux, or current, in the LC circuit; that is, (5.35) is appropriate if $\hat{f}(t)$ is proportional to a fluctuating current or magnetic flux. In the case of a fluctuating charge we should use $(a - a^\dagger)/i$ instead.[7] The only nonzero matrix elements will be

$$\langle n|\hat{B}|n+1\rangle = \langle n+1|\hat{B}|n\rangle = g\sqrt{n+1}; \quad n = 0, 1, \ldots, \qquad (5.36)$$

and, therefore, similar to (5.34),

$$\Gamma_{n,\uparrow} = \Gamma_{n\to n+1} = \frac{g^2}{\hbar^2}(n+1)S_f(-\omega_0); \quad \Gamma_{n,\downarrow} = \Gamma_{n+1\to n} = \frac{g^2}{\hbar^2}(n+1)S_f(\omega_0). \qquad (5.37)$$

A more detailed discussion of quantum spectrometers and their realizations can be found in Schoelkopf et al. (2003) and Clerk et al. (2010, Section II and Appendix B). Here it was sufficient to satisfy ourselves that both positive- and negative-frequency parts of the spectral density relate to the processes of emission (respectively, absorption) of quanta by the spectrometer and can be, in principle, directly measured.

5.1.3 *Quantum regression theorem

Direct calculation of quantum autocorrelators (5.1) is straightforward if the system is closed, and, therefore, its evolution is unitary. Then we can use the Heisenberg representation, in which the density matrix is time-independent, and the operators depend on time via

$$A(t) = U^\dagger(t)AU(t) = e^{iHt/\hbar}Ae^{-iHt/\hbar} \qquad (5.38)$$

---

[7] Of course, this is all purely a question of convention.

(if the full Hamiltonian is explicitly time-independent). Nevertheless, we are most interested in the evolution of open systems, where the system interacts with an uncontrollable environment (bath, reservoir), which needs to be traced out. As a result, the density matrix of the system itself, $\rho_S(t)$, will undergo non-unitary time evolution (see § 1.3). In order to obtain from (5.1) an expression which would contain $\rho_S = \text{tr}_E[\rho]$ instead of the full density matrix ρ of the system and its environment, we will write it as (Gardiner and Zoller, 2004, §5.2)

$$K_A(\tau) = \text{tr}\left[U^\dagger(t+\tau)AU(t+\tau)U^\dagger(t)A\left(U(t)\rho U^\dagger(t)\right)U(t)\right]$$

$$= \text{tr}\left[AU(\tau)A\rho(t)U^\dagger(\tau)\right] = \text{tr}_S\left[A\,\text{tr}_E\left(U(\tau)A\rho(t)U^\dagger(\tau)\right)\right]. \quad (5.39)$$

The last step is legitimate, since the operator A acts in the Hilbert space of the system, not of the environment. Now for the quantity $\mathcal{A}(\tau, t) \equiv \left(U(\tau)A\rho(t)U^\dagger(\tau)\right)$ we can write the equation of motion,

$$i\hbar\frac{\partial}{\partial\tau}\mathcal{A}(\tau, t) = [H, \mathcal{A}(\tau, t)], \quad (5.40)$$

and then, making the Markov approximation (as when deriving Eq. (1.53)), reduce it to the master equation for the quantity $\text{tr}_E[\mathcal{A}(\tau, t)]$, where the degrees of freedom of the environment have been traced out. The nonunitary evolution of an operator O in the Hilbert space of the system between the moments t_i and t_f is described by the operator $\mathcal{S}(t_f, t_i)[O]$ (1.75), which provides the solution to the master equation. Therefore, finally

$$K_A(\tau) = \text{tr}_S\left(A\mathcal{S}(t+\tau, t)[A\rho_S(t)]\right). \quad (5.41)$$

Note that the evolution operator applies to the whole combination $A\rho_S(t)$. If we wanted multipoint correlators of the kind $\langle A(t_1)B(t_2)\ldots C(t_n)\rangle$, we would find

$$\langle A(t_1)B(t_2)\ldots C(t_{n-1})D(t_n)\rangle$$

$$= \text{tr}_S\left(A\mathcal{S}(t_1, t_2)[B\mathcal{S}(t_2, t_3)[\ldots C\mathcal{S}(t_{n-1}, t_n)[D\rho_S(t_n)]\ldots]]\right). \quad (5.42)$$

The evolution operator $\mathcal{S}$ depends on the details of the environment, and its calculation can be rather involved. A welcome simplification is provided by the *quantum regression theorem* (Lax, 1963, 1968; Gardiner and Zoller, 2004, § 5.2.3), which allows us to find correlators from the known (e.g., phenomenological) equations for the expectation values of the observables. Suppose that the expectation values of the set of observables $A_1, A_2, \ldots$ satisfy *linear* equations

$$\frac{\text{d}}{\text{d}t}\langle A_i(t)\rangle = \sum_j G_{ij}(t)\langle A_j(t)\rangle. \quad (5.43)$$

Since

$$\langle A_i(t)\rangle = \text{tr}_S(A_i\rho_S(t)) = \text{tr}_S(A_i\mathcal{S}(t,t_0)[\rho_S(t_0)]), \qquad (5.44)$$

then

$$\frac{d}{dt}\langle A_i(t)\rangle = \text{tr}_S\left(A_i\frac{\partial}{\partial t}\mathcal{S}(t,t_0)[\rho_S(t_0)]\right), \quad t_0 < t. \qquad (5.45)$$

The trace is a linear operation, and

$$\frac{\partial}{\partial t}A_i\mathcal{S}(t,t_0)[\rho_S(t_0)] = \sum_j G_{ij}(t)A_i\mathcal{S}(t,t_0)[\rho_S(t_0)]. \qquad (5.46)$$

Therefore, for the correlator

$$K_{ij}(t+\tau,t) = \langle A_i(t+\tau)A_j(t)\rangle \equiv \text{tr}_S\left(A_i\mathcal{S}(t+\tau,t)[A_j\rho_S(t)]\right) \qquad (5.47)$$

we find

$$\frac{\partial}{\partial\tau}K_{ij}(t+\tau,t) = \text{tr}_S\left(\frac{\partial}{\partial\tau}A_i\mathcal{S}(t+\tau,t)[A_j\rho_S(t)]\right)$$

$$= \text{tr}_S\left(\sum_k G_{ik}(t+\tau)A_k\mathcal{S}(t+\tau,t)[A_j\rho_S(t)]\right)$$

$$= \sum_k G_{ik}(t+\tau)K_{kj}(t+\tau,t). \qquad (5.48)$$

The theorem can be applied as well when the nonlinear equations for the expectation values can be linearized. Note that the initial values for Eqs (5.48), the one-time correlators $K_{ij}(t,t)$, must be obtained from other considerations (for example, approximated by the equilibrium mean square values).

As a simple classical illustration, consider the electric charge on a leaky capacitor (or the velocity of a particle in a viscous liquid): $A(t) = Q(t)$ or $A(t) = v(t)$. The expectation value $\langle A(t)\rangle$ satisfies the phenomenological equation

$$\frac{d\langle A(t)\rangle}{dt} = -\gamma\langle A(t)\rangle. \qquad (5.49)$$

Therefore, from Eq. (5.48), we find for the correlator

$$\frac{dK_A(t)}{dt} = -\gamma K_A(t); \quad K_A(t) = K_A(0)e^{-\gamma|t|}, \qquad (5.50)$$

where we took into account the temporal symmetry of the correlator. The spectral density of fluctuations is

$$S_A(\omega) = K_A(0)\frac{2\gamma}{\omega^2+\gamma^2}. \qquad (5.51)$$

This is the typical *Lorentzian* spectral density, corresponding to the exponential decay of correlations. The factor $K_A(0)$ remains undetermined. For equilibrium charge fluctuations on a capacitor C at a high enough temperature T, it will follow from the equipartition theorem, $K_A(0) \equiv \overline{\delta Q^2} = 2C \times k_B T/2$. Since the relaxation rate in this case is $\gamma = 1/RC$, where R is the leakage resistance, we find for the spectral density of fluctuations of charge and of voltage ($V = Q/C$) on the capacitor, respectively

$$S_Q(\omega) = \frac{2k_B T R C^2}{1 + (\omega RC)^2}; \quad S_V(\omega) = \frac{2k_B T R}{1 + (\omega RC)^2}. \tag{5.52}$$

For small frequencies $\omega \ll 1/RC$ the frequency dependence can be neglected, and we arrive at $S_Q = 2k_B T R C^2$, and $S_V = 2k_B T R$. In the next section we will consider equilibrium thermal noise, represented by these expressions, in more detail. Here note that the equilibrium noise dependence on the relaxation rate in Eqs (5.51) and (5.52), following from the phenomenological equation (5.49), is actually the result of the general fluctuation–dissipation theorem (1.46).

5.2 Noise sources in solid-state systems

5.2.1 Thermal noise

In the previous section we have seen that since the relaxation and excitation rates of a qubit are proportional to the spectral density of noise, a qubit can be used as a quantum noise spectrometer. This also means, of course, that qubits subject to noise eventually lose their quantum coherence, and from this point of view it is very important to know what kinds of noise are prevalent in solid-state devices, in order to minimize their influence and extend the time for the quantum coherent evolution of the system.

Solid-state devices are many and various, and it is impossible to cover here in any detail this large subject, about which many books have been written (e.g., Buckingham, 1983; Van Der Ziel, 1986). We will concentrate on the three most important, most fundamental and most ubiquitous kinds of noise, viz. thermal noise, shot noise and 1/f-noise.

Thermal noise, also known as *Johnson* or *Johnson–Nyquist* noise, was first observed by Johnson (1927, 1928) as voltage fluctuations, proportional to the resistance and temperature, in a variety of resistors (metallic, electrolytic, "India ink on paper", carbon filaments, etc.). Nyquist (1928) explained the phenomenon as the manifestation of universal equilibrium fluctuations in the circuit. He derived an expression for its spectral density from thermodynamic considerations, proving the so-called *Nyquist's theorem*, which is actually a special case of the fluctuation–dissipation theorem, Eq. (1.46). We can, therefore, directly use (1.46)[8]:

$$S_{A,s}(\omega) = \hbar \operatorname{Im}\chi(\omega) \coth(\hbar\omega/2k_B T).$$ (5.53)

This is a rigorous quantum relation, which completely characterizes equilibrium noise. The antisymmetric part of the spectral density can always be restored from $S_{A,s}(\omega)$ using relation (1.44). The only remaining problem is to identify the appropriate generalized susceptibility $\chi(\omega)$.

We have introduced $\chi(\omega)$ (1.40) as the ratio of the change in the average value of the observable A at frequency ω to the (non-operator) generalized force[9] at the same frequency:

$$\chi(\omega) = \overline{\Delta A}(\omega)/f(\omega).$$ (5.54)

The average energy dissipation rate in the system due to the action of the generalized force is $Q(t) = \langle dH(t)/dt \rangle = -\langle B \rangle d f(t)/dt$; therefore, $f(\omega) = Q(\omega)/(i\omega\langle B \rangle)$, and

$$\chi(\omega) = \frac{i\omega \overline{\Delta A}(\omega)\langle B \rangle}{Q(\omega)}.$$ (5.55)

Consider, for example, equilibrium fluctuations of an electric current in a circuit. In the presence of an infinitesimal external voltage $V(t)$ the average value of current (a quantum operator) will be $I(t) \equiv \langle \hat{I} \rangle$, and its Fourier component

$$\overline{\Delta I}(\omega) \equiv I(\omega) = Y(\omega)V(\omega),$$ (5.56)

where $Y(\omega) = 1/Z(\omega)$ is the *admittance* (and $Z(\omega)$ the *impedance*) of the circuit. The energy dissipation rate $Q = IV$, so that the generalized susceptibility

$$\chi(\omega) = i\omega Y(\omega) \equiv i\omega(G(\omega) + iB(\omega)).$$ (5.57)

Finally, we obtain for the equilibrium current noise

$$S_{I,s}(\omega) = \hbar\omega G(\omega) \coth(\hbar\omega/2k_B T).$$ (5.58)

Here $G(\omega)$ is the conductance of the circuit. In the classical limit, $k_B T \gg \hbar\omega$, this expression reduces to the Nyquist formula

$$S_{I,s}(\omega) = 2k_B T G(\omega)$$ (5.59)

(the "electrical engineering" version of this expression, where only positive frequencies are used, is $S_I^{(EE)}(f) = 4k_B T G$).

---

[8] Nyquist's original derivation is presented, e.g., in Buckingham (1983, Appendix 2).
[9] Since we are considering the fluctuations of the observable $A(t)$, not of the generalized force $f(t)$, as in Eq. (5.25), we can choose the latter to be a c-number function: $H(t) = H_0 - f(t)B$, where B is a time-independent operator.

For voltage fluctuations, we should start from the reaction of the average voltage to an external current,

$$\overline{\Delta V}(\omega) \equiv V(\omega) = Z(\omega)I(\omega), \tag{5.60}$$

which would, of course, yield a different susceptibility,

$$\chi(\omega) = i\omega Z(\omega) \equiv i\omega(R(\omega) + iX(\omega)), \tag{5.61}$$

and the noise spectrum

$$S_{V,s}(\omega) = \hbar\omega R(\omega)\coth(\hbar\omega/2k_BT); \tag{5.62}$$

$$S_{V,s}(\omega) = 2k_BTR(\omega), \quad k_BT \gg \hbar\omega. \tag{5.63}$$

In the frequency range where the frequency dependence of the impedance or admittance can be neglected, and where $\hbar\omega \ll k_BT$, the thermal noise is white noise.

The above expressions for thermal noise do not depend on the origin of the admittance or impedance entering the formulas.[10] For example, a 2DEG point contact will demonstrate current or voltage noise (depending on the experimental setup), determined by its quantum conductance, nG_Q, or resistance, $1/nG_Q$, even though there is no relaxation or dissipation in the point contact itself. In the general case of a quantum conductance calculated within the scattering approach, the explicit calculation (Büttiker, 1990) also yields in the appropriate limit Eq. (5.59) with $G = (2e^2/h)\mathrm{tr}(t^\dagger t)$, where t is the submatrix of the scattering matrix (3.41). The seeming paradox is due to the fact that the transport properties of the system are determined by one part of it (the bottleneck), while the equilibration occurs in the other (the reservoirs), which are spatially separated.

5.2.2 Classical shot noise

Out of equilibrium, there appear additional sources of fluctuations (sometimes called *excess noise*). One of the most important of them is *shot noise*, which appears whenever there is a flow of discrete entities (e.g., an electric current in a conductor or in a vacuum, photons arriving at a detector, cars driving along a highway). This phenomenon was first identified in the electric current flowing through a vacuum tube (Schottky, 1918; see also Buckingham, 1983; Van Der Ziel, 1986) as due to random, independent events of electron emission by the cathode, much like shot being poured into a box. Correspondingly, a classical observable $A(t)$, subject to

---

[10] For reference, $Y = G + iB$ is admittance, G conductance, B susceptance (permittance); $Z = R + iX$ is impedance, R resistance, X reactance. Most of these terms were coined by Heaviside (Bolotovskii, 1985).

shot noise fluctuations, can be presented as a sequence of pulses of the same shape and unit area, starting at random moments of time:

$$A(t) = a \sum_n f(t - t_n). \tag{5.64}$$

Each pulse corresponds to, e.g., the detector current triggered by a photon, or a charge transferred by an electron. The shape of $f(t)$ depends on the physical situation and can be anything, except that $f(t < 0) = 0$ (due to causality), and $f(t) \to 0$ for large t sufficiently fast (otherwise, it will not have a unit – or finite – area). If the events are independent (there are no photon correlations at the source; electron–electron interactions can be neglected during their flight through the vacuum tube; etc.), the random times t_n in (5.64) are described by the Poisson distribution (e.g., Buckingham, 1983, Appendix 1):

$$P(m, t) = \frac{(vt)^m}{m!} e^{-vt}, \tag{5.65}$$

where $P(m, t)$ is the probability that during the interval $[0, t]$ there will occur exactly m pulses, and v is the average number of pulses per unit time. Indeed,

$$\langle m \rangle = \sum_{m=0}^{\infty} m P(m, t) = vt \left(\sum_{m=1}^{\infty} \frac{(vt)^{m-1}}{(m-1)!} \right) e^{-vt} = vt. \tag{5.66}$$

It is also straightforward to show that the dispersion of the number of events also equals vt:

$$\langle (m - \langle m \rangle)^2 \rangle = \langle m \rangle = vt. \tag{5.67}$$

If we introduce the *Fano factor*, the ratio of the dispersion to the average value,

$$F = \frac{\langle m^2 \rangle}{\langle m \rangle}, \tag{5.68}$$

we see that for the Poisson process $F = 1$.

Using these properties, we see that the average value of the shot noise is

$$\bar{A} \equiv \langle A(t) \rangle = va \int_{-\infty}^{\infty} dt f(t) = va. \tag{5.69}$$

Here it is more convenient to directly find the spectral density of noise using (5.12), and then determine its autocorrelator from the Wiener–Khintchin theorem. From (5.64),

$$\tilde{A}(\omega) = a \sum_n \tilde{f}(\omega) e^{i\omega t_n}, \tag{5.70}$$

and, therefore,

$$\langle \tilde{A}(\omega)\tilde{A}^*(\omega')\rangle = a^2 \tilde{f}(\omega)\tilde{f}^*(\omega') \left[\left\langle \sum_n e^{i(\omega-\omega')t_n} \right\rangle + \left\langle \sum_n e^{i\omega t_n} \right\rangle \left\langle \sum_m e^{-i\omega' t_m} \right\rangle \right]$$

$$= 2\pi \nu a^2 \left| \tilde{f}(\omega) \right|^2 \delta(\omega - \omega') + a^2 \left| \tilde{f}(0) \right|^2 (2\pi \nu)^2 \delta(\omega)\delta(\omega').$$

$$(5.71)$$

(Here we took into account that $\langle \sum_{n=-\infty}^{\infty} \exp(i\omega t_n) \rangle = \int_{-\infty}^{\infty} dt\, \nu \exp(i\omega t)$.) Substituting this into (5.12), we find for the spectral density of shot noise[11]

$$S_A(\omega) = \nu a^2 \left| \tilde{f}(\omega) \right|^2 + 2\pi \nu^2 a^2 \left| \tilde{f}(0) \right|^2 \delta(\omega). \qquad (5.72)$$

Eq. (5.72) is a special case of *Carson's theorem* for the spectral density of pulse processes (Buckingham, 1983). The second term here simply reflects the nonzero average of the noise variable $A(t)$. The autocorrelation function of shot noise is then

$$K_A(\tau) = \nu a^2 \int\limits_{-\infty}^{\infty} \frac{d\omega}{2\pi} e^{-i\omega\tau} \left| \tilde{f}(\omega) \right|^2 + \nu^2 a^2 \left| \tilde{f}(0) \right|^2. \qquad (5.73)$$

We see that the spectral density of shot noise is determined by the shape of individual pulses. For structureless pulses (i.e., when $f(t) = \delta(t)$ and $\tilde{f}(\omega) = 1$) the above formulas reduce to

$$S_A(\omega) = \nu a^2 + 2\pi \nu^2 a^2 \delta(\omega) \equiv a\overline{A} + 2\pi (\overline{A})^2 \delta(\omega); \qquad (5.74)$$

$$K_A(\tau) = \nu a^2 \delta(\tau) + \nu^2 a^2 \equiv a\overline{A}\delta(\tau) + (\overline{A})^2. \qquad (5.75)$$

Note that the nontrivial part of the spectral density of shot noise (i.e., its non-static component) in Eqs (5.72) and (5.74) is proportional to the average value of the noisy variable. For example, the spectral density of the current shot noise is proportional to the average electric current through the system, $\overline{I} = \nu e$:

$$S_I(\omega \neq 0) = e\overline{I}. \qquad (5.76)$$

This is a characteristic property of shot noise, unlike thermal noise (which is current-independent)[12] and 1/f-noise (which is usually proportional to the square of the average current and will be discussed later in this section).

Noise described by Eqs (5.74) and (5.76) or their equivalents is, for obvious reasons, called Poissonian noise. If there are negative or positive correlations between

---

[11] This expression differs by a factor of two from the "electrical engineering" formula, where only positive frequencies are considered.

[12] Their frequency dependence is often indistinguishable, since both (5.59) and (5.76) have flat, white noise spectra.

the pulses (i.e., whether the pulses tend to bunch together or to hold their distance), the distribution of pulses is described by what is called *sub-Poissonian* (respectively, *super-Poissonian*) statistics, leading to sub- or super-Poissonian shot noise. For the electric current, e.g., we will have

$$S_I(\omega \neq 0) = eF\bar{I}, \tag{5.77}$$

where F is the Fano factor.

5.2.3 *Quantum shot noise*

In the classical case the non-unity value of the Fano factor indicates the presence of interactions: e.g., for the shot noise of a vacuum diode this can be due to the non-negligible Coulomb repulsion between electrons; for the shot noise of photon detector output, it is due to the finite detector restoration time. In the quantum case there exist specific correlations due to quantum statistics, and one should expect deviations from $F = 1$ even in the absence of interactions. We will consider here, to be specific, quantum shot noise of an electric current (Khlus, 1987; Lesovik, 1989; Büttiker, 1990; Levitov and Lesovik, 1992).[13] First let us, after Lesovik (1989), consider shot noise in a 2DEG point contact. From our considerations in § 3.1 we can write the current operator as

$$\hat{I}(x,t) = \frac{ie\hbar}{2m^*} \int dy \left[\hat{\Psi}^\dagger(x,y,t) \frac{d}{dx} \hat{\Psi}(x,y,t) - \left(\frac{d}{dx} \hat{\Psi}^\dagger(x,y,t) \right) \hat{\Psi}(x,y,t) \right], \tag{5.78}$$

where the second-quantized electronic wave function,

$$\hat{\Psi}(x,y,t) = \sum_{\alpha=L,R} \sum_n \int dk \, \chi_{\alpha nk}(x,y,t) a_{\alpha nk};$$

$$\hat{\Psi}^\dagger(x,y,t) = \sum_{\alpha=L,R} \sum_n \int dk \, \chi^*_{\alpha nk}(x,y,t) a^\dagger_{\alpha nk} \tag{5.79}$$

is expressed through the complete orthogonal set of scattering states $\chi_{\alpha nk}(x,y,t)$, each corresponding to an electron wave incident from an equilibrium reservoir α in transverse mode n with the momentum $\hbar k$ (and energy $\varepsilon(k)$) at infinity (cf. Eq. (3.14)). Fermi operators $a^\dagger$ and a create and annihilate electrons in these states;

$$\langle a^\dagger_{\alpha nk} a_{\alpha nk} \rangle = f(\varepsilon(k) - \mu_\alpha), \tag{5.80}$$

---

[13] The corresponding effects in the statistics of photon counts (bunching and antibunching) have long been investigated in quantum optics; see, e.g., reviews (Smirnov and Troshin, 1987; Teich and Saleh, 1989; Dodonov, 2002) or textbooks (Scully and Zubairy, 1997; Orszag, 1999; Schleich, 2001).

where f is the Fermi distribution, and μ_α is the chemical potential of the αth bank; $\mu_L - \mu_R = eV$.

For a smooth confining potential there is no scattering between different transverse modes; therefore, the scattering states far from the bottleneck can be written as

$$\chi_{Lnk}(x, y, t) = \varphi_n(x, y)e^{-\frac{it}{\hbar}(\varepsilon(k)-\mu_L)} \begin{cases} e^{ikx} + r_{nn}e^{-ikx}, & x \ll 0; \\ t_{nn}e^{ikx}, & x \gg 0; \end{cases} \quad (5.81)$$

$$\chi_{Rnk}(x, y, t) = \varphi_n(x, y)e^{-\frac{it}{\hbar}(\varepsilon(k)-\mu_R)} \begin{cases} e^{-ikx} + \bar{r}_{nn}e^{ikx}, & x \gg 0; \\ \bar{t}_{nn}e^{-ikx}, & x \ll 0. \end{cases} \quad (5.82)$$

Here $\phi_n(x, y)$ is the eigenfunction (3.17) of the transverse part of the Hamiltonian, and the t's and r's are the ($\varepsilon(k)$-dependent) transmission and reflection coefficients, elements of the scattering matrix (Eq. (3.41)).

The symmetric part of the spectral density of the current noise is given by

$$S_I(\omega) = \int dt\, e^{i\omega t} \left(\frac{1}{2}\{\hat{I}(x, t), \hat{I}(x, 0)\} - \langle\hat{I}\rangle^2 \right). \quad (5.83)$$

Since (5.83) is the Fourier transform of noise autocovariance, this expression will not contain the static term, proportional to $\delta(\omega)$. On the other hand, it will automatically include equilibrium thermal noise. The shot noise is then given by $S_{\text{shot}}(\omega) = S_I(\omega) - S_{I,0}(\omega)$, where $S_{I,0}(\omega)$ is Eq. (5.83) calculated at $\mu_L = \mu_R = E_F$.

The current is conserved. Therefore, we can calculate the current and its correlators at any point along the channel, e.g., at $x \ll 0$, using the asymptotic expressions (5.81, 5.82). Substituting them in (5.78) one finds the expression (Lesovik, 1989), which at zero temperature and in the limit of low frequencies and small voltages (small enough to neglect the energy dependence of the transmission and reflection coefficients on a scale $eV \ll E_F$) simplifies to

$$S_{\text{shot}}(\omega) = \frac{2e^2}{h}eV \sum_n T_n(1 - T_n) = G_Q\, eV \sum_n T_n(1 - T_n), \quad (5.84)$$

where $T_n = |t_n(E_F)|^2 = |\bar{t}_n(E_F)|^2$ is the transparency of the point contact, and G_Q is the quantum conductance unit. In the limit of low transparency this expression reduces to

$$S_{\text{shot}}(\omega)|_{T_n \ll 1} = eV \left(G_Q \sum_n T_n \right) = e(VG) = e\bar{I}, \quad (5.85)$$

that is, the classical shot noise (5.76). However, if the channels are either open or closed, $T_n = 1$ or 0, the formula (5.84) predicts exactly *zero* shot noise. This remarkable result is readily understood in terms of quantum *anti*correlations between the electrons, due to the Pauli exclusion principle. Decreasing transparency makes these anticorrelations irrelevant (the electron flow becomes too sparse), and eventually the Poissonian shot noise is restored.

Levitov and Lesovik (1992, 1993) provided an illuminating statistical interpretation of Eq. (5.84). We will give here its simple intuitive explanation following Beenakker and Schönenberger (2003). Current noise can be considered as delta-correlated (see Eqs (5.59, 5.75)), so that

$$S_I(\omega) = \int dt e^{i\omega t} \langle \delta I(t) \delta I(0) \rangle = \overline{\delta I^2}. \tag{5.86}$$

However, for the charge Q transferred during the time interval τ we find the variance

$$\overline{\delta Q^2} = \int_0^\tau dt \int_0^\tau dt' \langle \delta I(t) \delta I(t') \rangle = \tau \overline{\delta I^2} = \tau S_I(\omega). \tag{5.87}$$

This expression is applicable for sufficiently low frequencies, $\omega \ll 1/\tau$. At zero temperature, only the electrons with energies within the band $[E_F - eV/2, E_F + eV/2]$ participate in the current. During the period τ the number of electrons attempting to pass between the reservoirs is of order[14] $N_\tau = \tau eV/h \gg 1$. At zero temperature each state contains one electron (i.e., they are perfectly anticorrelated). Therefore, the process of charge transfer can be considered as resulting from N_τ *independent* attempts to transfer charge e, each with the success probability T. This is a binomial distribution (see, e.g., Rytov et al., 1987, § 3), with the probability of transferring charge ne in N_τ attempts

$$P_n = \frac{N_\tau!}{n!(N_\tau - n)!} T^n (1 - T)^{N_\tau - n}, \tag{5.88}$$

the average charge transfer

$$\overline{Q} = \sum_n n e P_n = N_\tau e, \tag{5.89}$$

and the variance

$$\overline{\delta Q^2} = \overline{Q^2} - \overline{Q}^2 = e^2 N_\tau T(1 - T). \tag{5.90}$$

If there were several modes, each would contribute independently. Therefore, the spectral density of quantum shot noise:

$$S_I(\omega) = \overline{\delta Q^2}/\tau = \frac{e^2}{h} eV \sum_n T_n(1 - T_n) \tag{5.91}$$

(in our assumptions the equilibrium contribution to noise, Eq. (5.58), is negligible).

---

[14] The characteristic frequency eV/h is conceptually similar to the "attempt frequency" in the semiclassical expression for the tunnelling rate out of a potential well (see, e.g., Price, 1998).

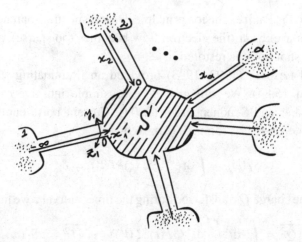

Fig. 5.1. Scattering theory description of quantum shot noise (see Eq. (5.92)).

The results obtained so far do not seem to depend on the system being a smooth 2DEG point contact. The handwaving argument, which brought us to Eq. (5.91), certainly did not use any details about the system, with the exception of independence of different "transport modes". As a matter of fact, this condition is also unnecessary. Following Büttiker (1990), let us represent the conductor as a scatterer, connected by leads to the external reservoirs kept at definite chemical potentials μ_α, $\alpha = 1, 2, \ldots$ (this is the same approach we used in § 3.1): see Fig. 5.1. The asymptotic behaviour of scattering states, each corresponding to an electron wave incident from the αth equilibrium reservoir in transverse mode n with wave vector k and energy $E = \varepsilon(k)$ at infinity (cf. Eqs (3.14, 5.81, 5.82)), can now be written as

$$\chi_{\alpha n E}(x, y, z, t) \sim e^{-\frac{it}{\hbar}(E - \mu_\alpha)} e^{ik_{\alpha n}(E)x} \varphi_{\alpha n}(y, z)$$

$$+ \sum_{\beta, m} \left[\frac{v_{\alpha n}(E)}{v_{\beta m}(E)} \right]^{-1/2} s_{\beta m, \alpha n}(E) e^{-ik_{\alpha m}(E)x} \varphi_{\beta m}(y, z), \quad |x| \gg 0. \qquad (5.92)$$

Here $\varphi_{\alpha n}(y, z)$ is the transverse mode wave function.[15] The longitudinal wave vector $k_{\alpha n}(E, x) = (1/\hbar)\sqrt{2m^*(E - \epsilon_{\alpha n}(x))}$, where $\epsilon_{\alpha n}(x)$ is the transverse mode energy (cf. Eqs (3.19, 3.20)). The longitudinal velocity in the mode (α, n) is given by $v_{\alpha n}(E) = \hbar[dk_{\alpha n}(E)/dE]$; the ratio of velocities in (5.92) is used to normalize the wave function to the unit probability flux. Finally, the scattering matrix elements

---

[15] To unclutter the formulas, we do not explicitly label the coordinates x_α, y_α, z_α in the αth lead. The coordinate x along each lead is measured from minus infinity (in the αth reservoir) to zero (inside the scattering region).

$s_{\beta m, \alpha n}(E)$ now include intermode scattering. (We use small, rather than capital, letters to denote the scattering matrix elements to avoid confusion with the noise spectral density.)

Expressing the second-quantized wave functions (5.79) through the scattering states (5.92), we can write the current operator in the γth lead as

$$\hat{I}_\gamma(x, t) = \frac{e\hbar}{2m^*i} \int_{(\gamma)} dy\,dz \left[\hat{\Psi}^\dagger(x, y, z, t) \frac{d}{dx} \hat{\Psi}(x, y, z, t) \right.$$
$$\left. - \left(\frac{d}{dx} \hat{\Psi}^\dagger(x, y, z, t) \right) \hat{\Psi}(x, y, z, t) \right]. \tag{5.93}$$

The integral is taken over the cross section of the γth lead far away from the scattering region, where the asymptotic condition (5.92) holds. The resulting expression for the shot noise in the low-frequency, zero-temperature limit is (Büttiker, 1990)

$$S_{\alpha\beta, \text{shot}}(\omega) = \frac{2e^2}{h} \sum_{\gamma \neq \delta} \int dE\, f(E - \mu_\gamma)(1 - f(E - \mu_\delta)) \, \text{tr}\left[s_{\alpha\gamma}^\dagger s_{\alpha\delta} s_{\beta\delta}^\dagger s_{\beta\gamma} \right] \tag{5.94}$$

(the summation over the transverse modes in each lead is implied; the velocities were cancelled by the 1D density of states, as in § 3.1). Here $S_{\alpha\beta}(\omega)$ is the Fourier transform of the symmetrized mutual covariance, $K_{\alpha\beta}(\tau) = (1/2)\left\langle \{\hat{I}_\alpha(\tau) - \langle \hat{I}_\alpha \rangle, \hat{I}_\beta(0) - \langle \hat{I}_\beta \rangle\} \right\rangle$. In a two-terminal conductor (5.94) reduces to

$$S_{\text{shot}}(\omega) = \frac{2e^2}{h} eV \, \text{tr}\left[r^\dagger r t^\dagger t \right]. \tag{5.95}$$

Here r, t are the submatrices of the scattering matrix, Eq. (3.41), which describe the transitions between different transverse modes. From the unitarity of the scattering matrix it follows that $r^\dagger r + t^\dagger t = \hat{1}$ (the unit $N_\perp \times N_\perp$-matrix, where $N_\perp$ is the number of transverse modes in each lead), and, therefore,

$$S_{\text{shot}}(\omega) = \frac{2e^2}{h} eV \, \text{tr}\left[t^\dagger t (1 - t^\dagger t) \right] = \frac{2e^2}{h} eV \sum_n T_n(1 - T_n), \tag{5.96}$$

where the T_n's are now, as in Eq. (3.45), the eigenvalues of the matrix $t^\dagger t$, and cannot be generally regarded as transparencies of a particular transverse channel.[16]

---

[16] The experimental evidence for the suppression of quantum shot noise in a 2DEG and atomic-size metallic point contacts is reviewed in van Ruitenbeek (2003) and Agraït et al. (2003).

5.2.4 1/f-noise

In addition to thermal and shot noise, practically every measured quantity in practically every system exhibits 1/f-noise.[17] This umbrella term covers all fluctuations with a spectral density that behaves approximately as $|\omega|^{-\alpha}$, with $\alpha \approx 1$, in an extremely broad range of frequencies (exceeding $10^{-6}-10^6$ Hz). Such fluctuations dominate in the low-frequency region and cause various problems, especially so when the quantum coherence is to be maintained, like in solid-state qubits.

In many (but far from all) cases the spectral density of the electrical (current or voltage) 1/f-noise can be approximated by the empirical formula (Hooge, 1969),

$$S_A(\omega) = \frac{\alpha_H \overline{A}^2}{N_c} \left| \frac{2\pi}{\omega} \right|^\alpha, \tag{5.97}$$

where $\alpha_H \approx 2 \times 10^{-3}$ is "Hooge's constant", weakly dependent on temperature, and N_c is the total number of charge carriers in the sample. The (approximately) quadratic dependence on the average signal is a qualitative feature helping to distinguish it from thermal and shot noise.

The most interesting feature of 1/f-noise is its scaling property: the contribution to the total power of fluctuations from each frequency decade,

$$\langle \delta A^2 \rangle_{[\omega_{\min}, 10\,\omega_{\min}]} \propto 2 \int_{\omega_{\min}}^{10\,\omega_{\min}} \frac{d\omega}{2\pi} \frac{1}{\omega} = \frac{1}{\pi} \ln \frac{10\,\omega_{\min}}{\omega_{\min}}, \tag{5.98}$$

is the same. The total noise power, $\langle \delta A^2 \rangle = \int_0^\infty (d\omega/2\pi) S_A(\omega)$, diverges, which should not bother us any more than the divergence of the total power of white noise at the upper integration limit, $\omega \to \infty$. In reality there always exists some high-frequency cutoff due to the minimal physical correlation time. For 1/f-noise, when the integral also diverges at the lower limit, $\omega = 0$, a low-frequency cutoff is in any case imposed by the finite observation time.

Fluctuations with 1/f-spectra appear in various, quite unrelated random processes and invite interesting mathematics (see, e.g., Mandelbrot, 1982). The topic remains very active, and it is way beyond the scope or purpose of this book to discuss the available experimental data and the numerous theories of its origin in different physical systems, whether it is stationary or not, whether there exists a universal mechanism of 1/f-noise, etc. These are reviewed in, e.g., Dutta and Horn (1981), Bochkov and Kuzovlev (1983), Buckingham (1983), Kogan (1985), Klimontovich (1986), Van Der Ziel (1986) and Weissman (1988). In solid-state systems of the type used for the realization of solid-state qubits, the 1/f-noise is predominantly due to the slow relaxation of tunnelling two-level systems (TLS) with an exponentially

---

[17] The effect was first reported by Johnson (1925). Sometimes it is also called *flicker noise* (Schottky, 1926).

wide range of relaxation times, which, as we shall see, is a theory consistent with all the available data and directly confirmed by experiments on phase qubits.

Formally, a 1/f-spectrum can be obtained from a weighted superposition of the Lorentzian spectra (5.51):

$$\int_{\gamma_{min}}^{\gamma_{max}} d\gamma \frac{2\gamma w(\gamma)}{\omega^2 + \gamma^2} = 2w_0 \frac{\arctan[\gamma_{max}/|\omega|] - \arctan[\gamma_{min}/|\omega|]}{|\omega|}, \tag{5.99}$$

if we assume that $w(\gamma) = w_0/\gamma$ in the interval $[\gamma_{min}, \gamma_{max}]$. For the frequencies $\gamma_{min} \ll |\omega| \ll \gamma_{max}$ the spectral density (5.99) indeed reduces to $\pi w_0/|\omega|$. In order to justify this model, we need a physical mechanism that would provide a $1/\gamma$-distribution of characteristic rates γ over an exponentially wide range $[\gamma_{min}, \gamma_{max}]$. Several such mechanisms exist (see, e.g., Kogan, 1985). The one relevant for us is based on the presence of TLS's, that is, defects, atoms or groups of atoms in disordered solids, which can occupy two almost degenerate states, represented by the bottoms of a two-well potential in the configuration space (Fig. 5.2). Spontaneous transitions between the wells, due to quantum tunnelling or thermal activation, lead to the fluctuations of the macroscopic parameters of the system (electrical resistivity, Josephson critical current density, etc.) and thus produce the slow fluctuations in current and voltage.[18]

The exponentially broad distribution of switching rates in this model appears in a very natural way (see, e.g., Kogan, 1985). For example, for thermally activated transitions between the wells the characteristic switching rate $\gamma \sim \omega_0 \exp[-V/k_B T]$ (Fig. 5.2). Plausibly assuming a flat probability distribution for the effective barrier heights in a sufficiently wide range $[V_{min}, V_{max}]$, we find

$$w(\gamma) = w_V(V) \left| \frac{dV}{d\gamma} \right| = \frac{k_B T \overline{w_V}}{\gamma} \bigg|_{\gamma_0 e^{-V_{max}/k_B T} \ll \gamma \ll \gamma_0 e^{-V_{min}/k_B T}}. \tag{5.102}$$

---

[18] A simple model of such a process in a single fluctuator is given by the *random telegraph signal* (RTS). RTS is a random function of time, $A(t)$, which can take two values, $A_1 = -a$ and $A_2 = a$. The switching rate between the states, v, is constant and the same for $A_1 \rightarrow A_2$ and $A_2 \rightarrow A_1$ transitions, and switching events are independent, so that the probability of m switches on the interval $[0, t]$ is given by the Poisson distribution $P(m, t)$ (5.65). (The generalizations to $v_{12} \neq v_{21}$ or $A_1 \neq -A_2$ are simple, but not essential here.) The autocorrelator of RTS can be calculated directly:

$$K_A^{(RTS)}(t) = \langle A(t)A(0) \rangle = a^2 P(m = \text{even}, t) - a^2 P(m = \text{odd}, t)$$

$$= a^2 \sum_{k=0}^{\infty} (P(2k, t) - P(2k+1, t)) = a^2 e^{-2v|t|}. \tag{5.100}$$

The spectral density is then a Lorentzian,

$$S_A^{(RTS)}(\omega) = \frac{4a^2 v}{\omega^2 + 4v^2}. \tag{5.101}$$

Comparing (5.100) and (5.101) with (5.50), (5.51), we see that two different random processes (here relaxation and RTS) can have the same correlators and spectral densities.

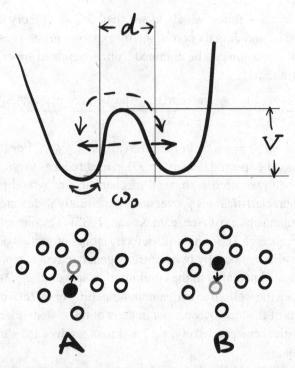

Fig. 5.2. Tunnelling two-level system (TLS) in a disordered solid: two atomic configurations (A, B) with close energies are separated by the potential barrier. The rate γ of either thermally activated or tunnelling transitions between A and B exponentially depends on the barrier parameters V, d. (Transitions can involve several atoms, defects, etc.) A plausible assumption of a uniform distribution of the TLS parameters within a certain range leads to a $1/\gamma$-distribution of transition rates in an exponentially wide range of frequencies.

At sufficiently low temperatures, $k_B T \ll V$, the thermally activated transitions are excluded and the TLS switching will be dominated by tunnelling with the rate

$$\gamma \sim \omega_0 e^{-\lambda}, \quad \lambda = d(2MV)^{1/2}/\hbar, \qquad (5.103)$$

where d is the effective barrier thickness, and M the (possibly very large on an atomic scale) mass of the tunnelling complex. Again, assuming a flat distribution of λ's, we find the switching rate probability $w(\gamma) = w_0/\gamma$ in an exponentially broad interval $[\omega_0 e^{-\lambda_{max}}, \omega_0 e^{-\lambda_{min}}]$, resulting in the effective 1/f-spectrum. However, when the number of active TLS's interacting with the system is reduced, the 1/f-spectrum will resolve into a superposition of a few Lorentzians, as has been observed, e.g., in charge qubits (Kafanov et al., 2008).

One of the direct confirmations of the TLS theory was provided by a series of experiments on phase qubits (TLS's were identified in all kinds of Josephson qubits; see Simmonds et al. (2009) and references therein), where the contributions from separate TLS's were spectroscopically resolved. It was shown that the TLS's reside in the tunneling barrier, and their number can be drastically reduced (with a corresponding improvement of the qubit decoherence time) by improving the barrier quality or changing its composition (Oh et al., 2006; Kline et al., 2009). Moreover, experiments (Simmonds et al., 2004; Martinis et al., 2005) have shown that some TLS's have long decoherence times (in excess of 1 μs) and, when in resonance with the qubit, undergo quantum Rabi oscillations. This is not surprising, since the TLS's are, after all, microscopic objects (e.g., the switching observed in phase qubits is consistent with a single electron charge moving by one atomic bond length), and their coupling to the environmental degrees of freedom, like lattice phonons, is small enough. But then such microscopic TLS's can themselves be used as qubits, accessible through the phase qubits in which they reside (Zagoskin et al., 2006).[19]

A quantum TLS in a Josephson qubit is an electric dipole inside the tunnelling barrier, which affects the qubit state either through a direct electrostatic interaction with the electric charge on the Josephson junction (Martinis et al., 2005) or by modulating the critical current (Simmonds et al., 2004). In either case the coupling is of the $\sigma_x \tilde{\sigma}_x$-type (where the tilde marks the TLS), and in the rotating wave approximation and neglecting the variation of the interlevel spacing in the phase qubit we obtain the Hamiltonian of the qubit-TLS's system (Zagoskin et al., 2006)

$$H = -\frac{\hbar \omega_{10}}{2} \sigma_z - \sum_j \left(\frac{\Delta_j}{2} \tilde{\sigma}_z^j + \lambda_j \sigma_x \tilde{\sigma}_x^j \right), \tag{5.104}$$

valid when the couplings $\lambda_j \ll \hbar \omega_{10}, \Delta_k$. The coupling term will only act in resonance, when $|\hbar \omega_{10} - \Delta_j| < \lambda_j$ (cf. § 4.1.4). Therefore, it can be switched on and off by changing the bias current (magnetic flux). The state of the qubit can thus be "stored" in the TLS ("quantum memory") for the duration of the latter's decoherence time, while other operations are performed on the qubit. The resonant character of qubit–TLS coupling also allows us to target a specific TLS inside the qubit as long as the tunnelling splits Δ_j, Δ_k are not too close together. We can also perform quantum manipulations between different TLS's inside the same phase qubit; in this case the latter will play the role of quantum bus (see Fig. 5.3). The possibility of using a TLS as quantum memory was confirmed in an experiment by Neeley et al. (2008) (Fig. 5.4), and if a controlled method of creating "good" TLS's in superconducting qubits were developed, they could add to our set of tools for building quantum devices.

---

[19] Such "naturally formed qubits" would be a quantum counterpart to the "cat's whisker detectors", the early semiconductor diodes.

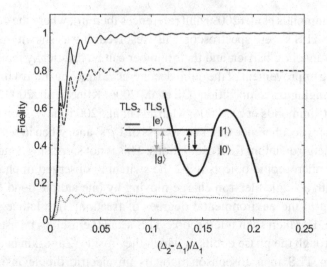

Fig. 5.3. Fidelity of the iSWAP gate on two TLS qubits in a phase qubit as a function of the TLS relative detuning. The fidelity F is defined as a minimum over the initial states $|\Psi_{\text{initial}}\rangle$ of the system of the matrix element $\langle\Psi_{\text{initial}}|U^{\dagger}\rho_{\text{final}}U|\Psi_{\text{initial}}\rangle$, where ρ_{final} is the final density matrix of the system, and $U = U_{\text{iSWAP}}$ is the desired unitary operation. Up to single qubit rotations, the iSWAP operation can be performed by bringing the qubit in resonance with TLS$_1$ for the time interval $\hbar/2\lambda_1$, then with TLS$_2$ for $\hbar/2\lambda_2$, and then again in resonance with TLS$_1$ for $\hbar/2\lambda_1$. In the end the qubit will be decoupled and disentangled from both TLS's. The calculations were performed taking the coupling $\lambda_1 = \lambda_2 = \lambda = 0.05\Delta_1$, and for the phase qubit decoherence rate $\Gamma_1 = \Gamma_2/2 = 0$ (solid line), $\lambda/20\pi$ (dashed line), and $\lambda/2\pi$ (dotted line). Decoherence in the TLS's is neglected. The inset shows quantum beats between the qubit and TLS$_1$ in resonance, while TLS$_2$ is effectively decoupled. (Reprinted with permission from Zagoskin et al., 2006, © American Physical Society.)

On balance, though, the naturally occurring TLS's in qubits are rather a nuisance, as is generally the 1/f-noise of whatever origin. They are especially unpleasant in charge qubits, since they are directly coupled to the relevant degree of freedom, and the system of TLS's acts as an additional source of decoherence (Martinis et al., 2005). We will discuss this matter in the next section.

5.3 Noise and decoherence

5.3.1 External noise and decoherence rates

We have seen that the relaxation and excitation rates of a qubit coupled to an external system (noise source), as in (5.25), are proportional to the spectral density of noise at the frequency of interlevel transitions, (5.31). This is reasonable, since these processes involve the exchange of quanta, $\hbar\Omega$, between the qubit and

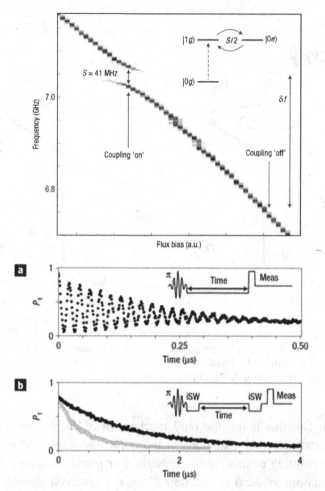

Fig. 5.4. Tunnelling two-level system in a phase qubit as quantum memory. (Left) Phase qubit excitation probability as a function of microwave frequency and flux bias. Coupling to the TLS at 7.05 GHz creates a splitting in the spectrum. The swap frequency (splitting) is $S = 41$ MHz. Inset: scheme of quantum coherent oscillations between the states of the system "qubit + TLS". (Right) Probability P_1 of the qubit to be found in the excited state as a function of time. (a) Free oscillations due to a resonant coupling to the TLS. The iSWAP operation time (position of the first minimum) is 12 ns. (b) Energy decay of the TLS. The qubit is initiated in the excited state, then its state is stored in the TLS (by performing the iSWAP operation), and after a period of time is restored from the TLS by another iSWAP operation, and measured. The results (black dots) correspond to a longer decay time ($T_{1,TLS} = 1.2\,\mu$s), than when the state of the qubit is not in the TLS (grey dots, $T_{1,qb} = 0.4\,\mu$s). (Reprinted by permission from *Nature Physics*, Neeley et al., 2008, © 2008 Macmillan Publishers Ltd.)

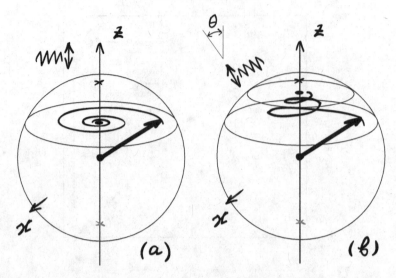

Fig. 5.5. Qubit dephasing by external noise. The basis is the eigenbasis of the unperturbed qubit: $H_0 = -\frac{\hbar}{2}\Omega\sigma_z$. The evolution of the Bloch vector is given by Eq. (1.88). (a) Pure dephasing: the external noise $-\frac{\hbar}{2}\delta\Omega(t)\sigma_z$ changes the absolute value, but not the direction, of the angular velocity with which the Bloch vector rotates, wiping out the coherences ($\propto \mathcal{R}_\perp$). (b) Dephasing and relaxation: the classical noise term in the Hamiltonian, $f(t)[\sigma_z\cos\theta + \sigma_x\sin\theta]$, is at an angle θ to the z-axis, eliminating $\mathcal{R}_\perp$ and affecting $\mathcal{R}_z$.

the reservoir. But this is not the only mechanism of decoherence. Coupling to the reservoir can induce fluctuations in the qubit interlevel spacing, Ω, without absorbing or emitting quanta (*puze dephasing*, or *phase damping*): the qubit (or a general quantum system) is "scattered" by the reservoir quanta, acquiring a random phase (Gardiner and Zoller, 2004, §6.1.2). This process will not affect the diagonal elements of the density matrix. However, the coherences depend on the corresponding interlevel spacings (e.g., $\rho_{12}(t) \propto \exp[i\Omega t + i\int_0^t dt'\,\delta\Omega(t')]$). (In terms of the Bloch vector representation of the density matrix, Fig. 5.5a, the fluctuations $\delta\Omega(t)$ are the fluctuations in the magnitude of the angular velocity with which the Bloch vector rotates around the z-axis; they produce random phase gains, which should eventually suppress its component $\mathcal{R}_\perp$ in the xy-plane.) Assuming (plausibly enough) that the fluctuations $\delta\Omega(t)$ are Gaussian with zero mean value, after averaging over them we will obtain for the damping of the off-diagonal terms

$$\left|\frac{\rho_{12}(t)}{\rho_{12}(0)}\right| = \left|\left\langle e^{i\int_0^t dt'\,\delta\Omega(t')}\right\rangle\right| = e^{-\frac{1}{2}\int\int_0^t dt_1 dt_2\,\langle\delta\Omega(t_1)\delta\Omega(t_2)\rangle} \equiv e^{-\eta(t)}. \qquad (5.105)$$

This standard result for the characteristic functional of a *Gaussian random process* (see, e.g., Rytov et al., 1989) can be easily obtained by elementary means.[20] The exponent is calculated explicitly (e.g., Makhlin and Shnirman, 2004),

$$\eta(t) = \frac{1}{2} \int \int_0^t dt_1 dt_2 \, \langle \delta\Omega(t_1)\delta\Omega(t_2) \rangle = \frac{1}{2} \int_{-\infty}^{\infty} \frac{d\omega}{2\pi} \, S_{\delta\Omega}(\omega) \int \int_0^t dt_1 dt_2 \, e^{-i\omega(t_1 - t_2)}$$

$$= \frac{1}{2} \int_{-\infty}^{\infty} \frac{d\omega}{2\pi} \, S_{\delta\Omega(\omega)}(\omega) \frac{\sin^2(\omega t/2)}{(\omega/2)^2} = \frac{|t|}{2} \int_{-\infty}^{\infty} \frac{d\omega}{2\pi} \, S_{\delta\Omega}(\omega) \frac{\sin^2(\omega t/2)}{|t|(\omega/2)^2},$$

$$(5.106)$$

and using the limit (e.g., Landau and Lifshitz, 2003, § 42, Eq. (42.2))[21]

$$\lim_{t \to \infty} \frac{\sin^2 \alpha t}{\alpha^2 t} = \pi \delta(\alpha), \qquad (5.107)$$

---

[20] Gaussian random processes arise as the result of the action of many independent small influences, and, therefore, describe many – but not all! – random processes in nature. As a counterexample, the evolution of stock prices is the result of many, probably small, but in no way independent events, and is, therefore, anything but Gaussian. From the mathematical point of view, the Gaussian random process is a random process where all correlators are reduced to the sums of products of first- and second-order correlators (that is, averages and covariances). Remarkably, the Marcinkiewicz theorem (see, e.g., Titulaer, 1975) states that this is also the *only* example of a random process with a finite number of nonzero irreducible nth-order correlators: either $n = 1, 2$, and the process is Gaussian, or there will exist irreducible correlators of arbitrarily large order. We assumed $\langle \delta\Omega \rangle = 0$, therefore, the average

$$\left\langle e^{i \int_0^t dt' \, \delta\Omega(t')} \right\rangle = \sum_{n=0}^{\infty} \frac{i^n}{n!} \int_0^t dt_1 \dots \int_0^t dt_n \, \langle \delta\Omega(t_1) \dots \delta\Omega(t_n) \rangle$$

can be written as

$$\sum_{n=0}^{\infty} \frac{i^{2n}}{(2n)!} \sum_{\text{pairs}} \int_0^t dt_1 \dots \int_0^t dt_{2n} \, \langle \delta\Omega(t_1)\delta\Omega(t_k) \rangle \dots \langle \delta\Omega(t_m)\delta\Omega(t_n) \rangle$$

$$= \sum_{n=0}^{\infty} \frac{i^{2n}}{(2n)!} N_{\text{pair}} \left(\int_0^t dt_1 \int_0^t dt_2 \langle \delta\Omega(t_1)\delta\Omega(t_2) \rangle \right)^n.$$

The number of ways in which $2n$ items can be split into pairs, N_{pair}, is counted as follows: we choose two $\delta\Omega$'s out of $2n$, then two more out of $2n - 2$, etc., which yields

$$N_{\text{pair}} = \frac{2n!}{(2n-2)!2!} \cdot \frac{(2n-2)!}{(2n-4)!2!} \dots \frac{4!}{2!2!} \cdot \frac{2!}{0!2!} \cdot \frac{1}{n!} = \frac{(2n)!}{2^n n!};$$

the last $1/n!$ factor compensates for $n!$ rearrangements of pairs, which we would otherwise count repeatedly. Substituting this in the above, we indeed find

$$\sum_{n=0}^{\infty} \frac{(-1)^n}{2^n n!} \left(\int_0^t dt_1 \int_0^t dt_2 \langle \delta\Omega(t_1)\delta\Omega(t_2) \rangle \right)^n = e^{-\frac{1}{2} \int_0^t dt_1 \int_0^t dt_2 \langle \delta\Omega(t_1)\delta\Omega(t_2) \rangle}.$$

[21] Recall that $\int_{-\infty}^{\infty} dz \frac{\sin^2 z}{z^2} = \pi$.

we see that the last integral in Eq. (5.106) indeed yields the exponential dephasing from the interlevel spacing fluctuations, with the rate

$$\Gamma_{\delta\Omega} = \frac{1}{2} S_{\delta\Omega}(0). \tag{5.108}$$

Here $S_{\delta\Omega}(0) = \int_{-\infty}^{\infty} d\tau \, \langle \delta\Omega(\tau)\delta\Omega(0)\rangle$ is the spectral density of these fluctuations *at zero frequency*. In other words, fluctuations on all timescales contribute *equally* to the dephasing.

While accounting for the influence of the environment in the general case requires such heavy artillery as path integrals (Feynman and Vernon, 1963; Feynman and Hibbs, 1965; Caldeira and Leggett, 1983; Leggett et al., 1987; Weiss, 1999) or diagrammatic techniques (Schoeller and Schön, 1994; Shnirman and Schön, 2003), in the limit of weak decoherence it can be treated in the second (i.e., lowest surviving) order of the perturbation theory. (This is what we actually need: there would be little sense to try and build a quantum device in the teeth of strong decoherence.) We start from the Hamiltonian

$$H = H_0 - \hat{f}(t)B + H_E \equiv -\frac{\hbar\Omega}{2}\sigma_z + (\sigma_x \sin\theta + \sigma_z \cos\theta)\hat{f}(t) + H_E. \tag{5.109}$$

Here the Pauli matrices are taken *in the eigenbasis of the qubit*, and we parameterize the coupling operator B by the mixing angle θ. The σ_x-component leads to transitions between the qubit levels, while the σ_z one will be responsible for pure dephasing. The "environmental" term H_E describes the dynamics of the reservoir variable $\hat{f}(t)$.[22] Now we can repeat, *mutatis mutandis*, our operations in Section 1.3 for the linear response theory. First we use the interaction representation with respect to H_0. The only operator affected will be B:

$$B \to \widetilde{B}(t) = e^{iH_0t/\hbar}Be^{-iH_0t/\hbar}$$
$$= \left(\sigma_+e^{-i\Omega t} + \sigma_-e^{i\Omega t}\right)\sin\theta + \sigma_z\cos\theta \equiv \sigma_\Omega(t)\sin\theta + \sigma_z\cos\theta. \tag{5.110}$$

Iterating the Liouville equation in the interaction representation (and dropping the tildes to unclutter the formulas), we find after averaging over the environment and

---

[22] If $\hat{f}(t)$ was a c-number (classical external noise), Eq. (5.109) would correspond to Fig. 5.5b.

assuming $\langle \hat{f}(t) \rangle_E = 0$:

$$\rho(t) = \rho(0) - \frac{1}{i\hbar} \int_0^t dt_1 \left\langle [\hat{f}(t_1)B(t_1), \rho(0)] \right\rangle_E$$

$$+ \frac{1}{(i\hbar)^2} \int_0^t dt_1 \int_0^{t_1} dt_2 \left\langle [\hat{f}(t_2)B(t_2), [\hat{f}(t_1)B(t_1), \rho(0)]] \right\rangle_E + \dots \quad (5.111)$$

$$= \rho(0) + \frac{1}{(i\hbar)^2} \int_0^t dt_1 \int_0^{t_1} dt_2 \left\langle [\hat{f}(t_2)B(t_2), [\hat{f}(t_1)B(t_1), \rho(0)]] \right\rangle_E + \dots.$$

Therefore, in the lowest order the decoherence rates will follow from

$$\dot{\rho}(t) \approx -\frac{1}{\hbar^2} \int_0^t dt' \left\langle [\hat{f}(t')B(t'), [\hat{f}(t)B(t), \rho(0)]] \right\rangle_E$$

$$\approx -\frac{1}{\hbar^2} \int_0^t dt' \left\{ \sin^2\theta \left\langle [\hat{f}(t')\sigma_\Omega(t'), [\hat{f}(t)\sigma_\Omega(t), \rho(t)]] \right\rangle_E \right.$$

$$+ \cos^2\theta \left\langle [\hat{f}(t')\sigma_z, [\hat{f}(t)\sigma_z, \rho(t)]] \right\rangle_E$$

$$+ \sin\theta\cos\theta \left(\left\langle [\hat{f}(t')\sigma_\Omega(t'), [\hat{f}(t)\sigma_z, \rho(t)]] \right\rangle_E \right.$$

$$\left. \left. + \left\langle [\hat{f}(t')\sigma_z, [\hat{f}(t)\sigma_\Omega(t), \rho(t)]] \right\rangle_E \right) \right\}. \quad (5.112)$$

(The replacement of $\rho(0)$ by $\rho(t)$ in the last stage produces the higher-order corrections only.) Expressing the density matrix through the Bloch vector components, $\rho = \frac{1}{2}[1 + \mathcal{R}_\alpha \sigma_\alpha]$, picking up the right components from the matrix equation (5.112), and essentially repeating the manipulations, which led to Eq. (5.30), we arrive at the relaxation and total dephasing (i.e., decoherence) rates

$$\Gamma = \frac{S_{f,s}(\Omega)}{\hbar^2} \sin^2\theta = \frac{1}{T_1}; \quad \Gamma_\phi = \frac{\Gamma}{2} + \gamma \equiv \frac{\Gamma}{2} + \frac{S_{f,s}(0)}{\hbar^2} \cos^2\theta = \frac{1}{T_2}, \quad (5.113)$$

where the difference between the excitation and relaxation rates is neglected. The times $T_{1,2}$ are those we have earlier introduced phenomenologically in the Bloch equations (1.90). The expressions (5.113) can be obtained in a more rigorous way, including the exactly solvable case $\theta = 0$ (Leggett et al., 1987; Weiss, 1999; Makhlin et al., 2001; Shnirman and Schön, 2003). They are also consistent with our earlier analysis, including the $\Gamma/2$-contribution from the relaxation rate to dephasing (which in Eq. (1.72) followed from our choice of Lindblad operators (1.64, 1.65)). We note here that, since the relaxation rate does depend on the qubit transition frequency Ω, while the pure dephasing rate does not, one can in principle influence the former, but not the latter, by tuning the qubit.

The optimistic reading of Eq. (5.113) is that in order to find the decoherence time of a quantum system one does not need to know any details of its environment beyond the noise spectrum of the generalized force operator $\hat{f}(t)$ (the coupling strength due to our choice of B in Eq. (5.109) is included in the definition of $\hat{f}(t)$). We shall see to what degree this optimism is justified.

5.3.2 Bosonic bath model of the environment

The general treatment of decoherence in quantum systems through the tracing out of the environment was realized by Feynman and Vernon (1963) (see also Feynman and Hibbs, 1965, §12.8). They used the path integral approach to show, in particular, that when the system is coupled to the environment sufficiently weakly, the latter can be represented by a set of harmonic oscillators (called the *oscillator*, or *bosonic bath*). Then tracing out the bosonic degrees of freedom leads to the non-unitary evolution of the system. We have seen an example of this earlier, when we derived the master equation (1.60) for a system coupled to a single bosonic mode.

This is a realistic enough situation, corresponding, e.g., to a qubit coupled to a single mode of an LC circuit. Nevertheless, in that case we could not simply trace out the mode of the oscillator: instead, we have seen that the qubit and resonator undergo quantum coherent Rabi oscillations, (4.61), the lifetime of which was determined by the intrinsic decoherence of the qubit and the oscillator losses, which must still, somehow, be brought into the picture.

Feynman and Vernon (1963) showed that losses appear when the spectrum of oscillator frequencies is *continuous*. Then there is no control over them, and moreover, due to the weakness of coupling between the system and the environment each of the environmental oscillator modes is only negligibly affected by the interaction with the system, while the opposite is not true. Therefore, tracing the environment out is justified.[23] We need also realize that the environment cannot be "completely" linear, since the non-interacting oscillator modes cannot exchange energy and cannot equilibrate. There must remain some infinitesimal interaction between them, somewhere, maybe infinitely far from our system, which ensures their equilibrium. (This is somewhat similar to the situation for the scattering theory of transport

---

[23] The same problem occurs also in classical statistical mechanics. To be consistent, one should treat thermal reservoirs as large, but finite, mechanical systems; the total energy of the system under investigation and the reservoirs is conserved. Then the Poincaré theorem (Mayer and Mayer, 1977) states that any mechanical system, starting from an arbitrary state, will always return to a state infinitesimally close to the initial state after a certain time T_0 (*Poincaré recurrence time*). This means, in particular, that whatever energy is lost by the system to the reservoirs will be returned after T_0. The amplitude of a damped oscillator coupled to a finite oscillator bath will behave as shown in Fig. 5.6, where the initial decay, well approximated by the irreversible Eq. (2.94), will eventually be followed by growth and will almost return to the initial state, to be repeated almost periodically (Tatarskii, 1987). For a large enough reservoir, T_0 is practically infinite: e.g., a mole of Ne at standard conditions has $T_0 \sim \exp[10^{25}]$ s, compared to the lifetime of the Universe $\sim \exp[40]$ s (Mayer and Mayer, 1977). This is, of course, excessive: a bath of 20 linear oscillators already has a "practically infinite" T_0 (Weiss, 1999, § 3.1.3).

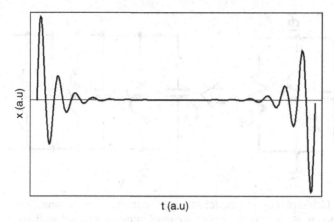

Fig. 5.6. The amplitude of a classical oscillator coupled to a bath of oscillators with a discrete equidistant set of frequencies, $\omega_n = n\Omega$, $n = 1, \ldots, \infty$: $x(t) \propto \{\cosh[\gamma(T_0/2 - t)]\sin\omega t\} / \{\omega\cosh(\gamma T_0/2)\}$, extended periodically outside the interval $[0, T_0]$ (Tatarskii, 1987). The Poincaré cycles are in this particular case *exactly* periodic.

we encountered in § 3.1.3, where the scattered electrons went to the equilibrium reservoirs at infinity and never – that is, not until the Poincaré recurrence time – came back.)

After these caveats, one can describe the two- (or many)-level quantum system ("spin") with decoherence by the *spin-boson model*, ascribing the decoherence of a quantum system to its interaction with a bath of linear oscillators with a continuous frequency spectrum. This model is often called the *Caldeira–Leggett model*, since Caldeira and Leggett (1983) (see also Leggett et al., 1987) developed and used it to describe quantum tunnelling in the presence of dissipation.[24]

Following Devoret (1997), let us show how this approach works in the case of a linear electrical circuit. Consider a classical lossy LC circuit (Fig. 5.7). There is only one variable, the nodal flux Φ, and the modified Lagrange equations of motion (2.94) are

$$\frac{C_0}{c^2}\ddot{\Phi} + \frac{1}{L_0}\Phi = -\frac{G}{c^2}\dot{\Phi}, \tag{5.114}$$

where G is the leakage conductance. Using Fourier components,

$$-\omega^2 \frac{C_0}{c^2}\Phi(\omega) + \frac{1}{L_0}\Phi(\omega) = i\omega\frac{Y(\omega)}{c^2}\Phi(\omega), \tag{5.115}$$

---

[24] When a distinction between the "spin-boson" and "Caldeira–Leggett" models is being made at all, it concerns the "central system", not the bath: the "spin-boson" then relates to the case of a quantum system with a finite-dimensional Hilbert space (like spin's, or a mesoscopic qubit's truncated to its lowest eigenstates).

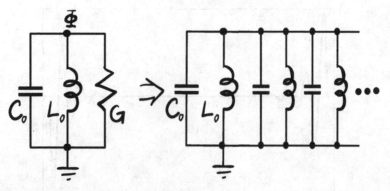

Fig. 5.7. Caldeira–Leggett model for an LC circuit (after Devoret, 1997): A dissipative element is replaced by a continuum of non-dissipative oscillator modes see Eq. (5.116).

where we have replaced the leakage conductance with a general, frequency-dependent admittance $Y(\omega) = G(\omega) + iB(\omega)$.

Let us now replace $Y(\omega)$ by an infinite number of purely reactive elements:

$$Y(\omega) = \sum_m Y_m(\omega) = -\sum_m \left(\frac{i\omega L_m}{c^2} + \frac{1}{i\omega C_m} \right)^{-1}. \qquad (5.116)$$

For real frequencies each of the terms is purely imaginary, as will be $Y(\omega) = iB(\omega)$. Let us now shift the frequency infinitesimally above the real axis: $\omega \to \omega + i\varepsilon$, where $\varepsilon \to 0$. We are allowed to do so by the definition of admittance (here $\widetilde{Y}(t) = \int (d\omega/2\pi) \exp[-i\omega t] Y(\omega)$),

$$I(t) = \int_{-\infty}^{\infty} d\tau \, \widetilde{Y}(t-\tau) V(\tau) \equiv \int_{-\infty}^{t} d\tau \, \widetilde{Y}(t-\tau) V(\tau), \qquad (5.117)$$

or $\widetilde{Y}(t) \equiv \widetilde{Y}(t)\theta(t)$. This is because causality does not allow the current at any time to be influenced by the voltage applied at later times.[25] Then the admittance

$$Y(\omega) = \int_{-\infty}^{\infty} dt \, e^{i\omega t} \widetilde{Y}(t) \equiv \int_{0}^{\infty} dt \, e^{i\omega t} \widetilde{Y}(t), \qquad (5.118)$$

remains well defined if ω has a *positive* imaginary part.[26] Using the Weierstrass formula (1.58), we find

---

[25] Compare with Eqs (1.37, 1.38) from the linear response theory.

[26] In other words, the admittance $Y(\omega)$ can be analytically continued to the upper complex halfplane. This and other useful mathematical consequences of causality, which impose restrictions on the physical response functions (the *dispersion relations*), are discussed in much detail by Nussenzveig (1972).

$$Y_m(\omega + i\varepsilon) = \frac{i\omega_m^2 C_m}{2}\left[\frac{1}{\omega + i\varepsilon - \omega_m} + \frac{1}{\omega + i\varepsilon + \omega_m}\right] \xrightarrow{\varepsilon \to 0}$$

$$\frac{\Upsilon_m}{2}\left[i\left(\mathcal{P}\frac{\omega_m}{\omega - \omega_m} + \mathcal{P}\frac{\omega_m}{\omega + \omega_m}\right) + \pi\omega_m\left(\delta(\omega - \omega_m) + \delta(\omega + \omega_m)\right)\right],$$

$$(5.119)$$

where $\omega_m = c/\sqrt{L_m C_m}$ is the resonant frequency of the mth constituent LC circuit, and $\Upsilon_m = c\sqrt{C_m/L_m}$ is its "nominal" admittance. Thus, we have managed to construct an active conductance out of purely reactive elements:

$$G(\omega) = \frac{\pi}{2}\sum_m \Upsilon_m \omega_m \left[\delta(\omega - \omega_m) + \delta(\omega + \omega_m)\right]$$

$$= \frac{\pi}{2}\sum_m \frac{1}{L_m}\left[\delta(\omega - \omega_m) + \delta(\omega + \omega_m)\right]. \qquad (5.120)$$

In order to mimic the conductance $G(\omega)$ it remains to replace the discrete set of oscillator modes by a continuum (i.e., by an oscillator bath).[27] The summation will become an integral over frequencies with the mode density $D(\omega)$:

$$\sum_m F_m\left[\delta(\omega - \omega_m) + \delta(\omega + \omega_m)\right] \to \int \frac{d\omega'}{2\pi} D(\omega') F(\omega'). \qquad (5.121)$$

Instead of using the modified Lagrange equations with the dissipative function, we can now add to the Lagrangian the terms representing the admittance,

$$\mathcal{L}_Y = \sum_m \left[\frac{C_m \dot{\Phi}_m^2}{2c^2} - \frac{(\Phi_m - \Phi)^2}{2L_m}\right]. \qquad (5.122)$$

The coupling between the oscillator bath and the system is due to the last term in the brackets and can be reduced to the contribution to the potential energy

$$U_{\text{int}} = -\left(\sum_m \frac{\Phi_m}{L_m}\right)\Phi. \qquad (5.123)$$

---

[27] If the nodal charges were chosen as canonical variables, instead of fluxes (e.g., Wells, 1967, §15.2.A), then we would have to deal with the circuit impedance $Z(\omega) = R(\omega) + iX(\omega) = \sum_m Z_m(\omega)$ (Fig. 5.7). Instead of Eq. (5.119) we would have

$$Z_m(\omega + i\varepsilon) \xrightarrow{\varepsilon \to 0} \frac{1}{2\Upsilon_m}\left[i\left(\mathcal{P}\frac{\omega_m}{\omega - \omega_m} + \mathcal{P}\frac{\omega_m}{\omega + \omega_m}\right) + \pi\omega_m\left(\delta(\omega - \omega_m) + \delta(\omega + \omega_m)\right)\right],$$

etc., etc.

Now it is possible to switch to the Hamiltonian formalism, quantize, go to the limit of a continuous oscillator distribution (5.121), average over the states of the oscillator, and in the end obtain non-unitary quantum dynamics for the observable Φ of our LC circuit. Note that the oscillator bath will also renormalize the non-dissipative dynamics of the system via the purely imaginary terms in (5.119) (as for the first line in Eq. (1.60)). Their contribution is negligible for sufficiently small leakage, and in any case can be compensated for by a shift in L_0 and C_0.

The Caldeira–Leggett model is usually presented in terms of a point mass M_0 (Caldeira and Leggett, 1983; Leggett et al., 1987; Weiss, 1999), the position q of which – assuming that it is in a quadratic potential – will satisfy the phenomenological equation of motion with the frequency-dependent viscous damping rate $\gamma(\omega)$,

$$-\omega^2 M_0 q(\omega) + M_0 \omega_0^2 q(\omega) = i\omega M_0 \gamma(\omega) q(\omega), \tag{5.124}$$

instead of (5.115). The classical Eq. (5.114) can be obtained by starting from the Hamiltonian system with

$$H = \frac{p^2}{2M_0} + \frac{M_0 \omega_0^2 q^2}{2} + \sum_m \left(\frac{p_m^2}{2M_m} + \frac{M_m \omega_m^2 q_m^2}{2} \right) - \left(\sum_m \lambda_m q_m \right) q. \tag{5.125}$$

Then, after calculations similar to those for the admittance, the damping rate turns out to be

$$\gamma(\omega) = \frac{1}{M_0} \frac{J(\omega)}{\omega}, \tag{5.126}$$

where the *spectral density of the environmental coupling* is (for positive frequencies ω)

$$J(\omega) = \frac{\pi}{2} \sum_m \frac{\lambda_m^2}{M_m \omega_m} \delta(\omega - \omega_m) \to \frac{\pi}{2} \int_0^\infty \frac{d\omega'}{2\pi} D(\omega') \frac{\lambda^2(\omega')}{M(\omega')\omega'}. \tag{5.127}$$

Comparing Eqs (5.114) and (5.124), we see the correspondences:

$$
\begin{array}{llll}
q_\alpha & \leftrightarrow \Phi_\alpha; & M_\alpha \leftrightarrow C_\alpha/c^2; & M_\alpha \omega_\alpha^2 \leftrightarrow 1/L_\alpha; \\
M_\alpha \omega_\alpha & \leftrightarrow \Upsilon_\alpha; & \lambda_\alpha \leftrightarrow 1/L_\alpha; & M_0 \gamma(\omega) \leftrightarrow G(\omega)/c^2
\end{array}
\tag{5.128}
$$

and it is possible to confirm that Eqs (5.126) and (5.127) translate into Eq. (5.120).

The spectral density $J(\omega)$ is determined by the distribution of the oscillator parameters and coupling coefficients – if we consider a reservoir of realistic oscillators (like phonon modes in a crystal), or vice versa, if we start from the irreversible phenomenological equations with damping and introduce a bath of fictitious oscillators to enable the use of the Hamiltonian formalism. In either case the behaviour of $J(\omega)$ critically affects the dynamics of the system interacting with the bath. In particular, the fluctuation–dissipation theorem relates $J(\omega)$ to the spectrum of equilibrium fluctuations of the quantized generalized force $\hat{f} = \sum_m \lambda_m \hat{q}_m$, which couples the bath to the system:

$$S_{f,s}(\omega) = \hbar J(\omega) \coth \frac{\hbar \omega}{2k_B T}, \tag{5.129}$$

where, as usual, $S_{f,s}(\omega)$ is the Fourier transform of the symmetrized correlator, $K_{f,s}(t) = \langle\{\hat{f}(t), \hat{f}(0)\}\rangle/2$. Therefore, $J(\omega)$ determines the relaxation and dephasing rates (as in (5.113)).

The behaviour of $J(\omega)$ at the high-frequency cutoff ω_c (which is much larger than all the characteristic frequencies in the problem; $J(\omega) \to 0$ for $\omega > \omega_c$) is not that important: the decoherence is determined by $S(\omega = 0)$ and $S(\omega = \Omega)$, way below the cutoff. For a bosonic bath $J(\omega)$ usually has power-law behaviour at low frequencies, which can be written as (Leggett et al., 1987; Weiss, 1999)

$$J(\omega) = \frac{\pi}{2} \alpha \hbar \omega_r^{1-s} \omega^s e^{-\omega/\omega_c}. \tag{5.130}$$

Here the dimensionless parameter α characterizes the strength of the system–bath interaction; a "reference frequency" ω_r is included to keep the proper dimensionality. The case $s = 1$ is called *ohmic* friction, because then the damping rate $\gamma \propto J(\omega)/\omega$ is frequency independent. In Eqs (5.114) and (5.115) this would yield a constant ohmic conductance $G = 1/R$. This is the most common case.[28] For example, the intrinsic decoherence in Josephson junctions due to thermally excited quasiparticles can be described by the ohmic bath with the noise spectrum (5.58), where the effective shunting conductance of the junction G drops exponentially at temperatures $k_B T$ below the superconducting gap Δ.

5.3.3 Spin bath

The bosonic bath model is built on the assumption that the quantum system we are investigating weakly interacts with each of a very large number of bath oscillators. For delocalized modes – like phonons – this is indeed the case; each oscillator mode is spread over the volume of the reservoir, and its interaction with the system drops as $\sim \mathcal{N}^{-1/2}$, where $\mathcal{N}$ is the total number of modes. This does *not* happen in the case when the environmental degrees of freedom are localized, like nuclear or

---

[28] The cases $s > 1$ $(0 < s < 1)$ are, of course, called *super-* *(sub-)*ohmic friction. The super-ohmic bath is less troublesome, than the ohmic one, because the bath influence at low frequencies is suppressed. A sub-ohmic bath corresponds, e.g., to the case of a system coupled to an RC-transmission line (Weiss, 1999, § 3.4.2) (with $s = 1/2$), and its decohering effects are qualitatively stronger than in the (ohmic) case of an LC-transmission line (see, e.g., Weiss, 1999, §20.3.2). Shnirman and Schön (2003) provide detailed analysis of decoherence and of the response functions of a two-level system interacting with sub-ohmic, ohmic and super-ohmic bosonic bath.

paramagnetic spins or TLS's. This kind of environment is called a *spin bath*. Since the coupling to each of the environmental "spins" is independent of their total number $\mathcal{N}$, there arise severe complications (Prokof'ev and Stamp, 2000; Stamp, 2008) unless the coupling is weak from the beginning (in which case the spin bath is equivalent to the oscillator bath, see Feynman and Vernon, 1963; Weiss, 1999, §3.5; Prokof'ev and Stamp, 2000, §4.2).

The complications arise because at least some spins in the spin bath are strongly coupled to the system, and, therefore, their evolution cannot be determined perturbatively. The state of the spin bath evolves with the state of the system, and their mutual influence is significant. This leads to distinct contributions to the decoherence (Weiss, 1999; Prokof'ev and Stamp, 2000): "topological decoherence" due to the randomization of the phase of the system caused by the reaction of the bath spins on the system's dynamics; "degeneracy blocking", when resonant tunnelling between the system's states is disrupted by the bath-induced energy mismatch between them; "orthogonality blocking" due to a vanishingly small overlap between the initial and final states of the spin bath; and the "bath fluctuations" contribution from its intrinsic dynamics. Fortunately, we do not need to go into the involved theory of these effects (which would require more powerful mathematical tools), since the types of mesoscopic qubit we consider here should be adequately described by the weak coupling limit, i.e., by the bosonic bath model.[29]

5.3.4 Decoherence from 1/f-noise

The contribution of 1/f-noise to the relaxation rate of Eq. (5.113) and to the relaxation-induced part of the dephasing rate can usually be safely neglected: its spectrum at the qubit transition frequency will be well below the level of thermal noise. The situation is drastically different for pure dephasing. It may seem from Eqs (5.108) and (5.113) that 1/f-noise sounds the death knell for any hope of achieving quantum coherence in a solid-state system: as the dephasing rate depends on the noise spectral density at zero frequency, $S(0)$, a noise with $S(\omega) \propto 1/|\omega|^{\alpha}$, $\alpha \approx 1$, will lead to instantaneous dephasing. Even if we accept the low-frequency cutoff $\omega_{\min}$ (e.g., supplied by the large, but finite, duration of any given experiment), the dephasing rate will be $\propto |\omega_{\min}|^{-\alpha}$, that is, very large. Given the ubiquity of

---

[29] The situation is different for such microscopic systems as "molecular magnets" – molecules with several magnetic ions, the spins of which are locked together (Gatteschi et al., 2006). Two different types of Rabi oscillation (with frequencies of 4.5 MHz and 18.5 MHz), corresponding to different resonant transitions, were observed in such a magnet (a vanadium cluster anion $\left[V_{15}^{IV} As_6^{III} O_{42} (H_2 O) \right]^{6-}$) by Bertaina et al. (2008) (see also Stamp, 2008). Their decoherence times at 4 K (respectively 800 ns and 340 ns) were predominantly due to the spin bath composed of the vanadium nuclear spins of the molecular magnet itself, with a smaller contribution from the hydrogen nuclear spins, and a negligible one (by two orders of magnitude) from the bosonic bath of lattice phonons.

1/f-noise, it is would appear strange that any of the experiments referred to in previous chapters could ever have succeeded.

A moment's reasoning about the physical meaning of 1/f-noise should allay these apprehensions, to a certain degree. If the noise is, e.g., produced by the overlapping random telegraph signals from a set of TLS's, during long periods between flips the qubit will remain in a *static* potential (if the TLS's are classical) or go through quantum beats with a few of them (which happen to be in resonance with the qubit) – which should not lead to any immediate loss of coherence.

Actually, Eqs (5.108) and (5.113) greatly overestimate the influence of low-frequency noise and even give the wrong functional form for decoherence as a function of time. Let us return to Eq. (5.106). We used there the limit $t \to \infty$, which in reality meant $|t| \gg 1/|\omega|$; it is for such times that the coherences decay as $\sim \exp[-\Gamma_{\delta\Omega}|t|]$. This is well and good for noise with a flat spectrum, but the condition $|\omega t| \gg 1$ clearly does not apply when the noise spectrum grows, or even diverges, at low frequencies. In order to analyse the decoherence effects of 1/f-noise we need, therefore, to use Eq. (5.106) directly. We will approximate the 1/f-noise spectral density with

$$S_{\delta\Omega}(\omega) = \left(\frac{\partial\Omega}{\partial A}\right)^2 C\overline{A}^2 \left|\frac{2\pi}{\omega}\right| \theta(\omega_{min} \leq |\omega| \leq \omega_{max}). \qquad (5.131)$$

Here A is the 1/f-fluctuating parameter, C a dimensionless constant, and the infrared and ultraviolet cutoff frequencies are chosen well below or above all the characteristic frequencies in the problem. Substituting this in (5.106), we find (Ithier et al., 2005)

$$\eta(t) = \frac{t^2}{2} \cdot \left(\frac{\partial\Omega}{\partial A}\right)^2 C\overline{A}^2 \cdot 2\int_{\omega_{min}}^{\omega_{max}} \frac{d\omega}{\omega} \frac{\sin^2(\omega t/2)}{(\omega t/2)^2} \approx t^2 \left(\frac{\partial\Omega}{\partial A}\right)^2 C\overline{A}^2 \ln\left(\frac{1}{\omega_{min}t}\right), \qquad (5.132)$$

since for the times $t \ll 1/\omega_{min}$ the low frequencies dominate the integral. We see that not only is the dependence of the decoherence on the infrared cutoff for such time intervals negligible,[30] but the dephasing is Gaussian instead of exponential: $\sim\exp[-at^2]$.[31]

5.4 Decoherence suppression

5.4.1 Dynamic suppression of decoherence

Dephasing due to 1/f-noise and other slow fluctuations in the qubit's environment can be significantly reduced using so-called *quantum bang-bang control*

---

[30] "Logarithm is not a function, it grows too slowly." (L. D. Landau).
[31] This is directly related to the *quantum Zeno effect*, which we will discuss in § 5.5.6.

(Viola and Lloyd, 1998; Viola et al., 1999a,b; Gutmann et al., 2005). Let us return to the representation of the qubit's density matrix by the Bloch vector and consider pure dephasing produced by a single classical TLS. Then the angular velocity of the Bloch vector precession around Oz is

$$\Omega(t) = \frac{2}{\hbar}|\mathcal{H}(t)| = \Omega + A(t), \tag{5.133}$$

where $A(t)$ is a random telegraph signal with amplitude a and the correlation function $a^2 \exp[-2\nu|t|]$ (see (5.100)). The precession angle (that is, the phase gain of the coherences) between flips of the TLS will be approximately $(\Omega \pm a)/2\nu$. Let us initially align the transverse component of the Bloch vector with, e.g., the x-axis, and after some time t_0 apply to the system a very large perturbation of very short duration t_π (the so-called π-*pulse*):

$$H_\pi(t) = \frac{\pi\hbar}{2t_\pi}\theta(t_0 < t < t_0 + t_\pi)\sigma_x \rightarrow \frac{\pi\hbar}{2}\delta(t - t_0)\sigma_x. \tag{5.134}$$

Its effect will be to flip the Bloch vector about the x-axis:

$$\begin{pmatrix} \mathcal{R}_x \\ \mathcal{R}_y \\ \mathcal{R}_z \end{pmatrix} \rightarrow \begin{pmatrix} \mathcal{R}_x \\ -\mathcal{R}_y \\ -\mathcal{R}_z \end{pmatrix}, \tag{5.135}$$

and, effectively, to reverse the direction of its precession around Oz (Fig. 5.8). If the TLS did not switch during the interval $[0, 2t_0]$, by time $2t_0$ the projection of the Bloch vector on the xy-plane, $\mathcal{R}_\perp$, will return to its initial position, and another π-pulse will also restore $\mathcal{R}_z$. The phase gain from the TLS during the period $2t_0$ is, thus, completely cancelled. If the TLS does switch between zero and $2t_0$, the precession speed will change, and there will be no full compensation of the random telegraph signal's contribution to the phase. Nevertheless, this extra phase will be, at worst, $\pm at_0$ instead of about $\pm a/2\nu$. Therefore, a periodic application of π-pulses suppresses dephasing as long as their period t_0 is less than the characteristic switching time, or, more generally, the correlation time of the longitudinal fluctuations. This is essentially the same idea as in the *refocusing technique* in NMR (e.g., Abraham et al., 1988, §6.6), which allows suppression of the effects of variations of the magnetic field (in space as well as in time), in which the nuclear spins rotate. If we had, instead of the Bloch vector, an ensemble of real spins precessing in the magnetic field, and at $t = 0$ they were all aligned, by $t = t_0$ they would turn by different angles, according to the local variation of the magnetic field. Nevertheless, after a π-pulse and then waiting for t_0, all of them would align again, producing the *spin echo* (Hahn, 1950; see also Abraham et al., 1988, Ch. 7; Blum, 2010, §8.4.2).

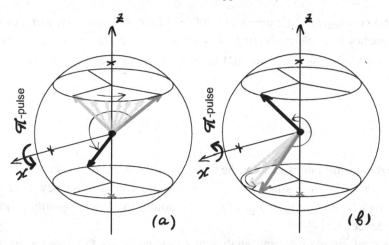

Fig. 5.8. Quantum bang-bang (spin echo) approach to the reduction of dephasing due to low-frequency fluctuations. (a) Due to the presence of random static external potential, the precession angle of the qubit's Bloch vector around the z-axis acquires a random term. After time t_0, a π-pulse flips the Bloch vector, and the direction of its precession is reversed. (b) As a result, after another interval t_0 the precession angle will be zero, and another π-pulse can restore the Bloch vector to its initial position. The random contribution to the qubit phase is completely cancelled (as well as the regular one). If the random potential changes in the interval $[0, 2t_0]$, thus changing the precession frequency, the cancellation of the random phase will be only partial.

We can easily enough calculate the dephasing in the echo experiment for Gaussian classical noise. Substituting in Eq. (5.105) the phase gain during the period $[0, 2t_0]$,

$$\Delta\phi_{\text{echo}}(2t_0) = \int_0^{t_0} dt' \delta\Omega(t') - \int_{t_0}^{2t_0} dt' \delta\Omega(t'), \qquad (5.136)$$

we find (Ithier et al., 2005):

$$\eta_{\text{echo}}(2t_0) = -\ln\left\langle e^{i\Delta\phi_{\text{echo}}(2t_0)} \right\rangle = \frac{(2t_0)^2}{2} \int_{-\infty}^{\infty} \frac{d\omega}{2\pi} S_{\delta\Omega}(\omega) \frac{\sin^4(\omega t_0/2)}{(\omega t_0/2)^2}. \qquad (5.137)$$

Comparing (5.137) with (5.106), we see that now the contribution from the low-frequency part of the noise spectrum (with $|\omega| < \pi/t_0$) is not stressed, but suppressed – as expected from physical considerations. To be more precise, it is necessary to know the noise spectrum. The integral in (5.137) will depend on the behaviour of $S_{\delta\Omega}(\omega)$ near the high-frequency cutoff ω_{max}. For example, 1/f-noise with a sharp cutoff at ω_{max} (i.e., $S_{\delta\Omega}(\omega) \propto |\omega|^{-1}\theta(\omega_{\text{max}} - \omega)$) produces $\eta_{\text{echo}}(2t_0) \propto (\omega_{\text{max}}t_0)^2 t_0^2$, while for $1/\omega^2$-decay at high frequencies one gets instead

$\eta_{\text{echo}}(2t_0) \propto (\omega_{\text{max}}t_0)t_0^2$ (Ithier et al., 2005). In any case, the decoherence due to low-frequency fluctuations is suppressed by the small parameter $\omega_{\text{max}}t_0$. Assuming instead a white noise spectrum and keeping in mind that $\int_{-\infty}^{\infty} dx \frac{\sin^4 x}{x^2} = \frac{\pi}{2}$, we see that

$$\eta_{\text{echo}}(2t_0) = S_{\delta\Omega}^{\text{white}} \cdot |t_0| = \frac{1}{2} S_{\delta\Omega}^{\text{white}} \cdot |2t_0|, \tag{5.138}$$

and the dephasing rate is exactly the same as without the refocusing, Eq. (5.108). This is not surprising, since white noise contains high-frequency fluctuations, against which refocusing is not effective (though the exact equality is more of a coincidence).

Numerical simulations and analytical calculations by Gutmann et al. (2005) showed that for large observation times the suppression of dephasing caused by a single classical fluctuator with the average flipping time τ_{bfl} by the application of the bang-bang protocol[32] (a train of π-pulses with period τ_{bb}) is linear in $\tau_{\text{bb}}/\tau_{\text{bfl}}$, at least in the range $10^1 < \tau_{\text{bb}}/\tau_{\text{bfl}} < 10^5$. In the more complex and interesting situation when the fluctuating environment has quantum dynamics of its own, *quantum bang-bang control* (alias *dynamic refocusing*) works as long as $\tau_{\text{bb}}/\tau_c \ll 1$, where τ_c is the correlation time for the environment; in the limit of continuous application of π-pulses ($\tau_{\text{bb}} \to 0$) not only the effects of noise would be totally eliminated, but also all the dynamics of the qubit would be completely frozen (Viola and Lloyd, 1998). This, again, is not surprising, since then all other terms in the Hamiltonian would be negligible compared to the bang-bang ones, which would keep flipping the Bloch vector about the x-axis.

Experimental confirmation of the approach comes, e.g., from the fact that qubit decoherence times measured using the spin-echo technique systematically exceed those measured directly. Byrd and Lidar (2002) give a general mathematical perspective of quantum bang-bang control.

5.4.2 Decoherence suppression by design: the optimal point

The drawback of dynamical refocusing is that it requires active interference in the dynamics of the system, inevitably bringing extra noise, dissipation and decoherence of its own. Instead, we could passively suppress the effects of decoherence by decoupling the qubits from noise sources. One way to do this is to ensure that in the first approximation small fluctuations of the environment do not affect the

---

[32] The term "bang-bang control" originates from classical control theory, where it means applying only maximum control, or none at all.

qubit. Consider a qubit Hamiltonian in the physical (diabatic) basis (using separate operators for the transverse ($\hat{f}_x$) and longitudinal ($\hat{f}_z$) noise):

$$H = -\frac{1}{2}(\Delta\sigma_x + \epsilon\sigma_z) - \hat{f}_x(t)\sigma_x - \hat{f}_z(t)\sigma_z$$

$$= -\frac{1}{2}\left((\Delta + 2\hat{f}_x(t))\sigma_x + (\epsilon + 2\hat{f}_z(t))\sigma_z\right). \tag{5.139}$$

We have left out the bath Hamiltonian H_E, assuming that the qubit's influence on the environment is negligible. In the adiabatic basis, (5.139) can be rewritten as

$$H = -\frac{1}{2}\left[(\Delta + 2\hat{f}_x(t))^2 + (\epsilon + 2\hat{f}_z(t))^2\right]^{1/2}\tilde{\sigma}_z$$

$$= -\frac{\hbar\Omega(\Delta,\epsilon)}{2}\tilde{\sigma}_z - \frac{\Delta\hat{f}_x(t) + \epsilon\hat{f}_z(t)}{\hbar\Omega(\Delta,\epsilon)}\tilde{\sigma}_z - \left[\frac{\hat{f}_x^2(t)}{\hbar\Omega(\Delta,\epsilon)}\left(1 - \frac{\Delta^2}{(\hbar\Omega(\Delta,\epsilon))^2}\right)\right.$$

$$\left. - \frac{\Delta\epsilon\{\hat{f}_x(t),\hat{f}_z(t)\}}{(\hbar\Omega(\Delta,\epsilon))^3} + \frac{\hat{f}_z^2(t)}{\hbar\Omega(\Delta,\epsilon)}\left(1 - \frac{\epsilon^2}{(\hbar\Omega(\Delta,\epsilon))^2}\right)\right]\tilde{\sigma}_z + o(\hat{f}^2) \tag{5.140}$$

where $\hbar\Omega(\Delta,\epsilon) = \sqrt{\Delta^2 + \epsilon^2}$ is the interlevel spacing in the unperturbed qubit, a function of an operator is understood, as usual, as its Taylor expansion (if it exists), and $o(\hat{f}^2)$ denotes terms of the third and higher orders in the noise variables. By tuning the bias, ϵ, to zero, we obtain

$$H = -\left[\frac{\Delta}{2} + \hat{f}_x(t) + \frac{\hat{f}_z^2(t)}{\Delta}\right]\tilde{\sigma}_z + o(\hat{f}^2). \tag{5.141}$$

This expression is valid for small noise, i.e., when $\langle\hat{f}_x\rangle \ll \Delta$ and $\langle\hat{f}_z^2\rangle \ll \Delta^2$. We see that by choosing $\epsilon = 0$ the influence of longitudinal noise is eliminated in the first order. If for some reason transverse noise is negligible, decoherence is only due to the second-order term and is, therefore, suppressed. This is the idea behind the operation of a qubit at the *optimal point* in the parameter space. It does not work for every type of qubit. For example, the Hamiltonian of a superconducting phase qubit (2.68) in the absence of an external ac field only contains the σ_z-term, and the optimal point simply does not exist. But it applies to qubits where the physical states are hybridized by tunnelling (like superconducting charge and flux qubits, or 2DEG quantum dot qubits).[33]

Makhlin and Shnirman (2004) investigated decoherence by longitudinal noise at the optimal point (putting $\hat{f}_x = 0$ in (5.141)), which then reduced to pure dephasing

[33] A flux qubit is coupled to external electromagnetic fields through the persistent current around its loop, the operator of which $\hat{I}_p \propto \sigma_z$ in the physical basis. Therefore, at the optimal point it is completely uncoupled from these fields, since there $\langle\sigma_z\rangle = 0$ either in the ground or excited state.

with quadratic coupling given by

$$e^{-\eta(t)} = \left\langle \widetilde{T} \exp\left[\frac{i}{\hbar} \int_0^t dt' \frac{\hat{f}_z^2(t')}{\Delta}\right] T \exp\left[\frac{i}{\hbar} \int_0^t dt' \frac{\hat{f}_z^2(t')}{\Delta}\right] \right\rangle, \qquad (5.142)$$

where T ($\widetilde{T}$) are the operators of (anti)chronological ordering (see, e.g., Zagoskin, 1998, §§ 1.3.1, 3.4.2). In the lowest-order perturbation theory, this reduces to Eq. (5.106) with $S_{\delta\Omega}$ replaced by

$$S_{f^2,s}(\omega) = \int_{-\infty}^{\infty} dt\, e^{i\omega t} \frac{1}{2} \left\langle \left[\frac{\hat{f}_z^2(t)}{\Delta}, \frac{\hat{f}_z^2(0)}{\Delta}\right] \right\rangle. \qquad (5.143)$$

For an ohmic bath, when the spectrum of fluctuations of $\hat{f}_z$ (not of $\hat{f}_z^2$!) $S_{f,s}(\omega) \propto \omega \coth(\hbar\omega/k_B T)$, the decay is exponential

$$\eta_{\text{ohmic}}(t) = \frac{t}{T_2} \propto t T^3, \qquad (5.144)$$

with the dephasing time growing at low temperatures faster ($\propto T^{-3}$) than for linear coupling (i.e., away from the optimal point), where it only grows as T^{-1}. For 1/f-noise with infrared cutoff $\omega_{\min}$, one gets instead

$$\eta_{1/f}(t) = \left(\frac{\gamma_f}{\pi} t \ln(\omega_{\min} t)\right)^2, \qquad (5.145)$$

where γ_f is proportional to the amplitude of the 1/f-noise. Beyond this approximation, the results for the ohmic bath do not change, while Eq. (5.145) changes to

$$\eta_{1/f}(t) = \begin{cases} \frac{1}{4} \ln\left[1 + \left(\frac{2}{\pi}\gamma_f t \ln(\omega_{\min} t)\right)^2\right], & \gamma_f t \ll 1 \\ \gamma_f/2, & \gamma_f t \ll 1 \end{cases} \qquad (5.146)$$

showing that in the long time limit, the dephasing is also exponential.

5.4.3 Transmon

Optimal point protection was used in the design of the quantronium qubit (§ 2.4.4) and especially for the so-called *transmon* qubit (Koch et al., 2007; Schreier et al., 2008; Houck et al., 2008, 2009). The transmon, like the "differential charge qubit" (DCQ) (Shnirman et al., 1997; Bladh et al., 2002; Shaw et al., 2007), consists of two superconducting islands, which are electrically *isolated* from the rest of the system. The Hamiltonian of the system is

$$H(n^*) = E_C(\hat{N} - n^*/2)^2 - E_J \cos\phi, \qquad (5.147)$$

where $\hat{N}$ is the operator of the number of Cooper pairs transferred between the islands, and ϕ is the gauge-invariant phase difference between them. In the case of the standard charge qubit, the Cooper pairs could tunnel between the SSET island and the bulk superconductor; the total charge of the system was not conserved, and this allowed us to fix the superconducting phase of the island. The conservation of the total charge on a DCQ on transmon does not make the phase difference between the islands indefinite, notwithstanding the number-phase uncertainty relation (2.47). This is because the operators of the total number of Cooper pairs on both islands, $\hat{N}_{tot}$, and of the phase difference between them, ϕ, commute. Indeed, since

$$\hat{N} = \frac{\hat{N}_1 - \hat{N}_2}{2}, \quad \hat{N}_{tot} = \hat{N}_1 + \hat{N}_2, \quad \phi = \phi_1 - \phi_2, \tag{5.148}$$

we see that

$$[\hat{N}_{tot}, \phi] = [\hat{N}_1, \phi_1] - [\hat{N}_2, \phi_2] = 0; \quad [\hat{N}, \phi] = \frac{[\hat{N}_1, \phi_1] + [\hat{N}_2, \phi_2]}{2} = -i. \tag{5.149}$$

The average phase of the islands, $\bar{\phi} = (\phi_1 + \phi_2)/2$, is indefinite, but it does not influence any observable and so is totally irrelevant.[34]

The crucial difference between the transmon and the ordinary SSET (as well as the DCQ) is that the transmon is designed to operate in the regime $E_J/E_C \gg 1$ (in practice, from a few dozen to several hundred), and not $E_J/E_C \lesssim 1$, typical for charge qubits. This is achieved by shunting the islands with a large capacitance C_B and similarly increasing the gate capacitance C_g (Fig. 5.9). In this regime we can no longer treat the Hamiltonian (5.147) the way we treated the Hamiltonian of an SSET, Eq. (2.122), restricting it to a subspace of states with N and $N+1$ Cooper pairs – which was essentially an expansion in E_J/E_C near the intersection of two charging energy parabolas (circled in Fig. 5.9). As we mentioned in Section 2.4 (see Eq. (2.134)), the eigenfunctions and eigenvalues of (5.147) can be found explicitly in terms of Mathieu functions (see Koch et al., 2007, where a detailed theory of the transmon is given). In the limit $E_J/E_C \gg 1$ the "charge dispersion" of the mth eigenstate (that is, the difference between the maximum and minimum of the mth eigenenergy, $E_m(n^*)$ ($m = 0, 1, \ldots$), as a function of the gate-induced charge n^*e, is given by (Koch et al., 2007)

$$\Delta E_m \approx (-1)^m \frac{E_C}{4} \frac{2^{4m+5}}{m!} \sqrt{\frac{2}{\pi}} \left(\frac{2E_J}{E_C}\right)^{\frac{m}{2}+\frac{3}{4}} e^{-4\sqrt{\frac{2E_J}{E_C}}}. \tag{5.150}$$

---

[34] This situation recalls Bohr's response (Bohr, 1935) to the EPR paradox, where he stressed that the full set of observables, which describe a two-partite (or, indeed, any) quantum system, is not fixed (any linearly independent combination will do), and the actual choice is dictated by the physical situation, i.e., by what can actually be measured in a given setup.

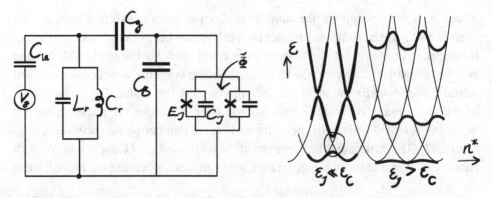

Fig. 5.9. (Left) Effective circuit diagram of the transmon qubit (after Koch et al., 2007). (Right) Scheme of energy levels of a charge qubit for different ratios of Josephson to charging energy.

This shows that the energy curve near the optimal points (e.g., $n^* = 1$) becomes exponentially flat as E_J/E_C grows. Therefore, the dephasing rate at the optimal point due to the charge fluctuations (induced by the external fields, or, more dangerously, local charged TLS's), being proportional to $\left(\partial^2(E_1 - E_0)/\partial n^{*2}\right)^2$, is exponentially suppressed. Moreover, due to the flatness of $E_{0,1}(n^*)$, dephasing is strongly suppressed even away from the optimal point.[35] However, even in the limit $E_J/E_C \gg 1$ the transmon remains sufficiently anharmonic:

$$E_{m+1} - E_m \approx \sqrt{2E_C E_J} - E_C(m+1), \tag{5.151}$$

so that its Hilbert space can be truncated to the envelope of the two lowest eigenstates of (5.147), and the transmon can be used as an effectively two-level system (a qubit). The best observed values of dephasing and relaxation times in transmons are $T_1 = 1.57\,\mu s$ and $T_2 = 2.94\,\mu s \approx 2T_1$. Therefore, pure dephasing must have been, indeed, strongly suppressed ($T_\phi > 35\,\mu s$), and the relaxation remains the main source of decoherence (Houck et al., 2009).

5.4.4 Decoherence suppression by detuning

Unlike pure dephasing, the relaxation rate of a qubit (5.31, 5.113) explicitly depends on its interlevel spacing. Therefore, as we have noted, it is possible to enhance or suppress the relaxation by tuning the qubit in or out of resonance with the prevalent noise source. This is the case when the external fields are filtered: if the noise

---

[35] Theoretical predictions for the transmon dephasing time by 1/f-fluctuations of charge run as high as $T_\phi \sim 25$ ms, leaving the 1/f-fluctuations of the Josephson critical current ($T_\phi \sim 35\,\mu s$) and the contribution from the relaxation as the leading sources of decoherence (Houck et al., 2009).

spectral density is concentrated in a band of width $\Delta\omega$, then tuning the qubit away from this band would suppress the relaxation. Such a situation arises for flux qubits placed in a high-quality LC tank circuit (Fig. 4.4); or in the circuit QED setup, with charge qubits placed in a high-quality strip-line resonator (Fig. 4.11). However, if the qubit and the leaky resonator (with finite exponential decay or leakage rate κ) are in resonance, the qubit relaxation rate, and with it the decoherence rate, will be increased by the quality factor of the resonator, Q (see, e.g., Blais et al., 2004). In quantum optics this is known as the *Purcell effect* (Purcell, 1946).[36] The resonator must be leaky enough ($\kappa > g$, where g is the qubit-resonator coupling constant in Eq. (4.46)); otherwise the qubit and the oscillator states will instead hybridize and form dressed states (4.61), the relaxation rate of which for precise resonance will be $(\kappa + \Gamma_i)/2$ (where Γ_i is the intrinsic qubit relaxation rate due to everything except the relaxation by emitting photons into the resonator; see Blais et al. (2004)).

For a resonator with frequency ω_0 and leakage rate κ, the spectral density of fluctuations is given by Eq. (5.18) with the delta-functions being replaced by the Lorentzians:

$$S_X(|\omega|) = [n_B(\hbar\omega_0/k_BT) + 1]\frac{\kappa/2}{(\omega - \omega_0)^2 + (\kappa/2)^2};$$

$$S_X(-|\omega|) = n_B(\hbar\omega_0/k_BT)\frac{\kappa/2}{(\omega + \omega_0)^2 + (\kappa/2)^2}. \tag{5.152}$$

At zero temperature and at large detuning (dispersive regime, $|\Omega - \omega_0| \equiv |\delta| \gg \kappa, g$, where $\hbar\Omega$ is the qubit level splitting), this leads to the qubit relaxation rate due to the spontaneous emission into the resonator (Blais et al., 2004)

$$\Gamma_\kappa \approx \frac{g^2}{\delta^2}\kappa \ll \kappa. \tag{5.153}$$

The total relaxation rate of the qubit and of the cavity are, respectively (Blais et al., 2004; Houck et al., 2007)

$$\Gamma_{tot} \approx \Gamma_i + \frac{g^2}{\delta^2}\kappa; \quad \kappa_{tot} \approx \kappa + \frac{g^2}{\delta^2}\Gamma_i. \tag{5.154}$$

Another way to look at (5.153) and (5.154) is to recall that every dressed state (4.61) is part-qubit excitation and part-resonator photon, with the mixing angle θ_n

---

[36] Purcell (1946) noted that the rate of spontaneous emission at the wavelength λ in a resonator attuned to it is increased, compared to free space, by a factor $f = 3Q\lambda^3/4\pi^2 V$, where Q is the quality of the resonator and V its volume. The physical reason is that there is precisely one oscillator in the corresponding frequency range v/Q. Purcell was interested in increasing the emission rate in order to speed up the time required to reach thermal equilibrium of a spin subsystem: he pointed out that in free space the relaxation time of a nuclear magneton at room temperature and at 10 MHz would be 5×10^{21} s.

dependent on the detuning. When $\delta = 0$, $\theta_n = \pi/4$, and the mixing is maximal (hence, the $(\kappa + \Gamma_i)/2$ relaxation rate mentioned above). In the dispersive regime, $\theta_n \approx g\sqrt{n+1}/\delta \ll 1$; therefore, in the second-order perturbation theory we obtain mutual contributions to the relaxation rates of the qubit and resonator proportional to $(g/\delta)^2$.

5.5 Measurements and decoherence

5.5.1 Quantum description of measurement

Measurement is a source of decoherence by definition. It is, after all, a process of interaction of a quantum system with a special kind of environment, a detector, which erases the coherences in the system's density matrix, leading to the reduction, or collapse, of the quantum state, i.e., its projection of the quantum state onto an eigenstate of the measured observable. In Section 1.1 we included the collapse of the wave function among the axioms of quantum mechanics. The term "collapse" is in itself deceptive, since it implies an instantaneous measurement. Such language is well suited to describe, e.g., a flash of light when ionizing radiation hits a scintillation detector. But there is actually nothing in nature or in the rules of quantum mechanics which would demand the process of measurement to be instantaneous. The experimentally realized qubit measurement procedures cover the whole range, from an instantaneous collapse ("strong" or "projection measurement") to a protracted process ("weak continuous measurement", Braginsky and Khalili (1992)). The decohering effect ("back-action") of the detector on the system depends on the character of the measurement of the quantum state. The loss of coherence in itself is a necessary, but not sufficient, condition of the measurement. For an ideal (i.e., quantum limited) detector, the only decoherence introduced by the detector is due to the measurement.

Unlike the decohering environments discussed before (except the spin bath in a strong coupling regime), the state of the detector must also be affected by the state of the measured system – otherwise it would not be much of a detector – and affected in a way that allows the determination of the state of the system. Therefore, our previous treatment of decoherence requires some refinement.[37]

Let us first consider a strong measurement. We saw earlier an example, when the "which-state"-measurement of a two-level quantum system was modelled by a simple Lindblad operator (1.63), leading to the evolution equation (1.67). The rate γ (the pure dephasing rate) at which the state was reduced to the sum of projectors, $\rho_0|0\rangle\langle 0| + \rho_1|1\rangle\langle 1|$, could be chosen at will. Of course, it is only a model

---

[37] A detailed exposition of measurement theory is given by Braginsky and Khalili (1992); see also Kholevo (1982), Namiki et al. (1997), Gardiner and Zoller (2004), Chapter 2, Clerk et al. (2010) and references therein.

(Chapman et al., 1975). Whether it is applicable to a specific situation remains to be seen. Generally, one cannot exclude relaxation and excitation of the system in the process of measurement. If we use the decoherence time as an approximation of the measurement time, τ_m, we must conclude from Eq. (5.113) that the measurement of a qubit will induce the noise acting on the system (detector back-action) given by

$$\frac{1}{2} S_{f,s}(\Omega) \sin^2 \theta + S_{f,s}(0) \cos^2 \theta \approx \frac{\hbar^2}{\tau_m}. \tag{5.155}$$

It will grow indefinitely as we attempt to shorten the measurement time to zero. Unless $\theta = 0$ exactly, the detector back-action noise will have a transverse component, which will induce transitions between the qubit levels and disrupt the very state we were going to measure. Suppose, e.g., that we want to determine the state of a phase qubit (Fig. 2.4). This is equivalent to the measurement of its energy with the error $\Delta E < \hbar \omega_{01}$, the interlevel spacing, so from general considerations (see, e.g., Landau and Lifshitz, 2003, § 44) the measurement cannot take less than $\tau_m \sim 1/\omega_{01}$. However, there are other sources of decoherence besides the detector, with some decoherence rate Γ_e, and it is necessary to keep $\tau_m \ll 1/\Gamma_e$, which does not allow too slow a measurement, or too weak a coupling between the detector and the system.

To be more specific, in order to read out a phase qubit, we can increase the bias current $I(t)$ so that the state $|1\rangle$ would rapidly tunnel out into the continuum and produce the dissipative voltage pulse, which is detected. Alternatively, if there are three levels in the potential well, we can apply an rf pulse at the frequency ω_{12} (or ω_{02}) in order to induce the Rabi transition from the state $|1\rangle$ (respectively $|0\rangle$) to the state $|2\rangle$, with the same outcome. In either case the current pulse must change on the scale of τ_m, and the Fourier transform of $I(t)$ will have a nonvanishing magnitude at frequencies up to $\sim 1/\tau_m$. Therefore, the effective Hamiltonian (2.68) will contain, along with the longitudinal σ_z-term, the transverse $\sigma_{x,y}$-terms, which can change the state of the qubit during its measurement. The resulting "imprecise", or "approximate", measurement (as opposed to the "precise", ideal projection measurement) yields the correct information about the state of a quantum system only with some probability, and the cycle of quantum state preparation and measurement must be repeated several times. The realization of a single-shot measurement, where the qubit state can be determined with a probability close to unity after one such cycle, is a nontrivial problem, which has been, nevertheless, solved for several qubit types.[38]

---

[38] For example, electron spins in 2DEG quantum dots (Hanson et al., 2005); phase (Cooper et al., 2004), charge (Astafiev et al., 2004) and flux qubits (Lupascu et al., 2006); transmons (Mallet et al., 2009); for a review, see Johansson et al. (2006b).

5.5.2 *Quantum nondemolition (QND) measurement*

In the simple model of measurement, Eq. (5.155), the dangerous detector back-action is due to the transverse terms, which can induce transitions between the qubit levels. This is because the qubit state measurement is actually the measurement of the observable σ_z (in the energy basis); longitudinal noise, also proportional to σ_z, commutes with this observable and, therefore, cannot change it. More generally, if the measured observable A commutes with the full Hamiltonian $H = H_S + H_D + H_I$ (where H_S, H_D and H_I are the Hamiltonians of the system, of the detector and of the system-detector interaction, respectively), and is not explicitly time-dependent, then the Heisenberg equations of motion for it,

$$i\hbar \dot{A} = [A, H] = 0, \tag{5.156}$$

show that it will be constant during and after the measurement, and, therefore, its measurement will always give the same outcome: the state of the system will not be "demolished" by the measurement. To look at this from another angle, $[A, H] = 0$ guarantees that a Hermitian operator A and the full Hamiltonian have a complete set of common eigenstates (Landau and Lifshitz, 2003, §§4, 11), $\{|a_j\rangle\}$, $A|a_j\rangle = a_j|a_j\rangle$, and a projective measurement – according to the projection postulate – after collapsing the initial state ρ of the system into one of these states (with probability $p_j = \langle a_j|\rho|a_j\rangle$), will *always* return the same value a_j of A and leave the system in the same state $|a_j\rangle$. The restriction

$$[A, H_S + H_I] = 0 \tag{5.157}$$

is called the *strong* QND condition (we excluded H_D, because, of course, $[A, H_D] = 0$ by the very definition of the detector Hamiltonian). As a matter of fact, this is a sufficient, but not necessary condition: if U_m is the evolution operator, which describes the evolution of the observable A during the measurement, then (Braginsky and Khalili, 1992, § 4.4*)

$$\left[U_m^\dagger A U_m - A\right]|\Psi_{D,0}\rangle = 0 \tag{5.158}$$

is the necessary and sufficient condition of a QND measurement. Here $|\Psi_{D,0}\rangle$ is the state of the detector before the measurement. In principle, (5.158) can be satisfied for a special choice of $|\Psi_{D,0}\rangle$, while $\left[U_m^\dagger A U_m - A\right] \neq 0$.[39] As Braginsky and Khalili (1992) pointed out, the realization of a weak QND condition

---

[39] Such equalities, which hold for certain projections, but not for the operators themselves, are called in quantum field theory *weak* equalities (Dirac, 2001), not to be confused with weak measurements, which will be considered later.

(5.158) would allow interesting possibilities, e.g., using the *same* detector with the *same* coupling to the system in order to realize QND measurements of *different* noncommuting observables; but the required engineering of the quantum state of the detector made this impractical at that time. The possibilities provided by the current state of qubit design and manipulation mean that it is worthwhile revisiting the subject. Here we will consider only the strong QND condition (5.157), which, unless we deliberately choose a pathological case, implies $[A, H_S] = 0$ and $[A, H_I] = 0$ independently. The first of these means that only the integrals of motion of the unperturbed system can be QND-measured, and so circumscribes the information about the system which can be obtained this way. (We have tacitly omitted all the sources of decoherence except the detector.) The second restricts the possible detector-system interaction; in terms of Eq. (5.109), A and B must commute.

The obvious candidate observable for a QND measurement is the energy, and the readout of a qubit state can usually be reduced to an energy measurement. Consider, e.g., a qubit coupled to a resonator in the dispersive regime (large detuning δ), with the Hamiltonian (4.69),

$$
H = \underbrace{-\frac{\hbar\Omega}{2}\sigma_z}_{H_S} + \underbrace{\hbar\omega\left(a^\dagger a + \frac{1}{2}\right)}_{H_D} + \underbrace{\frac{\hbar g^2}{\delta}\sigma_z\left(a^\dagger a + \frac{1}{2}\right)}_{H_I} \tag{5.159}
$$

(plus corrections of higher order in $|g/\delta| \ll 1$). The interaction term can be considered both as the renormalization of the qubit energy by the cavity (the ac Stark shift) and as the cavity frequency shift dependent on the qubit's state. It is the latter that we will use for the measurement, the resonator and its environment playing the role of the detector. Since $[H_I, H_S] = 0$, this will be a QND measurement.

Looking at (5.159) from the qubit's perspective, we see that H_I describes detector noise. Measuring the frequency of the resonator will always lead to fluctuations of the photon number about its average value, $\delta\hat{n} = a^\dagger a - \langle a^\dagger a\rangle$. The resulting dephasing term $\hbar(g^2/\delta)\sigma_z\delta\hat{n}$ will produce dephasing, which – assuming a flat noise spectrum – is given by

$$
\gamma = 2\left(\frac{g^2}{\delta}\right)^2 S_{\delta\hat{n}}(0). \tag{5.160}
$$

In the best-case scenario, when the resonator becomes a quantum limited detector, all of the dephasing will be due to the measurement: the dephasing rate and appropriately defined information acquisition rate coincide (Clerk et al., 2010, Section III.B).

5.5.3 Weak continuous measurement

The energy of an energy eigenstate can be measured with arbitrary precision, if
we do not limit the measurement time (Landau and Lifshitz, 2003, § 44). This can
be done by means of *weak continuous measurement*, WCM, when the state of the
detector, which is weakly coupled to the system, is monitored continually. As a
result, the information about the system's state is acquired not instantaneously, like
in a strong projective measurement, but little by little over a period of time. Like
in the strong case, the state of the system after the measurement is projected onto
an eigenstate of the measured observable, in accordance with the measurement
postulate. The advantage of WCM is that due to the small coupling to the detector
the back-action is minimized. We have already encountered one example of WCM:
the measurement of the charge state of a quantum dot by the quantum point contact
detector (Sections 3.1.4 and 5). The state of the dot is determined by measuring
the QPC conductance, i.e., averaging over the outcome of many single-electron
scattering events by the dot potential, each only weakly affecting the quantum
state of the dot and gradually dephasing it (see Eq. (3.56)) and contributing to the
quantum state-dependent average QPC current.

The WCM approach is especially convenient for solid-state-based quantum
devices: the spatial proximity of the device and the control and readout circuit
makes it desirable to minimize the coupling strength between the measured system
and the detector, in order to preserve the quantum coherence of the system.[40] One
WCM scheme in particular proved useful, i.e., the measurement of a superconducting qubit state by monitoring the frequency shift of the resonator to which the
qubit is coupled (e.g., Il'ichev et al., 2003; Izmalkov et al., 2004b; Blais et al., 2004;
Lupascu et al., 2004; Wallraff et al., 2004). It is instructive to consider its pre-cQED
version, the *impedance measurement technique*, IMT (Il'ichev et al., 2003, 2004),
where a flux qubit is inductively coupled to a high-quality LC circuit (tank circuit)
(Fig. 4.4), the resonance frequency of which is significantly lower than the qubit
splitting (typically ~ 50 MHz vs. 1 GHz), which in terms of cQED means that we are
deep in the dispersive regime. The resonance frequency of the tank $\omega_T = c/\sqrt{LC}$
depends on the total inductance of the circuit, which includes the contribution from
the qubit (or qubits). As we have seen in Section 2.5, the inductance of the latter
contains a quantum inductance contribution, (2.142, 2.143), which depends on the
qubit state. Therefore, a measurement of the resonance frequency of the LC circuit
is a measurement of the qubit state. In IMT, the resonance frequency of the circuit
is determined by monitoring its response to the continuous ac signal at a frequency
close to ω_T.

---

[40] WCM is discussed in detail in Braginsky and Khalili (1992), Chapters VI and VII, Makhlin et al. (2001),
Section V, Clerk et al. (2010), Section III.

In the quasiclassical limit (Greenberg et al., 2002) the voltage in the tank will satisfy

$$\ddot{V} + \frac{\omega_T}{Q}\dot{V} + \omega_T^2 V = -\frac{M\omega_T^2}{c^2}\dot{I}_q + \frac{1}{C}\dot{I}_{ac}. \tag{5.161}$$

Here $Q = \omega_T/\kappa$ is the quality factor of the tank circuit (with decay rate κ), M is the mutual inductance of the tank and the qubit, $I_q(t) = \langle \hat{I}_q \rangle$ is the expectation value of the current operator in the qubit loop (which in the two-state approximation is given by $\hat{I}_q = I_p \sigma_z$, with I_p being the amplitude of the persistent current in the flux qubit), and $I_{ac}(t) = I_b \cos\omega t$; $\omega \approx \omega_T$, is the small harmonic forcing term. The superconducting current in the qubit loop can be obtained directly from the qubit Hamiltonian,

$$I_q(t) = c\left\langle \frac{\partial H_{qb}}{\partial \Phi} \right\rangle = c\frac{\partial E_\alpha}{\partial \Phi}, \tag{5.162}$$

by differentiating the expectation value of its energy over the magnetic flux Φ through the qubit loop (we consider here the persistent current qubit and can, therefore, neglect its self inductance L_q when calculating Φ). In the last step we assume that the qubit is in its energy eigenstate $|\alpha\rangle$ (otherwise a WCM would have been impossible). Since the external flux through the qubit loop $\Phi = MI_T/c$, we arrive at

$$\ddot{V} + \frac{\omega_T}{Q}\dot{V} + \omega_T^2 V = -k^2 L_q \omega_T^2 \frac{\partial^2 E_\alpha}{\partial \Phi^2}V + \frac{1}{C}\dot{I}_{ac}, \tag{5.163}$$

where the dimensionless qubit-tank coupling coefficient $k = M/\sqrt{LL_q}$. The resonance frequency of the tank circuit is shifted by

$$\Delta\omega_T = \omega_T \cdot k\sqrt{L_q \frac{\partial^2 E_\alpha}{\partial \Phi^2}}\bigg|_{\Phi=\Phi_{dc}}. \tag{5.164}$$

Here Φ_{dc} is the dc flux bias through the qubit loop; degeneracy between the physical states (with clock- and anticlockwise currents around the loop) is achieved at $\Phi_{dc} = \Phi_0/2$, the half-flux quantum. While away from the degeneracy point this contribution is zero, close enough to it the second derivative has a sharp dip (in the ground state) or peak (in the excited state), with the characteristic width determined by the tunnelling between the physical qubit states.

For the forcing frequency $\omega \approx \omega_T$ the approximate solution of (5.163) can be found using an ansatz $V(t) = V_T(t)\cos(\omega t + \chi(t))$, where $V_T(t)$ and $\chi(t)$ are slow functions of time (on the scale of $1/\omega_T$), and finding the stationary solutions. By measuring the impedance of the tank, or more specifically, the phase angle χ between the forcing current and the response voltage, we measure the frequency shift, and, therefore, the qubit state. For small L_q and forcing amplitude I_b the result is (Greenberg et al., 2002; Il'ichev et al., 2004)

$$\tan \chi = k^2 Q L_q \left. \frac{\partial^2 E_\alpha}{\partial \Phi^2} \right|_{\Phi = \Phi_{dc}}. \tag{5.165}$$

A consistent quantum-mechanical treatment starts from the Hamiltonian, which in the diabatic basis can be written as

$$H = -\frac{1}{2} (\epsilon \sigma_z + \Delta \sigma_x) + \hbar \omega_T \left(a^\dagger a + \frac{1}{2} \right) - \sigma_z \left(\gamma \hat{I}_T + f(t) \right) + H_{dec}, \tag{5.166}$$

where the coefficient $\gamma = M I_p / c^2$ describes the qubit-tank coupling, $\hat{I}_T = c\sqrt{\hbar \omega_T / 2L}(a + a^\dagger)$ is the tank current operator, $f(t)$ is a small c-number ac drive, and all the terms responsible for the decoherence in the qubit and tank are swept into H_{dec}. This Hamiltonian is equivalent to (5.159) only sufficiently far from the qubit degeneracy point, $\epsilon = 0$ (which corresponds to $\Phi_{dc} = \Phi_0/2$). Near it the σ_z-coupling term no longer commutes with the qubit Hamiltonian ($\sim \sigma_x$), so that IMT is not a QND measurement in the strict sense of (5.157). Nevertheless, for a small enough drive amplitude the transition probability between the qubit states remains negligible even at $\epsilon \ll \Delta$, and the result (5.165) stands (Smirnov, 2003; Il'ichev et al., 2004).

5.5.4 Detector back-action and measurement rate

In order to obtain general relations for the detector back-action during continuous measurement, which follow from the quantum uncertainty relations, we will limit ourselves to the case of linear detection, i.e., we will develop the linear response theory for a detector coupled to the measured system (Braginsky and Khalili, 1992, Chapter VI; Averin, 2003; Clerk et al., 2010, Section IV). We will denote the qubit-detector coupling term by

$$H_I = \lambda B \hat{f}, \tag{5.167}$$

where for the sake of convenience we choose dimensionless system (B) and detector ($\hat{f}$) operators and introduce the coupling strength λ, which must be much less than the characteristic system and detector energy scales (otherwise the linear approximation would fail). The detector will be modelled by a black box with input $\hat{f}$ and output $\hat{q}$. The observable $\hat{q}$ reflects the state of the detector when it is measuring the system (like the position of a needle), and we can always choose it to have a zero expectation value when the detector is decoupled from the system. Then using (5.167) in the linear response theory we will obtain Eq. (1.38) in the form

$$q(t) = \lambda \int_{-\infty}^{\infty} dt' \ll q(t) f(t') \gg^R \langle B(t') \rangle, \tag{5.168}$$

where the retarded Green's function[41] of the detector input/output observables $\hat{f}, \hat{q}$ is

$$\ll q(t)f(t') \gg^R = \frac{1}{i\hbar} \langle [q(t), f(t')] \rangle_0 \theta(t - t'). \tag{5.169}$$

The average is taken over the (stationary) unperturbed state of the detector, and it will, therefore, depend only on the difference $t - t'$. We can also introduce the output/input Green's function (reverse gain),

$$\ll f(t)q(t') \gg^R = \frac{1}{i\hbar} \langle [f(t), q(t')] \rangle_0 \theta(t - t'), \tag{5.170}$$

linking the expectation value of $\hat{f}$ to that of the system's observable B via

$$f(t) = \lambda \int_{-\infty}^{\infty} dt' \ll f(t)q(t') \gg^R \langle B(t') \rangle. \tag{5.171}$$

If we take the operator $B = \sigma_z$, that is, assume weak continuous QND measurement, the spectral density of fluctuations of $\hat{f}$ at zero frequency, $S_{\hat{f},s}(0)$, will determine the dephasing rate

$$\gamma = 2\frac{\lambda^2}{\hbar^2} S_{\hat{f},s}(0). \tag{5.172}$$

In order to compare it to the measurement rate, we need to introduce a definition of the latter. Following Clerk et al. (2010), let us consider the integrated detector output,

$$\hat{Q}(t) = \int_0^t dt' \, \hat{q}(t'). \tag{5.173}$$

If the duration of the measurement is much longer than the detector correlation time, then we can assume that the statistics of $\hat{Q}(t)$ is Gaussian, and for the qubit state $|\alpha = 0, 1\rangle$ the expectation value of $\hat{Q}(t)$ is

$$\langle \hat{Q}(t) \rangle_{|0\rangle} = \lambda \ll qf \gg^R_{\omega=0} t; \quad \langle \hat{Q}(t) \rangle_{|1\rangle} = -\lambda \ll qf \gg^R_{\omega=0} t, \tag{5.174}$$

while its variance is, in the first approximation, independent of the qubit state:

$$\langle \delta \hat{Q}^2 \rangle_{|0\rangle} = \langle \delta \hat{Q}^2 \rangle_{|1\rangle} = S_{\hat{q},s}(0)t, \tag{5.175}$$

where $S_{\hat{q},s}(\omega)$ is the symmetrized spectral density of the fluctuations of $\hat{q}(t)$. Actually Eqs (5.174) and (5.175) contain the essence of the WCM approach. The actions of individual "measurement acts" (e.g., the passage of a single electron through a quantum point contact) on the detector do not differ much for different states of the measured system, and the expectation values of the integrated signal for the $|0\rangle$ and

---

[41] Alias the generalized susceptibility, alias the detector gain, $\chi_{qf}(t - t')$.

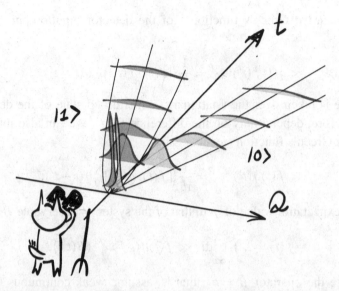

Fig. 5.10. Probability distribution for the expectation value of the integrated detector output under weak continuous measurement (WCM) (see Eq. (5.173)). The separation between the average values for two different states of the measured system (qubit) grows linearly with time, while the distribution width only grows as $\sqrt{t}$, which eventually allows us to resolve the qubit states.

$|1\rangle$ qubit states separate linearly in time; but the width of the outcome distribution function only grows as a square root of the measurement, which makes it possible to distinguish between the two cases, once you have waited long enough (Fig. 5.10). We can, therefore, define the measurement rate as the normalized rate at which the two $|0\rangle$- and $|1\rangle$-distributions of the integrated detector output separate

$$\Gamma_m t = \frac{\left[\left(\langle\hat{Q}(t)\rangle_{|0\rangle} - \langle\hat{Q}(t)\rangle_{|1\rangle}\right)/2\right]^2}{\langle\delta\hat{Q}^2\rangle_{|0\rangle} + \langle\delta\hat{Q}^2\rangle_{|1\rangle}}, \tag{5.176}$$

so that

$$\Gamma_m = \frac{\lambda^2 \left(\ll qf \gg_{\omega=0}^R\right)^2}{2S_{\hat{q},s}(0)}, \tag{5.177}$$

and the ratio of the measurement and back-action dephasing rates for weak continuous QND measurement is, as one would expect, independent of the qubit-detector coupling strength:

$$\frac{\Gamma_m}{\gamma} = \frac{\hbar^2 \left(\ll qf \gg_{\omega=0}^R\right)^2}{4S_{\hat{q},s}(0)S_{\hat{f},s}(0)}. \tag{5.178}$$

Now we will use a general inequality (see Clerk et al., 2010, Section IV.3, and references therein)

$$S_{\hat{q},s}(0)S_{\hat{f},s}(0) - \left|S_{\hat{q}\hat{f},s}(0)\right|^2 \geq \frac{\hbar^2}{4}\left|\tilde{\chi}_{\hat{q}\hat{p}}(\omega = 0)\right|^2, \qquad (5.179)$$

where the cross spectral density

$$S_{\hat{q}\hat{f},s}(\omega) = \frac{1}{2}\int_{-\infty}^{\infty} dt\, e^{i\omega t}\left\langle\{\hat{q}(t), \hat{f}(0)\}\right\rangle_0, \qquad (5.180)$$

and the cross gain

$$\tilde{\chi}_{\hat{q}\hat{p}}(\omega) \equiv \ll qf \gg_\omega^R - \left(\ll fq \gg_\omega^R\right)^*. \qquad (5.181)$$

If the detector has zero reverse gain (which is desirable and possible), then (5.179) becomes

$$S_{\hat{q},s}(0)S_{\hat{f},s}(0) - \left|S_{\hat{q}\hat{f},s}(0)\right|^2 \geq \frac{\hbar^2}{4}\left|\ll qf \gg_{\omega=0}^R\right|^2. \qquad (5.182)$$

Substituting this into (5.178), we obtain

$$\frac{\Gamma_m}{\gamma} \leq 1 - \frac{\left|S_{\hat{q}\hat{f},s}(0)\right|^2}{S_{\hat{q},s}(0)S_{\hat{f},s}(0)} \leq 1. \qquad (5.183)$$

We see that, at best, the measurement rate can be the same as the back-action decoherence rate. When this condition is achieved, we have a quantum-limited detector, since the constraint (5.179) eventually follows from the quantum-mechanical uncertainty relations.[42] For classical noise the Schwarz inequality would only require

$$S_q(0)S_f(0) - \left|S_{qf}(0)\right|^2 \geq 0, \qquad (5.184)$$

instead of (5.179), (5.182).

5.5.5 *Rabi spectroscopy*

Rabi oscillations arising when a qubit is subjected to a resonant drive will eventually decay with a rate of $1/\tau \sim \max(\Gamma, \Gamma_\phi)$; after that, as we have seen in § 1.4.4, the qubit will follow the drive. Nevertheless, quantum correlations on the characteristic scale $1/\omega_R$ are manifest in the stationary fluctuations of its parameters as long as the drive is present. This provides a rather counterintuitive way of observing quantum correlations indefinitely, even though the corresponding decoherence time is finite

---

[42] See Braginsky and Khalili (1992) and Clerk et al. (2010) for discussions of the strategies needed to achieve quantum-limited detection, and relevant references.

and may be relatively short, and using for this purpose the stationary noise in the observed system. On a closer inspection, the possibility of noise Rabi spectroscopy is not that strange. In the absence of fluctuations the Bloch vector of the qubit will follow the drive (see Eq. (1.130)). But the reaction of a qubit to a random jolt, produced by external noise, will be quantum coherent on the scale of τ. The Bloch vector, tilted away from the established path (i.e., from the stationary position in the rotating frame of reference), will return to it through decaying Rabi oscillations. The behaviour of the qubit observables (e.g., $\langle \sigma_z(t) \rangle$) can, thus, be looked at as a series of overlapping, random Rabi oscillation trains, decaying on the scale of τ; these oscillations will average to zero in the average behaviour of $\langle \sigma_z(t) \rangle$, but will survive in the correlators, $\langle \sigma_z(t)\sigma_z(0) \rangle$, and will show up in the spectral density of qubit noise. A formal, but direct, way of looking at the situation is through the quantum regression theorem (5.48): since the expectation values $\langle \sigma_i(t) \rangle$, $i = x, y, z$, undergo Rabi oscillations in the presence of the resonant drive, so will the stationary correlators $\langle \sigma_i(t)\sigma_j(0) \rangle$.

In an experiment (Il'ichev et al., 2003), an aluminium persistent current qubit was inductively coupled to a high-quality, low-frequency niobium tank circuit and driven at the qubit's resonance frequency Ω (Fig. 5.11), and the stationary noise spectrum of the tank voltage was measured as a function of the drive power. The long lifetime of excitations in the tank allows the qubit to produce a significant fluctuating voltage, while being only weakly coupled to the tank (which allows us to neglect the back-action of the tank on the qubit compared to the resonant drive). The situation can be described by an equation similar to (5.161),

$$\ddot{V} + \frac{\omega_T}{Q}\dot{V} + \omega_T^2 V = -\frac{M\omega_T^2}{c^2}I_p\langle \dot{\sigma}_z \rangle, \tag{5.185}$$

where the expectation value $\langle \sigma_z(t) \rangle$ will undergo decaying Rabi oscillations, $\sim \cos(\omega_R t)e^{-|t|/\tau}$. By the quantum regression theorem, the same equations will hold for the correlators. It is more convenient, instead of finding the voltage correlator $K_V(t)$ and then taking its Fourier transform, to find the spectral density directly from the Fourier transform of the voltage, $\widetilde{V}(\omega)$, using the relation (5.12). Rewriting (5.185) in the Fourier representation, we can express $\widetilde{V}(\omega)$ through $\widetilde{\langle \sigma_z \rangle}(\omega)$, which immediately yields an expression for $S_{V,\text{Rabi}}(\omega)$ (this will be the part of the fluctuating voltage due to qubit fluctuations):

$$S_{V,\text{Rabi}}(\omega) = 2k^2 \frac{\epsilon^2}{\hbar^2\Omega^2} \frac{L_q I_p^2}{c^2 C} \cdot \frac{\omega^2}{\tau} \cdot \left[\frac{\omega_R^2}{(\omega^2 - \omega_R^2)^2 + (\omega/\tau)^2} \right]$$

$$\cdot \left[\frac{\omega_T^2}{(\omega^2 - \omega_T^2)^2 + (\kappa\omega)^2} \right]. \tag{5.186}$$

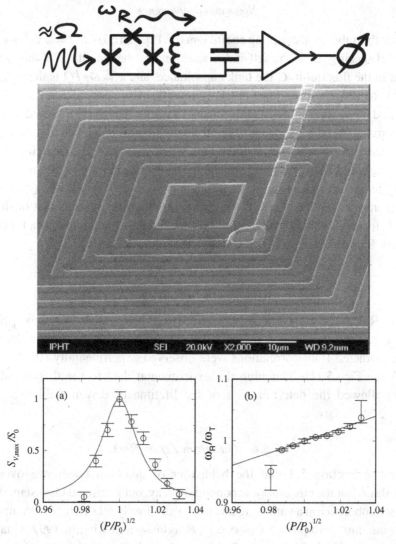

Fig. 5.11. Rabi spectroscopy. (Top) Schematics of the experimental setup and the micrograph of the aluminium persistent current flux qubit inside the surrounding niobium pickup coil (part of the resonant tank circuit) (reprinted with permission from Il'ichev et al., 2003, © 2003 American Physical Society). A resonant HF signal sent to the qubit translates into a noise signal at the much lower qubit Rabi frequency. When the latter is close to the resonance frequency of the LC tank circuit, it contributes to the increased noise signal in the tank. The qubit has loop inductance 24 pH, and the ratio of smaller-to-larger junction critical currents $\alpha = 0.8$, with $I_c = 600$ nA and $C_J = 3.9$ fF for the larger Josephson junctions. The measured resonance frequency of the qubit $\Omega/2\pi = 868$ MHz. The tank circuit resonance frequency $\omega_T/2\pi = 6.284$ MHz and quality factor $Q = 1850$. Estimated tank parameters $L_T = 0.2\,\mu$H, $C_T = 3$ nF, and the mutual inductance $M = 70$ pH. The nominal temperature of the system was 10 mK, while the noise temperature of the amplifier was 300 mK. (Bottom) The experimental data (reprinted with permission from Il'ichev et al., 2003, © 2003 American Physical Society). (a) Normalized spectral density of the noise signal in the tank as a function of the normalized HF signal amplitude (circles) and the theoretical curve (5.187). (b) Same data replotted for the normalized Rabi frequency.

Here k is the qubit-tank coupling coefficient (cf. Eq. (5.163)), $\hbar\Omega = \sqrt{\Delta^2 + \epsilon^2}$ is the interlevel spacing, L_q is the self-inductance, and I_p the amplitude of the persistent current in the flux qubit, C the tank capacitance, and $\kappa = \omega_T/Q$ is the tank decay rate.[43] In (5.186) the first square bracket comes from the spectral density of the qubit noise, and the second shows its filtering by the tank. We see that, indeed, due to the Rabi correlations, the spectrum of qubit fluctuations is enhanced at $\omega \approx \omega_R \ll \Omega$, which leads to increased noise in the tank when $\omega_R \approx \omega_T$. By changing the drive power one could change ω_R and so scan the Rabi contribution to the qubit noise by moving its spectrum past the static frequency window of width ω_T/Q around the tank resonance frequency ω_T. If the qubit damping rate exceeds that of the tank, $\kappa\tau \ll 1$, the peak value of $S_{V,\text{Rabi}}(\omega)$ is achieved at $\omega = \omega_T$, and, as a function of the qubit Rabi frequency (Il'ichev et al., 2003)

$$\frac{S_{V,\text{Rabi}}^{\max}(w)}{S_0} = \frac{w^2 u^2}{(w^2-1)^2 + u^2} \approx \frac{(u/2)^2}{(w-1)^2 + (u/2)^2}. \tag{5.187}$$

Here the Rabi noise in the tank is normalized to its peak value, $S_0 = \max_{\omega_R} S_{V,\text{Rabi}}^{\max}$; $w = \omega_R/\omega_T$ and $u = 1/\omega_T\tau$ are the reduced Rabi frequency and qubit decay rate.

The predicted Rabi correlations were observed experimentally (Il'ichev et al., 2003) (see Fig. 5.11). Matching the experimental data to the theoretical curve (5.187) allowed the determination of the lifetime of driven Rabi oscillations, $\tau_{\text{Rabi}} = 2\tau \approx 2.5\,\mu\text{s}$.

5.5.6 *Quantum Zeno effect

We saw in Section 5.3 that the behaviour of quantum coherence over short timescales is not the customary exponential decay, but is instead Gaussian: the system loses coherence at a vanishing rate as $t \to 0$. Closely related to this phenomenon is the *quantum Zeno paradox* (see, e.g., Braginsky and Khalili, 1992, Chapter 7; Namiki et al., 1997, Chapter 8). Its simple formulation is as follows. Suppose a quantum system, starting from an eigenstate $|n\rangle$ of some observable, undergoes a series of periods of free evolution of length Δt, governed by a Hamiltonian H, interrupted by precise strong projective measurements of this observable.[44] Then

---

[43] If we take into account the additional qubit damping due to the tank back-action, $1/\tau$ must be replaced with

$$\Gamma(\omega) = \frac{1}{\tau_0} + 4k^2 \frac{\epsilon^2}{\hbar^2\Omega^2} \frac{L_q I_p^2}{\hbar^2 c^2} \frac{\omega_T^2 \kappa k_B T}{(\omega^2 - \omega_R^2)^2 + (\kappa\omega)^2},$$

which follows from the Bloch equations for the σ-operators of a qubit acted upon by the strong resonant drive and the magnetic flux produced by the tank at temperature T. In the limit we are considering here, this result, obtained by the method of quantum Langevin equations (Smirnov, 2003), reduces to (5.186), with $1/\tau \to \Gamma(\omega)$ of Eq. (43). Note also that the effect disappears at the degeneracy point, $\epsilon = 0$, since there the matrix element of the qubit current, which couples it to the tank, is zero.

[44] For the more general case of imprecise measurement see Braginsky and Khalili, 1992, Chapter 7.

in the limit of continuous monitoring ($\Delta t \to 0$) the probability p_{nn} to remain in the initial state tends to unity.[45]

Indeed, the state of the system at time Δt is given by

$$|\Psi(\Delta t)\rangle = e^{-\frac{i}{\hbar}H\Delta t}|n\rangle = \left[1 - \frac{i}{\hbar}H\Delta t - \frac{1}{2\hbar^2}H^2\Delta t^2 + \ldots\right]|n\rangle, \qquad (5.188)$$

and the probability to be projected back to the initial state after the measurement is

$$p_{nn}(\Delta t) = |\langle n|\Psi(\Delta t)\rangle|^2 = \left|1 - \frac{i}{\hbar}\Delta t\langle n|H|n\rangle - \frac{1}{2\hbar^2}\Delta t^2\langle n|H^2|n\rangle + \ldots\right|^2$$

$$= 1 - \frac{\Delta t^2}{\hbar^2}\left[\langle n|H^2|n\rangle - (\langle n|H|n\rangle)^2\right] + \cdots \equiv 1 - \frac{\Delta t^2(\delta E_n)^2}{\hbar^2} + \ldots$$

$$(5.189)$$

After time $N\Delta t$, the probability of staying in the initial state $p_{nn}(N\Delta t) = p_{nn}(\Delta t)^n$, and in the continuous limit[46]

$$p_{nn}(t) = \lim_{\Delta t \to 0}\left(1 - \frac{\Delta t^2(\delta E_n)^2}{\hbar^2}\right)^{\frac{t}{\Delta t}} = \lim_{\Delta t \to 0} e^{-\frac{t\Delta t}{\hbar^2}(\delta E_n)^2} = 1. \qquad (5.190)$$

---

[45] The closest thing Zeno of Elea (Fifth century BCE) had to say about this situation was quoted by Aristotle (2008) as: "if everything when it occupies an equal space is at rest, and if that which is in locomotion is always occupying such a space at any moment, the flying arrow is therefore motionless."

[46] It is worth noting here that at *very* long times the probability $p_{nn}(t)$ also does not decay exponentially, but as some inverse power of time. This unexpected result follows mathematically from the fact that the spectrum of the Hamiltonian is limited from below (see Namiki et al., 1997, Appendix A.2 and references therein).

6

Applications and speculations

It is known that dragons do not exist ... but each kind does so in a completely different manner. Imaginary and null dragons do not exist in a much less interesting way, than the negative ones.

Stanisław Lem, *Dragons of Probability*,
translated from Polish by the author

6.1 Quantum metamaterials

6.1.1 A qubit in a transmission line

Now we can finally discuss what kind of structures can be formed using qubits as basic units, and what would be their properties. Consider, for example, the optical properties of such a structure. For wavelengths much larger than the size of a single element, or the distance between elements, the structure will appear to be a continuous material with, possibly, strange properties. Such materials built out of artificial unit blocks are known as metamaterials (e.g., Saleh and Teich, 2007, 5.7) and, indeed, demonstrate such exotic behaviour as having a negative refractive index. Compared to them, *quantum metamaterials* (Rakhmanov et al., 2008), where the building blocks are qubits (or more complex artificial atoms), which maintain quantum coherence during the characteristic time of the signal propagation through them, and whose quantum state can be, in principle, selectively controlled, promise a wider and even more interesting range of behaviour. Such a device could be called an extended quantum coherent system, and our investigations will take us from the field of already fabricated structures, already conducted experiments and already verified theoretical models and into the even more interesting here-be-dragons territory of suppositions, suggestions and open questions.

We will start from secure ground and begin with a single artificial atom (qubit). We have already seen in Section 4.3 that a qubit in a resonator behaves as an atom in a cavity. We obtain the simplest analogue to a real atom interacting with light (microwaves, X-rays, etc.) in space by replacing a resonator, where only standing

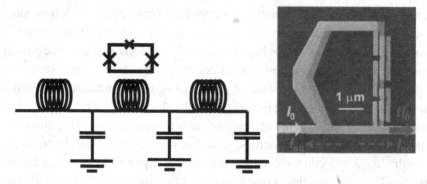

Fig. 6.1. A flux qubit inductively coupled to a section of a transmission line: lumped-elements circuit (left) and a micrograph (right, from Astafiev et al. (2010a) reprinted with permission from AAAS).

electromagnetic waves are allowed, with a transmission line, along which the waves can propagate. Then, for a qubit one can consistently introduce such "atomic" properties as scattering amplitude, transmission and reflection coefficients, etc. After that, a number of qubits can be put in a transmission line, forming a 1D "optical crystal". This is also a practical way of bringing an electromagnetic wave to solid-state qubits, which sit on a substrate in a cryostat, surrounded by control and readout circuitry, and are not that easily accessible from the outside. This is just a slight modification of the transmission line resonators employed in circuit QED experiments, which we encountered in § 4.3.1 (Fig. 4.11).

Using the Lagrangian (4.70) for an infinite stretch of a transmission line, we find that in the absence of dissipation the flux variable, quite predictably, satisfies the wave equation

$$\ddot{\Phi}(x,t) - s^2 \frac{\partial^2}{\partial x^2} \Phi(x,t) = 0, \tag{6.1}$$

and so do the current and voltage in the line,

$$I(x,t) = \frac{c}{\widetilde{L}} \frac{\partial}{\partial x} \Phi(x,t); \quad V(x,t) = \frac{1}{c} \dot{\Phi}(x,t). \tag{6.2}$$

Here $s = c/\sqrt{\widetilde{L}\widetilde{C}}$ is the phase velocity of the electromagnetic wave in the line, and $\widetilde{L}$ and $\widetilde{C}$ are the inductance and capacitance per unit line length. The dispersion relation is, of course, $\omega = sk$.

The wave equations for the current and voltage could be obtained directly from the *telegraph equations*,

$$\begin{cases} \frac{\partial}{\partial x} V(x,t) = \frac{\widetilde{L}}{c^2} \frac{\partial}{\partial t} I(x,t); \\ \\ \frac{\partial}{\partial x} I(x,t) = \widetilde{C} \frac{\partial}{\partial t} V(x,t), \end{cases} \tag{6.3}$$

which express Faraday's law and the conservation of current (to check their validity, consider a segment of the equivalent lumped-elements scheme of the transmission line, or simply look attentively at Eqs (6.1, 6.2)). These equations have the advantage that we can immediately include the effect of a qubit. Following Astafiev et al. (2010a), we will consider the case of a flux qubit, which induces in the adjacent section of the transmission line the magnetic flux $\Phi_q = (M/c)\langle\hat{I}_q\rangle$, where M is the mutual inductance and $\hat{I}_q$ is the operator of the persistent current in the qubit. Given that the qubit is only $\sim 10\ \mu$m across, while the wavelength of the electromagnetic field in the line $\lambda \sim 1$ cm, we can treat the qubit as a point-like scatterer located at the origin. Then, considering a unit segment of the line at $x = 0$ (Fig. 6.1), we see that[1]

$$\begin{cases} V(+0, t) = V(-0, t) - \frac{M}{c}\frac{\partial}{\partial t}\langle\hat{I}_q(t)\rangle \\[2mm] I(+0, t) = I(-0, t). \end{cases} \tag{6.4}$$

For a monochromatic current wave incident from left infinity, $I_0(x, t) = \mathrm{Re}\{I_0\exp[ikx - i\omega t]\}$, we can introduce transition/reflection amplitudes t, r via

$$I(x < 0, t) = \mathrm{Re}\left\{I_0 e^{ikx - i\omega t} - \mathsf{r}\, I_0 e^{-ikx - i\omega t}\right\}; \quad I(x > 0, t) = \mathrm{Re}\left\{\mathsf{t}\, I_0 e^{ikx - i\omega t}\right\}. \tag{6.5}$$

We are interested in a stationary solution; therefore, the qubit current will oscillate with the frequency of the wave in the line:

$$\langle\hat{I}_q(t)\rangle \equiv \mathrm{Re}\left\{\overline{I_{q,\omega}}e^{-i\omega t}\right\} = \{\mathrm{Re}\overline{I_{q,\omega}}\}\cos\omega t + \{\mathrm{Im}\overline{I_{q,\omega}}\}\sin\omega t. \tag{6.6}$$

Substituting this and (6.5) into the telegraph equations (6.3) and with the matching conditions at $x = 0$, Eq. (6.4), we find:

$$\mathsf{t} + \mathsf{r} = 1; \quad \mathsf{r} = \frac{i\omega M\overline{I_{q,\omega}}}{2c^2 Z I_0} = -\frac{\omega M\,\mathrm{Im}\overline{I_{q,\omega}}}{2c^2 Z I_0} + \frac{i\omega M\,\mathrm{Re}\overline{I_{q,\omega}}}{2c^2 Z I_0}, \tag{6.7}$$

where $Z = \sqrt{\widetilde{L}/c^2\widetilde{C}}$ is the impedance of the transmission line.

The quadratures of $\langle\hat{I}_q(t)\rangle$ must be determined from the driven qubit dynamics. Consider for simplicity the case when the qubit is at its degeneracy point. Then in the energy eigenbasis, where $|0\rangle$ ($|1\rangle$) is the ground (excited) state, the qubit current operator $\hat{I}_q = I_p\sigma_x$. Its expectation value

$$\langle\hat{I}_q(t)\rangle = \mathrm{tr}\left[\rho(t)\cdot I_p\sigma_x\right] = I_p\mathcal{R}_x(t), \tag{6.8}$$

---

[1] Of course, the same result will follow from the general formalism of Section 2.3, if we use the lumped-elements model for the transmission line and then put the length of each unit segment to zero. Zhou et al. (2008) developed a theory for a charge qubit, where the transmission line was modelled by a series of coupled LC circuits, and the electromagnetic field was treated quantum-mechanically.

where $\mathcal{R}_x$ is a component of the Bloch vector. We can now use the same approach that we applied to the Rabi oscillations in Sections 1.4.3 and 1.4.4, with the role of the driving field played by the current in the incident wave at the position of the qubit, $I_0(0, t)$.

The Hamiltonian of the qubit is

$$H = -\frac{\hbar\Delta}{2}\sigma_z - \frac{\hbar\eta}{2}\sigma_x\left(e^{i\omega t} + e^{-i\omega t}\right),\tag{6.9}$$

where $\hbar\eta = M I_p I_0/c^2$. Introducing the detuning $\delta = \omega - \Delta$ and the Rabi frequency $\omega_R = \sqrt{\eta^2 + \delta^2}$, we find that the transformation of the Bloch vector from the laboratory frame (energy eigenbasis) to the rotating frame of reference (interaction representation) is given by the matrix (cf. Eq. (1.116))

$$\mathbf{W}(t) = \begin{pmatrix} (\eta/\omega_R)\cos\omega t & -(\eta/\omega_R)\sin\omega t & \delta/\omega_R \\ \sin\omega t & \cos\omega t & 0 \\ -(\delta/\omega_R)\cos\omega t & (\delta/\omega_R)\sin\omega t & \eta/\omega_R \end{pmatrix}; \quad \mathbf{W}^{-1}(t) = \mathbf{W}^{\mathrm{T}}(t).\tag{6.10}$$

In the rotating frame, the Bloch equations in the rotating-wave approximation, with the relaxation and dephasing rates Γ_1 and Γ_2, are given by (1.124). Their static solution, $\mathcal{R}_{\mathrm{stat}}$, is immediately found:

$$\mathcal{R}_{\mathrm{stat}} = \frac{1}{1 + (\eta^2/\Gamma_1\Gamma_2) + (\delta/\Gamma_2)^2}\begin{pmatrix} (\delta/\omega_R)((\omega_R/\Gamma_2)^2 + 1) \\ -\eta/\Gamma_2 \\ -\eta/\omega_R \end{pmatrix}.\tag{6.11}$$

After transforming it back to the laboratory frame,

$$\mathcal{R}_{\mathrm{lab}}(t) = \mathbf{W}^{\mathrm{T}}(t)\mathcal{R}_{\mathrm{stat}},\tag{6.12}$$

and substituting the x-component of $\mathcal{R}_{\mathrm{lab}}(t)$ in (6.8), we obtain

$$\langle \hat{I}_q(t)\rangle = \frac{I_p\eta/\Gamma_2}{1 + (\eta^2/\Gamma_1\Gamma_2) + (\delta/\Gamma_2)^2}\left(\frac{\delta}{\Gamma_2}\cos\omega t - \sin\omega t\right).\tag{6.13}$$

Comparing this to (6.6) and substituting into (6.7), we eventually find (Astafiev et al., 2010a)

$$r = \frac{\Gamma_1}{2\Gamma_2}\frac{1 + i(\delta/\Gamma_2)}{1 + (\delta/\Gamma_2)^2 + \eta^2/(\Gamma_1\Gamma_2)}.\tag{6.14}$$

The maximal reflection amplitude is achieved in exact resonance and, as the field amplitude I_0 drops ($\eta \to 0$), it approaches the limit $r_0 = \Gamma_1/2\Gamma_2$. If there is no pure dephasing, that is, $\Gamma_2 = \Gamma_1/2$, this limit is unity: a vanishingly weak wave is totally reflected.

Eq. (6.14) shows that by measuring the complex reflection amplitude as a function of the field amplitude one can establish the relation between the relaxation and dephasing rates in the qubit. We can also return to Eq. (6.7) and notice that the ratio (ω/Z) in the expression for r can be rewritten as

$$\frac{\omega}{Z} = \frac{1}{\hbar}\left[\hbar\omega Z^{-1}\coth(\hbar\omega/2k_B T)\right]_{T=0} = \frac{1}{\hbar}S_{I,s}(\omega), \tag{6.15}$$

the spectral density of zero-point current fluctuations in the transmission line.[2] If we assume that *all* relaxation in the qubit is due to its coupling to the line, then (see Eq. (5.34))

$$\frac{\omega}{Z} = \frac{1}{\hbar}S_{I,s}(\omega) = \frac{\hbar}{g^2}\Gamma_1, \tag{6.16}$$

where the coupling coefficient $g = MI_p/c$. Therefore, for the amplitude of the scattered current wave we find in this case (Astafiev et al., 2010a)

$$I_{sc} \equiv rI_0 = \overline{I_{q,\omega}}\frac{i\hbar\Gamma_1}{2MI_p^2/c^2}. \tag{6.17}$$

In an experiment carried out by Astafiev et al. (2010a), where a flux qubit was placed in a 1D transmission line, the measured reflection amplitude was in very good agreement with the formula (6.14) (Fig. 6.2). A remarkable feature of this formula is that when deriving it we assumed that the qubit is driven by the incident wave only, $I_0(0,t)$. A naïve attempt at a "self-consistent" solution by replacing $I_0(0,t)$ with the total current at $x=0$, $I(0,t) = tI_0(0,t)$, yields results that are totally irreconcilable with the observed data. This circumstance confirms the fact that the qubit in the experiment did behave as a point-like quantum scatterer: in order for it to be driven by the *total* current, it must repeatedly interact with the wave it just scattered, which – in the absence of other scatterers in the line – simply never returns to the qubit.

The qubit in the transmission line demonstrates other "atom-like" properties, revealed, for example, by the inelastic scattering by the qubit, i.e., for the scattering when the electromagnetic wave changes its frequency (for which the relative power $(1 - |r|^2 - |t|^2)$ is still available). It is a known result from quantum optics that the spectrum of the scattered light will contain, in addition to the central peak at the incident frequency, ω, two side peaks, at frequencies $\omega \pm \omega_R$ (known as the *resonance fluorescence triplet*, a.k.a. the Mollow triplet; see, e.g., Scully and

---

[2] The fact that our idealized transmission line does not contain any active elements does not invalidate this consequence of the fluctuation–dissipation theorem. All the dissipation and equilibration happens somewhere in infinity, from where the electromagnetic waves arrive and where they go after being scattered. The situation is again similar to the ballistic transport in a point contact.

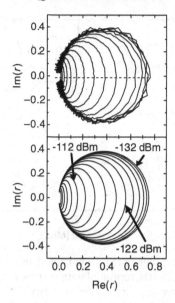

Fig. 6.2. Reflection amplitude for a qubit in a transmission line (from Astafiev et al., 2010a, reprinted with permission from AAAS). The experimental data (top) are in good agreement with the formula (6.14) (bottom). Different curves correspond to the driving power changing from -132 dBm to -102 dBm in steps of 2 dBm. (dBm, sometimes written as dBmW, is an abbreviation for the power ratio in decibels (dB) of the measured power with reference to one milliwatt (mW). In other words, power (mW) $= 10^{\text{power(dBm)}/10}$.)

Zubairy (1997), Chapter 10). They appear due to the modulation of the atomic dipole moment (in our case, $\langle\sigma_x\rangle$) by the Rabi oscillations. The effect can be readily understood if we recall the treatment of quantum Rabi oscillations in § 4.2.3. There we saw that due to the Jaynes–Cummings coupling $\hbar g(a\sigma_- + a^\dagger\sigma_+)$ between the qubit and the field mode the degenerate states $|0, n+1\rangle$ and $|1, n\rangle$ undergo dynamic Stark splitting into doublets of dressed states, $|n\pm\rangle$, with the interlevel distance $\omega_R = 2g\sqrt{n+1}$ (see (4.59, 4.61)). It is immaterial here whether the field mode is a standing wave, as in an oscillator, or a propagating wave, as in a transmission line. For a large enough field amplitude, $n \gg 1$, the splitting for n and $n+1$ will be practically the same, and there will be now three resonant transition frequencies: ω (between $|n+1, -\rangle$ and $|n-\rangle$, and between $|n+1, +\rangle$ and $|n+\rangle$); $\omega + \omega_R$ (between $|n+1, +\rangle$ and $|n-\rangle$); and $\omega - \omega_R$ (between $|n+1, -\rangle$ and $|n+\rangle$); see Fig. 6.3. The central peak, to which two transitions contribute, should be twice as high as the side ones. Without going further into the effect's theory, which was, of course, initially developed for real atoms in free space, we just say here that the triplet observed by Astafiev et al. (2010a) is in quantitative agreement with the theoretical predictions for qubit "atom" in a transmission line "space".

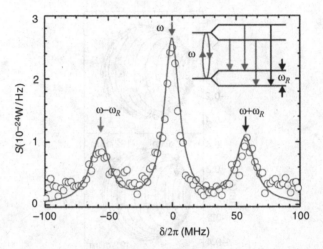

Fig. 6.3. Resonance fluorescence (Mollow triplet) in a qubit in the transmission line (from Astafiev et al., 2010a, reprinted with permission from AAAS). The solid line is the theoretical curve (e.g., Scully and Zubairy, 1997, Chapter 10). The side peaks appear due to the splitting of the "atomic" levels by Rabi oscillations in the presence of a strong driving (inset).

Now let us recall that the two-state ("qubit") approximation, with its Pauli matrices, etc., is only valid when transitions to the higher levels (which are generally present in mesoscopic devices – such as a persistent current flux qubit) can be neglected because of the system's nonlinearity. But they can be, instead, intentionally exploited. For example, the use of the three lowest levels, with the transition frequencies ω_{10}, ω_{21}, ω_{20}, allows the realization of a wider variety of "artificial-atomic" effects, such as quantum amplification (Astafiev et al., 2010b) or electromagnetically-induced transparency (Abdumalikov et al., 2010). In the latter case, in addition to the probe wave with frequency $\omega \equiv \omega_p = \omega_{10} + \delta_p$, the artificial atom was illuminated by the control wave with $\omega_c = \omega_{21} + \delta_c$. In the RWA (assuming that both the detuning and the anharmonicity, $\omega_{21} - \omega_{10}$, are small compared to ω_p, ω_c) the Hamiltonian of the qubit (or, more accurately, "qutrit", since it has three states) is

$$H = -\hbar\left(\delta_p \varsigma_{11} + (\delta_p + \delta_c)\varsigma_{22}\right) - \frac{\hbar}{2}\left(\eta_p(\varsigma_{01} + \varsigma_{10}) + \eta_c(\varsigma_{12} + \varsigma_{21})\right). \quad (6.18)$$

Here $\varsigma_{ij} = |i\rangle\langle j|$ is the transition operator between the "atom"'s states. We could use a spin-1 object instead of the Bloch vector in order to parameterize the qutrit, but the game is not really worth the candle. It is more straightforward to solve the equations of motion for the qubit density matrix in the energy representation, with the Lindblad terms describing the dephasing, with dephasing rates γ_{ij}, $i \neq j$, and relaxation with rates Γ_{21} and Γ_{10}; because of the symmetry of the eigenstate wave functions the transitions $|0\rangle \leftrightarrow |2\rangle$ are prohibited at the qubit's degeneracy point (Astafiev et al., 2010b; Abdumalikov et al., 2010). In the limit of weak probe drive, $\eta_p \ll \gamma_{10}$, and at zero temperature, the transmission amplitude for the probe signal turns out to be (Abdumalikov et al., 2010)

$$t = 1 - \frac{\Gamma_{10}}{2(\gamma_{10} - i\delta_p) + \eta_c^2[2(\gamma_{20} - i\delta_p - i\delta_c)]^{-1}}, \qquad (6.19)$$

and at zero pure dephasing in the (01)-channel ($2\gamma_{10} = \Gamma_{10}$) and in exact resonance it will change from zero at zero control η_c to unity when $\eta \to \infty$. The transmission coefficient, $T = |t|^2$, is then given by

$$T = \left(\frac{\eta_c^2}{2\Gamma_{10}\gamma_{20} + \eta_c^2}\right)^2. \qquad (6.20)$$

The physical reason for this effect lies in the depopulation of the lower two states by the action of the strong control field, which "coherently captures" the "atom" in the state $|2\rangle$ (Scully and Zubairy, 1997, Section 7.3). The transmission coefficient T observed by Abdumalikov et al. (2010) changed between zero and 96%. This high switching efficiency is due to the strong coupling of the qubit to the field, confined to a 1D transmission line, and can be further improved by reducing the dephasing and the non-radiative decay of the qubit states and the signal leakage from the line.

6.1.2 *Classical wave propagation along a multiqubit chain in a transmission line*

Since, with respect to the scattering of electromagnetic waves, a single qubit does behave as an atom in space, it is logical to expect that a spatially periodic multiqubit structure can demonstrate the properties of a transparent (in an appropriate range) material.

As we have already mentioned, the idea of building optical (in the broad sense) media – metamaterials – from artificial elements with prescribed properties in place of atoms originated in optics, and is now under active development and is included in textbooks (see, e.g., Saleh and Teich, 2007, 5.7). The specifics of using qubits is, thus, neither in their being artificial, nor being quantum (since natural atoms in normal optical media are as quantum as it gets), but in the possibility – in principle – of individual access to and control of the quantum state of each constituent qubit, and of the controlled quantum coherence of the whole structure, and this is where some novel properties specific to quantum metamaterials are expected.[3]

---

[3] This is also where currently we have to rely on theory alone, and not yet a completely developed theory at that.

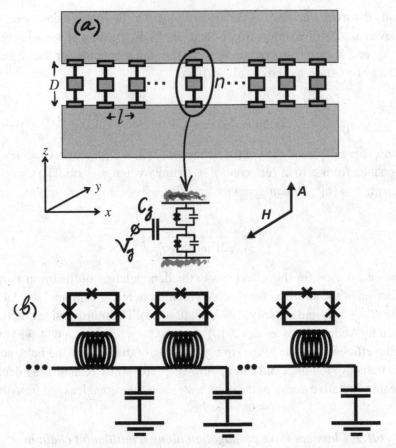

Fig. 6.4. Quantum metamaterial based on (a) charge qubits (after Rakhmanov et al., 2008); (b) flux qubits (after Zagoskin et al., 2009).

For a change, consider a periodic structure of superconducting charge qubits in a transmission line, formed by two bulk superconductors (Fig. 6.4a). We can somewhat simplify our task and, instead of starting from the Lagrangian of the equivalent lumped-elements circuit, follow the approach of Rakhmanov et al. (2008) and directly write down the energy of the system in terms of the superconducting phases on the qubit islands, ϕ_n, and the vector potential of the electromagnetic field, A:

$$E = \sum_n \left\{ \frac{E_J}{2\omega_J^2} \left[\left(\frac{2\pi D \dot{A}_{zn}}{\Phi_0} + \dot{\varphi}_n \right)^2 + \left(\frac{2\pi D \dot{A}_{zn}}{\Phi_0} - \dot{\varphi}_n \right)^2 \right] + \frac{Dl}{8\pi} \left(\frac{A_{z,n+1} - A_{z,n}}{l} \right)^2 \right.$$
$$\left. - E_J \left[\cos \left(\frac{2\pi D A_{zn}}{\Phi_0} + \varphi_n \right) + \cos \left(\frac{2\pi D A_{zn}}{\Phi_0} - \varphi_n \right) \right] \right\}. \tag{6.21}$$

The magnetic field in the gap between superconducting banks $\mathbf{H} = H\mathbf{e}_y = \nabla \times \mathbf{A}$, and we can choose $\mathbf{A} = A_z(x)\mathbf{e}_z$. Since we are considering wavelengths $\lambda \gg l, D$ (the unit cell size), we can assume that the vector potential is approximately constant within the cell, $A_z(x) \approx A_{zn}$. In (6.21) we have also taken into account that in the presence of a vector potential the superconducting phase acquires a gauge term, $\alpha_n = 2\pi D A_{zn}/\Phi_0$. For simplicity, we assume that all qubits are identical, with two Josephson junctions with critical current I_c and capacitance C each, and $\omega_J^2 = eI_c/\hbar C; E_J = I_c\Phi_0/2\pi c$.

Introducing dimensionless energy, time and vector potential, respectively, as $\mathcal{E} = E/E_J, \tau = \omega_J t$ and α_n, we can rewrite (6.21) as

$$E = \sum_n \left[\left(\frac{\partial \varphi_n}{\partial \tau}\right)^2 + \left(\frac{\partial \alpha_n}{\partial \tau}\right)^2 - 2\cos\alpha_n \cos\varphi_n + \beta^2(\alpha_{n+1} - \alpha_n)^2 \right], \quad (6.22)$$

where

$$\beta^2 = \left[(\Phi_0/2\pi)^2/8\pi l D E_J\right] \quad (6.23)$$

characterizes the ratio of electromagnetic to Josephson energy in the system. We consider a classical field, therefore the (dimensionless) Hamiltonian of the system is obtained simply by replacing in (6.22) the nth qubit charging energy term with the operator $-(\partial_{\varphi_n})^2$, to yield

$$H = \sum_n \left[H_n + \left(\frac{\partial \alpha_n}{\partial \tau}\right)^2 + \beta^2(\alpha_{n+1} - \alpha_n)^2 \right]. \quad (6.24)$$

The expression

$$H_n = -\left(\frac{\partial}{\partial \varphi_n}\right)^2 - 2\cos\alpha_n \cos\varphi_n \rightarrow -\left(\frac{\partial}{\partial \varphi_n}\right)^2 + \alpha_n^2 \cos\varphi_n \quad (6.25)$$

is the Hamiltonian of the nth qubit in the limit of a weak classical electromagnetic field, $\alpha_n \ll 1$. The qubits in this scheme only interact through the electromagnetic wave, and we can use perturbation theory, assuming that initially the state of the qubits is factorized, so that each of them can be characterized by a separate density matrix, or even a wave function, $|\Psi_n(\tau)\rangle$. Then we can take the expectation value of (6.24) over $|\Psi(\tau)\rangle = |\Psi_n(\tau)\rangle \otimes |\Psi_n(\tau)\rangle \otimes \dots$ and vary it to find the equations of motion for the field (Rakhmanov et al., 2008):

$$\frac{\partial^2 \alpha_n}{\partial \tau^2} - \beta^2(\alpha_{n+1} - 2\alpha_n + \alpha_{n-1}) + \alpha_n\langle\Psi_n|\cos\varphi_n|\Psi_n\rangle = 0. \quad (6.26)$$

In the continuous limit, with n replaced by the dimensionless continuous variable ξ, and the second difference by a second derivative, this becomes a wave equation, with β playing the role of dimensionless phase velocity:

$$\frac{\partial^2 \alpha}{\partial \tau^2} - \beta^2 \frac{\partial^2 \alpha}{\partial \xi^2} + \alpha \langle \Psi(\xi) | \cos \varphi(\xi) | \Psi(\xi) \rangle = 0. \qquad (6.27)$$

For a typical charge qubit in a strip-line resonator (e.g., Gambetta et al., 2006) the Josephson energy is $\sim h \times 6$ GHz, three orders of magnitude larger than the qubit dephasing rate (~ 5 MHz); the resonator leakage rate is ~ 0.5 MHz. Taking the unit cell size $D \sim l \sim 10\,\mu m$, we get $\beta \sim 30$ (i.e., during a period of oscillation the wave will travel across ~ 30 unit cells). Therefore, we are justified in using in the first approximation a continuous wave equation and neglecting the decoherence effects.

For a chain of flux qubits inductively coupled to a transmission line (Fig. 6.4b) the straightforward use of the standard circuit approach is more convenient. It will give the discrete,

$$\frac{\partial^2 \Phi_n}{\partial t^2} - \Omega^2(\Phi_{n+1} - 2\Phi_n + \Phi_{n-1}) - \Omega^2 \langle (\Psi_n | \hat{\phi}_n - \hat{\phi}_{n-1}) | \Psi_n \rangle = 0, \qquad (6.28)$$

and continuous,

$$\frac{\partial^2 \Phi}{\partial t^2} - s^2 \frac{\partial^2 \Phi}{\partial x^2} - s\Omega \langle \Psi(\xi) | \frac{\partial \hat{\phi}}{\partial x} | \Psi(\xi) \rangle = 0, \qquad (6.29)$$

dimensioned counterparts to Equations (6.26) and (6.27). Here $\Phi_n \to \Phi(x)$ is the node flux, $\Omega = c/\sqrt{\Delta L\,\Delta C}$, $\Delta L, \Delta C$ are the inductance and capacitance of the transmission line within a unit cell, $s = l\Omega \equiv c/\sqrt{\widetilde{L}\widetilde{C}}$ is the phase velocity, and $\hat{\phi}_n$ is the operator of the qubit-induced flux, $\hat{\phi}_n = M\hat{I}_{q,n} \to \hat{\phi}(x)$ (Zagoskin et al., 2009).

The differences between the two cases (e.g., the unperturbed spectrum of (6.29) is gapless, unlike that of (6.27)) do not affect their essential equivalence. In both cases a classical electromagnetic wave propagates through the line of qubits, which to it looks like a continuous medium characterized by the "local quantum state", $|\Psi(\xi)\rangle$ (or $|\Psi(x)\rangle$), which can be specified beforehand. The situation is, in a sense, complementary to the classic double- or multiple-slit experiment, when a quantum particle is scattered by an extended classical object. Now we scatter a *classical* wave off an extended *quantum* object with controllable properties (including its quantum state).

Let us return to the charge-qubit metamaterial of Eq. (6.27), for which some theoretical predictions have been made by Rakhmanov et al. (2008). Expanding the "local" wave function over the qubit basis states in the Heisenberg representation,

$$|\Psi(\xi, \tau)\rangle = C_0(\xi, \tau)|0\rangle_\xi e^{-i\epsilon_q \tau/2} + C_1(\xi, \tau)|1\rangle_\xi e^{i\epsilon_q \tau/2}, \qquad (6.30)$$

where ϵ_q is the interlevel spacing in units of $\hbar\omega_J$, and using the standard time-dependent perturbation theory (e.g., Landau and Lifshitz, 2003, § 40), we find

$$i\frac{dC_j(\xi,\tau)}{d\tau} = \alpha^2(\xi,\tau)\sum_{k=0,1} V_{jk}(\xi,\tau)C_k(\xi,\tau), \tag{6.31}$$

where $V_{jk}(\xi,\tau) \equiv \langle j|\cos\varphi(\xi,\tau)|k\rangle$. Expanding the wave amplitude and wave function coefficients in the perturbation theory series $\alpha(\xi,\tau) = \alpha^0(\xi,\tau) + \alpha^1(\xi,\tau) + \ldots$, $C_j(\xi,\tau) = C_j^0(\xi) + C_j^1(\xi,\tau) + \ldots$, we find

$$iC_0^1(\xi,\tau) = \int_0^\tau d\tau'\,(\alpha^0(\xi,\tau'))^2\left(V_{00}C_0^0(\xi) + V_{01}C_1^0(\xi)e^{-i\epsilon_q\tau'}\right);$$

$$iC_1^1(\xi,\tau) = \int_0^\tau d\tau'\,(\alpha^0(\xi,\tau'))^2\left(V_{11}C_1^0(\xi) + V_{10}^*C_0^0(\xi)e^{i\epsilon_q\tau'}\right). \tag{6.32}$$

As a result, we obtain for the unperturbed wave

$$\frac{\partial^2\alpha^0}{\partial\tau^2} - \beta^2\frac{\partial^2\alpha^0}{\partial\xi^2} + V^0\alpha^0 = 0. \tag{6.33}$$

The specifics of the quantum metamaterial are hidden in the coefficient

$$V^0(\xi,\tau) = |C_0^0(\xi)|^2V_{00} + |C_1^0(\xi)|^2V_{11} + \left(C_0^0(\xi)C_1^{0*}(\xi)e^{i\epsilon_q\tau}V_{10} + H.c.\right), \tag{6.34}$$

where *H.c.* stands for a Hermitian conjugate. This is explicitly dependent on the initial state of the system, which – at least in theory – can be externally controlled. In the next order we find:

$$\frac{\partial^2\alpha^1}{\partial\tau^2} - \beta^2\frac{\partial^2\alpha^1}{\partial\xi^2} + V^0\alpha^1 + V^1\alpha^0 = 0, \tag{6.35}$$

where V^1 is the first-order correction to V^0 of (6.34) due to changes in $C_{0,1}$ (6.32) induced by the initial wave.

If all qubits are initially in the ground (excited) state[4] ($C_0^0 = 1, C_1^0 = 0$ or, respectively, $C_0^0 = 0, C_1^0 = 1$), from (6.33) follows the dispersion law

$$k_{\text{ground}}(\omega) = \frac{1}{\beta}\sqrt{\omega^2 - V_{00}}; \quad k_{\text{excited}}(\omega) = \frac{1}{\beta}\sqrt{\omega^2 - V_{11}}. \tag{6.36}$$

---

[4] In the latter case the qubit line is expected to start lasing, since it is in state with a complete population inversion. Unfortunately, perturbation theory cannot be used to investigate the lasing regime, since it is valid only as long as the field and the corrections to the qubit wave functions are small. While lasing in a single-qubit system was experimentally realized (Astafiev et al., 2007) and theoretically investigated (Hauss et al., 2008a,b; Ashhab et al., 2009; André et al., 2009a,b), neither has yet been achieved for a quantum-coherent multiqubit structure.

This is more or less routine, though it already shows how to adjust the refractive index of a medium by tuning the quantum states of its constituent units. The spectrum is gapped: for sufficiently low frequencies (6.36) yields imaginary wave vectors, corresponding not to propagation, but to extinction of the wave inside the medium. More interesting is the case when the qubits are initially in a superposition state, $C_0^0 = C_1^0 = 1/\sqrt{2}$. Then due to coherent quantum beats (ζ is some irrelevant phase shift)

$$V^0(\tau) = \frac{1}{4}\left[V_{00} + V_{11} + 2|V_{01}|\cos(\epsilon_p\tau + \zeta)\right], \qquad (6.37)$$

and

$$k(\omega, \tau) \approx \sqrt{\omega^2 - \frac{V_{00} + V_{11} + 2|V_{01}|\cos(\epsilon_p\tau + \zeta)}{4\beta^2}}. \qquad (6.38)$$

Therefore, if the incident wave frequency ω is close to the threshold, $\omega_c = \sqrt{V_{00} + V_{11}}/2\beta$, the metamaterial will alternate between the transparent and reflecting states with the frequency of quantum beats ϵ_p (and will also generate a signal at frequencies ϵ_p and $\omega \pm \epsilon_p$). Such a system, if realized, would provide a medium that can exist in a superposition of states with different refractive indices – an "optical Schrödinger's cat". It would be really interesting to see what it looks like, even if only in the microwave range.

Suppose now that we managed to create a spatially periodic initial qubit state, e.g., the simplest one, with alternate sections of equal length Λ, which are in state $|A\rangle$ or $|B\rangle$. These can be eigenstates of the unperturbed qubit Hamiltonian or their superpositions. (In the latter case the following discussion only makes sense if the quantum beat frequency is small enough, i.e., $\epsilon_q^2 \ll V_{00}, V_{11}$.) In the corresponding pieces of the structure, which we recognize as the quantum analogue of a 1D photonic crystal, the wave will satisfy the equations

$$\frac{\partial^2\alpha}{\partial\tau^2} - \beta^2\frac{\partial^2\alpha}{\partial\xi^2} + V_{AA}\alpha = 0 \text{ or } \frac{\partial^2\alpha}{\partial\tau^2} - \beta^2\frac{\partial^2\alpha}{\partial\xi^2} + V_{BB}\alpha = 0. \qquad (6.39)$$

We will, as usual in such cases, look for the solution in the form of a Bloch wave, $\alpha(\xi, \tau) = u(\xi)\exp[ik\xi - i\omega\tau]$, with a 2Λ-periodic function $u(\xi)$ and the quasi-momentum k in the first Brillouin zone, $-\pi/\Lambda \le k \le \pi/\Lambda$ (see, e.g., Saleh and Teich, 2007, 7.2). However, the solutions of Eqs (6.39) in either the A- or B-region can be written as sums of $\exp[\pm ik\xi]$-terms with constant coefficients. Using the continuity of α and $\partial_\xi\alpha$ and the periodicity of $u(\xi)$, we obtain the set of linear algebraic equations for these coefficients, the solvability condition for which leads to the implicit dispersion relation $k(\omega)$ (Rakhmanov et al., 2008):

$$\cos \kappa_A \Lambda \cos \kappa_B \Lambda - \frac{\kappa_A^2 + \kappa_B^2}{2\kappa_A \kappa_B} \sin \kappa_A \Lambda \sin \kappa_B \Lambda = \cos 2k\Lambda; \qquad (6.40)$$

$$\kappa_A^2 = \frac{\omega^2 - V_{AA}}{\beta^2}; \quad \kappa_B^2 = \frac{\omega^2 - V_{BB}}{\beta^2}.$$

From Eq. (6.40) it follows that the spectrum $\omega(k)$ will contain gaps if the difference between κ_A and κ_B is large enough (i.e., $|V_{AA} - V_{BB}| > \sim \beta^2$, which according to (6.23) means that the electromagnetic energy of a unit cell must be much less than its Josephson energy, consistent with the conditions of applicability of our perturbative approach). An example of such a spectrum is shown in Fig. 6.5: choosing as one of the states $|A\rangle$, $|B\rangle$ a superposition of qubit eigenstates would produce a bandgap structure, which "breathes" with the frequency of quantum beats, ϵ_q.

6.1.3 Ambidextrous metamaterials

Arguably, the most interesting classical metamaterials are the left-handed meta-materials (LHMs), or negative-index metamaterials (NIMs). Without going into details, while in conventional media the electric and magnetic fields of the electro-magnetic wave and the wave vector form a right-handed set, in left-handed media they form a left-handed set. Therefore, the wave vector and the Poynting vector $\mathbf{S} = \frac{c}{4\pi} \mathbf{E} \times \mathbf{B}$ have opposite senses: the energy in the wave flows opposite to the direction of the phase propagation. The same holds, of course, for the group and phase velocities. This corresponds to a negative sign for the refractive index (since the phase velocity in a medium $s = c/n$).[5] The light propagation through such media is quite counterintuitive. From Snell's law, for example, it follows that the refracted beam will lie on the same side of the normal to the boundary as the inci-dent one (which, as is easy to see, makes a flat slab of a left-handed material into a lens). The media also have reversed Doppler and Vavilov–Cherenkov effects and negative light pressure. The theoretical possibility of such materials was predicted by Veselago (1964, 1968), but their first realization had to wait until 2000 (Smith et al., 2000; Engheta and Ziolkowski, 2006; see also Saleh and Teich, 2007, 5.7).

---

[5] The left-handedness can only be achieved if the medium has both negative permittivity, ϵ, and permeability, μ, hence another name for the LHMs, double-negative metamaterials (DNGs). As was pointed out by Veselago (1964, 1968), ϵ and μ can be negative simultaneously only if both are frequency-dependent. This follows directly from the requirement that the period-averaged energy density of an (almost) monochromatic wave (see, e.g., Landau et al., 1984, § 80)

$$\overline{W}_\omega = \frac{\partial(\omega\epsilon)}{\partial\omega}\overline{|E_\omega|^2} + \frac{\partial(\omega\mu)}{\partial\omega}\overline{|H_\omega|^2},$$

must be nonnegative.

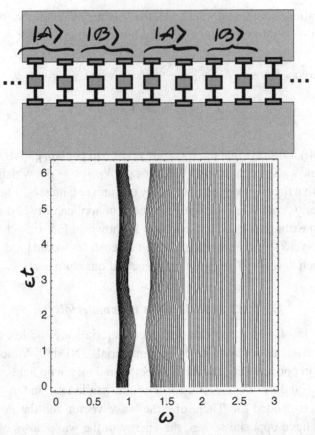

Fig. 6.5. (Top) "Quantum photonic crystal" – a simple periodic arrangement of the qubits' quantum states in a 1D quantum metamaterial. Here $|A\rangle = |0\rangle$, $|B\rangle = (|0\rangle e^{-i\epsilon_q \tau} + |1\rangle e^{i\epsilon_q \tau})/\sqrt{2}$. (Bottom) Breathing quantum photonic crystal (reprinted with permission from Rakhmanov et al., 2008, © 2008 American Physical Society): the structure of the bandgaps undergoes periodic oscillations due to the quantum beats between the qubits' states; see Eq. (6.40).

In the 1D case the left-handedness means simply that the product $kv_g(k)$, where the group velocity $v_g(k) = \partial\omega/\partial k$, is negative. This can be achieved in left-handed transmission lines (LHTLs, Fig. 6.6), which have been especially useful in combining left-handedness with nonlinearity (see Kozyrev and van der Weide, 2008, and references therein). Let us write the Lagrange equations for the equivalent lumped-element circuits of a usual (right-handed) transmission line and an LHTL:

$$\mathcal{L}_{\text{RHTL}} = \sum_n \left[\frac{C\dot{\Phi}_n^2}{2c^2} - \frac{(\Phi_n - \Phi_{n-1})^2}{2L} \right]; \quad \ddot{\Phi}_n - \omega_0^2(\Phi_{n+1} + \Phi_{n-1} - 2\Phi_n) = 0,$$

$$(6.41)$$

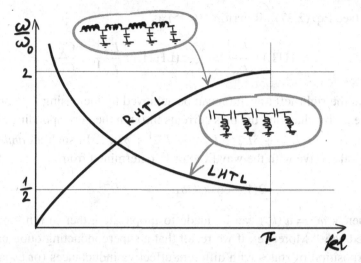

Fig. 6.6. Schematic representation of dispersion relations (6.43) in a right-handed (RHTL) and left-handed (LHTL) transmission line.

$$\mathcal{L}_{\text{LHTL}} = \sum_n \left[\frac{C\left(\dot{\Phi}_n - \dot{\Phi}_{n-1}\right)^2}{2c^2} - \frac{\Phi_n^2}{2L} \right]; \quad \Phi_n - \frac{1}{\omega_0^2}\left(\ddot{\Phi}_{n+1} + \ddot{\Phi}_{n-1} - 2\ddot{\Phi}_n\right) = 0,$$

$$(6.42)$$

where $\omega_0^2 = c^2/LC$. Substituting in (6.41) and (6.42) $\Phi_n(t) = F\exp[ikln - i\omega t]$, l being the unit segment length, we find the dispersion relations ($-\pi \leq kl \leq \pi$):

$$\omega_{\text{RHTL}}(k) = 2\omega_0 \left| \sin\left(\frac{kl}{2}\right) \right|; \quad \omega_{\text{LHTL}}(k) = \frac{\omega_0}{2\left|\sin\left(\frac{kl}{2}\right)\right|}, \quad (6.43)$$

and in the LHTL $kv_g(k)$ is indeed negative.

In a more general case, when each branch contains a capacitance and an inductance in parallel, the dispersion relation is

$$\omega^2(k) = \frac{2L_x^{-1}(1 - \cos kl) + L_y^{-1}}{2(C_x/c^2)(1 - \cos kl) + C_y/c^2}, \quad (6.44)$$

which reduces to Eqs (6.43) in the limits $L_y^{-1}, C_x \to 0$ or $L_x^{-1}, C_y \to 0$. Whether a transmission line will be right- or left-handed depends on its parameters: it follows from (6.44), that

$$\text{sgn}\left[k\frac{\partial\omega}{\partial k}\right] = \text{sgn}(L_x^{-1}C_y - L_y^{-1}C_x). \quad (6.45)$$

Therefore, this transmission line can be switched between right- and left-handedness, if it incorporates tunable capacitances or inductances, e.g., Josephson junctions (see Eq. (2.27)). Rewriting (6.45) as

$$(RHTL): \frac{L_x^{-1}}{L_y^{-1}} > \frac{C_x}{C_y}; (LHTL): \frac{L_x^{-1}}{L_y^{-1}} < \frac{C_x}{C_y}, \tag{6.46}$$

we see that the right-left transition can be achieved by increasing L_x^{-1} or decreasing L_y^{-1} (e.g., by changing the bias currents through the corresponding Josephson junctions): $L_x^{-1} \rightarrow L_x^{-1} + \delta L_x^{-1}$; $L_y^{-1} \rightarrow L_y^{-1} + \delta L_y^{-1}$. In such an *ambidextrous* metamaterial a wave with the wave vector k_c determined from

$$2(1 - \cos k_c l)\delta L_x^{-1} = -\delta L_y^{-1} \tag{6.47}$$

and frequency $\omega_c = \omega(k_c)$ can be made to propagate either in a normal or in a left-handed way.[6] Moreover, if we recall that a superconducting qubit can be put in a superposition of states with different effective inductances (or capacitances) (Section 2.5), there appears to be a possibility of fabricating an ambidextrous *quantum* metamaterial (Zagoskin and Saveliev, 2010) that is in a superposition of right-handed and left-handed states. How an electromagnetic wave will actually propagate through such a medium is still an open, and very interesting, question.

6.2 Quantum slide rules

6.2.1 Adiabatic quantum computing

The main problem of "standard" quantum computing, based on the consecutive application of unitary transformations (quantum gates) to qubits and pairs of qubits, remains the need to maintain quantum coherence in the system for long enough either to finish the computation, or to realize one of the quantum error-correction schemes (see, e.g., Le Bellac, 2006, 7.4). The latter requires many extra qubits and operations, which in turn tend to shorten the decoherence time in the system. These, probably not fundamental, but still formidable, difficulties motivate the search for less demanding approaches to quantum computing. One of these is *adiabatic quantum computing* (AQC) (Farhi et al., 2000, 2001).

The idea of the approach is as follows. Suppose we know how to encode the solution to the problem in the ground state $|\Psi_0\rangle$ of some Hamiltonian H_0, and we can build a physical system, which has such a Hamiltonian. This is, admittedly, a tall order, but, at least in principle and with a number of caveats, it can be done for an arbitrary quantum algorithm (Aharonov et al., 2004; Mizel et al., 2007). Then, if only we could put the system in its ground state, the problem would be solved. A straightforward cooling down would not be efficient, since for any

---

[6] Note that (6.47) has a nontrivial solution only if *both* inductances are changed.

sufficiently complex problem the corresponding H_0 is likely to have a large number of metastable minima, in which the system will freeze long before its temperature is anything close to the energy of the first excited state (*glassy behaviour*, Santoro et al., 2002; Das and Chakrabarti, 2008). However, if we apply to the system a strong perturbation, H_1 (of course, $[H_0, H_1] \neq 0$), such that the ground state of $H_0 + H_1$, $|\Psi_{0+1}\rangle$, is easily achieved (e.g., a system of interacting static spins can be placed in a *very* strong magnetic field), wait until the system relaxes into $|\Psi_{0+1}\rangle$, and *very* slowly remove the perturbation (in the absence of other interactions with the outside world), then, by virtue of the adiabatic theorem of § 1.4.5, in the end the system will be found in the desired state $|\Psi_0\rangle$.

AQC has some "built-in" advantages over "standard", or "circuit model" quantum computing. First, the ground state of a quantum system is naturally protected from decoherence. It has nowhere to relax to, and cannot be excited either, if only the temperature is low enough:

$$k_B T \ll \min_{\lambda \in [0,1]} [E_1(\lambda) - E_0(\lambda)], \qquad (6.48)$$

the minimum gap between the ground and first excited state of the adiabatic Hamiltonian

$$H(\lambda) = H_0 + \lambda H_1, \ 0 \leq \lambda(t) \leq 1. \qquad (6.49)$$

(Condition (6.48) should hold as well for the effective noise temperatures of whatever noise sources are present in the system.) The requirement of "slow change", which restricts $\lambda(t)$ so as to avoid exciting the system from its ground state, was considered in detail in § 1.4.5, Eqs (1.144, 1.154). Dephasing is of no concern, since it affects the relative phases of different state components in the energy eigenbasis, and the wave function of the system has only one such component, the instantaneous ground eigenstate of the Hamiltonian. This is in sharp contrast with circuit model quantum computing, which requires that the system is maintained in a generally fragile superposition of the Hamiltonian's eigenstates. While quantum error-correction schemes make this possible in principle, they are very costly in terms of extra physical qubits and operations (e.g., Le Bellac, 2006, 7.4).[7] Second, there is no need to perform precisely timed unitary operations (quantum gates) on

---

[7] This is, of course, an idealized picture. A real adiabatic quantum computer will always be an open system, interacting with the rest of the universe, and satisfying a nonunitary master equation for its density matrix rather than a unitary Schrödinger equation for its wave function. As a result, using the instantaneous eigenstates of the isolated system's Hamiltonian $H(\lambda)$ as the preferred basis (as we did in § 1.4.5) eventually stops being a good approximation; e.g., dephasing can no longer be ignored. After some critical time τ_c, dependent on the details of the system, the evolution becomes nonadiabatic. In other words, there is some nongeneric optimal speed at which AQC should be run (Sarandy and Lidar, 2005a,b). We will assume that τ_c is large enough, and limit our further discussion to times $t < \tau_c$.

qubits and pairs of qubits. This means no control signals, which eventually contribute to decoherence, and no sophisticated control circuitry, which would serve as an additional noise source and make the task of maintaining quantum coherence in the system so much harder – because of noise as well as simply because the additional elements make the device larger and harder to cool down and keep cold.

There is a price to pay, of course. First of all, universality: an adiabatic quantum computer essentially encodes the solution in its hardware (qubit–qubit couplings), and will require retuning or even rebuilding in order to solve a different problem. This is not surprising, since AQC is an analogue counterpart to "digital" circuit quantum computing. Analogue computers (see, e.g., Jackson, 1960; Blum, 1969; Dewdney, 1984) simulate, rather than calculate, the result. Therefore, they can do it very fast and pretty cheaply in terms of the resources involved. The simplest example is a slide rule; more sophisticated devices were used to aim anti-aircraft guns, to calculate the seepage of water under dams, etc., long before digital computers could do anything comparable, or even existed.[8] Indeed, Robert Hooke used an analogue computer to verify that the law of inverse squares leads to approximately elliptical trajectories of celestial bodies.[9] Another problem, next to nonuniversality, is that of accuracy: unlike a digital computer, the accuracy of an analogue one is limited by the accuracy of its hardware and cannot be controlled at will. Still, better a not-so-accurate answer than none at all. And, since from the point of view of hardware the current state of quantum computing is closer to that of classical computing in the 1880s rather than in the 1990s, it would be unwise to neglect slide rules in the expectation of the eventual advent of Pentiums.

6.2.2 Adiabatic algorithms

In many cases the form of the Hamiltonian H_0 directly follows from the problem to be solved. Consider, for example, the maxcut problem.[10] The problem is as follows. Let us assign to every edge of an N-node graph some weight. A "cut" partitions this graph into two sets, and its payoff is the sum of the weights of all the edges that cross the cut (i.e., connect nodes, which belong to the two sets). The "maxcut" is the

---

[8] An example given by Dewdney (1984) is the spaghetti analogue computer for sorting N numbers using exactly $N + 1$ operations (which is better than any known digital sorting algorithm). The first N operations consist of taking a piece of spaghetti from the bag and marking and cutting it to the length given by the corresponding entry in our number list. The $(N + 1)$th operation is holding them in your fist against a flat surface, and letting the gravity do the rest: the pieces of spaghetti (and the numbers from the list) will be sorted according to their size.

[9] After a series of experiments conducted together with Henry Hunt in January 1681 (Jardine, 2003, Chapter 8).

[10] This was the example used by Steffen et al. (2003) for an experimental demonstration of an AQC with three qubits, based on the NMR approach to quantum computing (for a brief introduction to the latter see, e.g., Le Bellac, 2006, 6.1).

cut with the biggest payoff. It is easy to persuade oneself that at least one maxcut should exist: if one set ("A") is empty and the other ("B") contains the whole graph, the payoff is zero; adding one node to set A will increase (or decrease) the payoff by the sum of the weights of all the edges connected to it; keep adding nodes to A, and eventually, when all the nodes belong to A, the payoff is zero again. Therefore, unless all possible cuts have zero weight (a trivial case), at least one maxcut exists. It is hard to find a maxcut in a given graph: this is an NP-complete problem[11] and is, therefore, a good testing problem for quantum computing. Each cut can be uniquely described by an N-component vector $\vec{s} = (s_1, s_2, \ldots)^{\mathrm{T}}$, where $s_j = 0$ if the jth node belongs to A, and $s_j = 1$ if it belongs to B. The payoff function is then

$$P(\vec{s}) = \sum_{j,k=1}^{N} w_{jk} s_j (1 - s_k) + \sum_{j=1}^{N} w_j s_j. \tag{6.50}$$

Here w_{jk} are the edge weights; the second term in (6.50) breaks the symmetry between the sets A and B. This payoff function can be identified with the ground state energy of the Hamiltonian of N coupled qubits:

$$H = -\frac{1}{2} \sum_{j} \left[\epsilon_j \sigma_z^j + \Delta_j \sigma_x^j \right] + \sum_{j<k} J_{jk} \sigma_z^j \sigma_z^k, \tag{6.51}$$

if we choose $J_{jk} = w_{jk}/2$, $\epsilon_j = w_j$ and $\Delta_j / J_{jk} \ll 1$. The solution will be encoded in the ground-state wave function $|0\rangle = |s_1 s_2 \ldots s_N\rangle$. One natural way of implementing such a Hamiltonian is using superconducting flux qubits. We have seen that all the parameters of (6.51), including the couplings, can be tuned by the external magnetic fluxes. A straightforward realization of the Hamiltonian (6.51) with inductively coupled flux qubits is restricted to a planar graph with a rather small number of nearest-neighbour or next-nearest-neighbour couplings, but this is mostly a technical problem. At any rate, these restrictions are immaterial at the level of the smallest nontrivial maxcut problem ($N = 3$), which has been attempted (Grajcar et al., 2005; Izmalkov et al., 2006; van der Ploeg et al., 2006) as a test case (Fig. 6.7). Choosing $J_{12} = J_{23} = J_{13} = 300$ mK, $\Delta_1 = \Delta_2 = \Delta_3 = 96$ mK, the adiabatic evolution from the initial polarized state to the biases $\epsilon_1 = 0.315$ K, $\epsilon_2 = 0.252$ K and $\epsilon_3 = 0.525$ K (which is achieved by the tuning fluxes of order $0.01\ \Phi_0$) should bring the system to ground state $|0\rangle = |\uparrow_1 \downarrow_2 \uparrow_3\rangle$, which encodes the maxcut problem solution (for the payoff function given by the ground state energy of (6.51) with $\Delta_j = 0$). A local minimum, $|\downarrow_1 \uparrow_2 \uparrow_3\rangle$, has a similar energy and could trap the system if it were classically annealed. The experiments showed

---

[11] Loosely, a problem is NP-complete if it is the hardest amongst those problems whose solution can be verified in polynomial time (as a function of the input size).

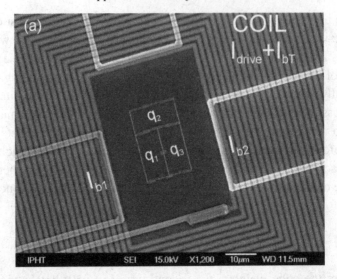

Fig. 6.7. An experimental circuit for the realization of a ($N = 3$) maxcut adiabatic quantum algorithm (from van der Ploeg et al., 2006, © 2007 IEEE, with permission). Aluminium persistent current qubits are placed inside a niobium pickup coil; the tuning fluxes, f_{qj}, $j = 1, 2, 3$ (in units of Φ_0), are induced by the currents in Π-shaped bias lines. The qubits are antiferromagnetically coupled through shared Josephson junctions and (to a lesser degree) mutual inductances. The quantum state of the qubits was determined using the impedance measurement technique (IMT). The circuit parameters, $J_{12} = J_{23} = J_{13} = 610$ mK, $\Delta_1 = \Delta_2 = \Delta_3 = 70$ mK, as well as its effective temperature, were determined from fitting the IMT data to the theoretical predictions. The effective temperature of the device, $T_{\text{eff}} = 70$ mK (as opposed to the nominal mixing chamber temperature $T = 10$ mK), must be further lowered to allow a demonstration of AQC in this device.

that a lower effective temperature of the system or a larger gap between the ground and excited states is still required in order to run an AQC algorithm.

AQC is naturally well suited for solving optimization problems on graphs; another important and actively studied example is the travelling salesman problem (TSP, see, e.g., Johnson and McGeoch, 1997). In its canonical formulation, it is necessary to find the shortest path on a graph, which passes through all the nodes and returns to the starting point. The payoff function is obviously the length of the path. There are different ways of writing the adiabatic TSP Hamiltonian, but usually it is cast as that of an Ising spin system, with the potential energy term representing the path length (e.g., Martonák et al., 2004); such a system can also be, in principle, realized as a set of qubits.

What should we do if the form of the Hamiltonian H_0 cannot be so easily guessed? An elegant way of finding the Hamiltonian H_0 for *any* standard (i.e., circuit model) quantum algorithm is called *ground state quantum computing* (GSQC, Mizel et al.,

2001) and proceeds as follows. We know that any quantum algorithm can be reduced to a series of one-qubit unitary rotations and two-qubit CNOT gates (e.g., Le Bellac, 2006, 5.3). It is, therefore, sufficient to consider only the one-qubit and two-qubit cases. Start with a single qubit, on which we would want to perform M gates, $U_1, \ldots, U_M$, where U_j is a 2×2-unitary matrix. We will model this qubit by an array of $2(M + 1)$ quantum dots (e.g., 2DEG quantum dots), containing a single electron (Fig. 6.8). The top (bottom) dot in the $(m + 1)$th column represents the qubit in state $|0\rangle$ ($|1\rangle$) at the mth step of the algorithm. If we denote by $c_{m,0(1)}^\dagger$ the creation operator of an electron on the corresponding quantum dot, any wave function can be written as

$$|\Psi\rangle = \sum_{m=0}^{M} \sum_{j=0}^{1} A_{mj} c_{mj}^\dagger |\text{vac}\rangle, \tag{6.52}$$

where $|\text{vac}\rangle$ is the vacuum state in the Fock space (to distinguish it from the state $|0\rangle$ of a qubit). It is convenient to introduce "vector operators", $C_m^\dagger = (c_{m0}^\dagger, c_{m1}^\dagger)$ and $C_m = \left[C_m^\dagger \right]^\dagger$. It is straightforward to check that the operator $P_m = C_m^\dagger C_m$ projects the state $|\Psi\rangle$ onto the state in the mth column (so that if for this particular value of m the coefficients A_{m0}, A_{m1} in (6.52) are zero, then $P_m|\Psi\rangle = |\text{vac}\rangle$). Similarly, the operator $T_{m,m-1} = C_m^\dagger C_{m-1}$ transfers the electron from the $(m-1)$th to the mth column.

By initializing a single qubit in some state $|\psi_0\rangle$ and then acting upon it consecutively by the operators U_j, $j = 1, \ldots, M$, we obtain

$$|\psi_1\rangle = U_1|\psi_0\rangle, |\psi_2\rangle = U_2|\psi_1\rangle, \ldots$$
$$|\psi_M\rangle \equiv |0\rangle\langle 0|\psi_M\rangle + |1\rangle\langle 1|\psi_M\rangle = U_M|\psi_{M-1}\rangle. \tag{6.53}$$

Therefore, if we could find a wave function to satisfy the equations

$$P_m|\Psi\rangle = U_m T_{m,m-1} P_{m-1}|\Psi\rangle \ (m = 1, \ldots, M);$$
$$P_0|\Psi\rangle = \left[A_{00} c_{00}^\dagger + A_{01} c_{01}^\dagger \right] |\text{vac}\rangle, \tag{6.54}$$

we would have modelled the time evolution (6.53) and could obtain the desired final state of the qubit by projecting the state of the system on the last column:

$$P_M|\Psi\rangle = A_{M0} c_{M0}^\dagger |\text{vac}\rangle + A_{M1} c_{M1}^\dagger |\text{vac}\rangle = \mathcal{C} \left\{ \langle 0|\psi_M\rangle c_{M0}^\dagger |\text{vac}\rangle + \langle 1|\psi_M\rangle c_{M1}^\dagger |\text{vac}\rangle \right\}. \tag{6.55}$$

In Eqs (6.54) the column number, m, represents the number of the unitary gate applied to the single qubit. The normalization constant $\mathcal{C}$ accounts for the fact that the electron state is spread over $2(M+1)$ quantum dots. The whole scheme nicely illustrates the strategy of exchanging space for time, common for AQC, chess and the art of war.

It is straightforward to check that the state $|\Psi\rangle$, satisfying (6.54), is the ground state of the Hamiltonian (Mizel et al., 2001)

$$H_0 = E \sum_{j=1}^{M} \left[P_{j-1} + P_j - C_j^\dagger U_j C_{j-1} - C_{j-1}^\dagger U_j^\dagger C_j \right]. \tag{6.56}$$

Actually there are *two* ground states, one corresponding to the initial qubit wave function $|\psi_0\rangle = |0\rangle$ (i.e., $A_{00} \neq 0$, $A_{01} = 0$), and the other with $|\psi_0\rangle = |1\rangle$ ($A_{00} = 0$, $A_{01} \neq 0$). This degeneracy is easily lifted by adding to H_0 a small term (e.g., $-\epsilon c_{00}^\dagger c_{00}$) favouring one state over the other. Expressed directly through the electron creation–annihilation operators, the Hamiltonian (6.56) has the form

$$H_0 = \sum_{m=0}^{M} \sum_{j=0}^{1} \epsilon_{mj} c_{mj}^\dagger c_{mj} + \sum_{\langle mj, m'j' \rangle} \left(\Delta_{mj,m'j'} c_{mj}^\dagger c_{m'j'} + H.c. \right), \tag{6.57}$$

where *H.c.* stands for a Hermitian conjugate and the second sum is only extended over the nearest neighbours. Therefore, it does not seem impossible to implement H_0 using, e.g., 2DEG quantum dots, since all we need is to control on-dot potentials, ϵ, and interdot tunnelling matrix elements, Δ, of the electron. (The practicality of such an implementation is a different question!) For example, a phase shift gate,

$$U_\theta = \begin{pmatrix} \exp(i\theta) & 0 \\ 0 & 1 \end{pmatrix}, \tag{6.58}$$

is a one-step operation ($M=2$) and can be realized by choosing

$$\epsilon_{00} = \epsilon_{11} = \epsilon_{01} = \epsilon_{10} = E;$$
$$\Delta_{00,10} = E e^{i\theta}; \ \Delta_{01,11} = E; \ \Delta_{00,11} = \Delta_{01,10} = 0. \tag{6.59}$$

While tuning the amplitude of the tunnelling matrix elements is straightforward, tuning their phases may require, e.g., using the magnetic or electrostatic Aharonov–Bohm effect in order to produce the appropriate phase shift in the electron wave during tunnelling between the quantum dots.

Consider now a universal two-qubit gate, CNOT. (Since a sequence of CNOT gates and single-qubit rotations allows, in particular, the swapping of the states of any two qubits, we only need to perform CNOT operations on the nearest neighbours.) This is again a one-step operation, and it requires $2 \times 2 \times 2$ quantum dots

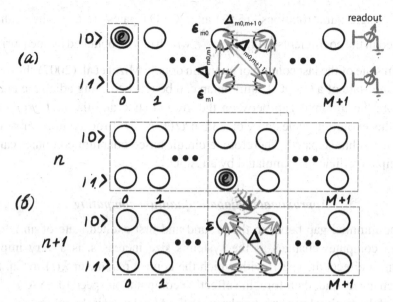

Fig. 6.8. Ground-state quantum computing (GSQC). (a) A single qubit undergoing M unitary operations is represented by $2(M+1)$ quantum dots with a single, trapped electron. The dot potentials and tunnelling matrix elements are controllable. (b) A two-qubit gate in GSQC. The tunnelling matrix elements and dot potentials in "qubit $n+1$" are affected by the presence of an electron on the corresponding $|1\rangle$-dot of "qubit n".

(Fig. 6.8). The corresponding Hamiltonian turns out to be

$$H_{\text{CNOT}} = \left(c_{10}^{(1)\dagger} C_1^{(2)\dagger} - c_{00}^{(1)\dagger} C_0^{(2)\dagger} \right) \left(C_1^{(2)} c_{10}^{(1)} - C_0^{(2)} c_{00}^{(1)} \right)$$

$$+ \left(c_{11}^{(1)\dagger} C_1^{(2)\dagger} - c_{01}^{(1)\dagger} C_0^{(2)\dagger} \sigma_x \right) \left(C_1^{(2)} c_{11}^{(1)} - \sigma_x C_0^{(2)} c_{01}^{(1)} \right)$$

$$+ C_0^{(1)\dagger} C_0^{(1)} C_1^{(2)\dagger} C_1^{(2)} + C_1^{(1)\dagger} C_1^{(1)} C_0^{(2)\dagger} C_0^{(2)}. \tag{6.60}$$

Here the upper index labels the qubit-representing row. If we look at the structure of H_{CNOT}, we see that the physics of the situation is that the quantum state of one qubit – i.e., the charge on the quantum dots of one row – influences (e.g., electrostatically) the evolution of the other, i.e., the tunnelling between the quantum dots in the other row.

We see now that the solution of an arbitrary problem solvable by "standard" quantum computing can be encoded into the ground state, $|\Psi\rangle$, of a system of $2N(M+1)$ quantum dots, containing N electrons (where N is the number of qubits, and M is the number of steps of the algorithm), which mimics by its spatial structure the application of the corresponding quantum algorithm in time (and is, therefore, called the "universal history state"). The Hamiltonian of the system is the sum of "single-qubit rotations" (6.57) and "CNOT gates" (6.60). The solution is contained in the coefficients $\left\{ A^n_{Mj} \right\}^{n=1..N}_{j=0,1}$, which can be obtained by measuring the quantum state of the last column of quantum dots.[12] Mizel et al. (2007) showed that the ground state of a GSQC Hamiltonian can be reached using adiabatic quantum evolution; the minimal gap between the ground state and the first excited state during this evolution scales no worse than $O(1/N^4 M^2)$ for the number of qubits N and algorithm steps M. Therefore, a circuit-based quantum computer can be – in principle – efficiently simulated by an AQC.[13]

Approximate adiabatic quantum computing

How the minimal gap between the ground and first excited state of an adiabatic quantum computer changes as the system's size increases, is a very important question: it affects the speed with which the control parameter $\lambda(t)$ in Eq. (6.49) can be changed, and, therefore, the effective computation speed of an AQC.[14] The scaling of the minimal gap and its relation to the AQC speed is the subject of ongoing research and much debate (e.g., Cao and Elgart, 2010; Mizel, 2010, and references therein). What is certain is that the spectrum of $H(\lambda)$ (and so the minimal gap) is sensitive to the specific quantum algorithm it realizes, and the values of the control parameter, where the gap becomes dangerously narrow, are not known beforehand (see Fig. 6.9).

Let us, therefore, ask not how to stay in the ground state, but how far we can deviate from it and still obtain useful results. This may not work for every algorithm, but it certainly makes sense for optimization problems, like TSP. In the TSP the energy of the solution state directly yields the route length, so the first excited state will give the best possible approximation to the optimal solution. If in the process of lifting the perturbation we manage not to excite the system too far from the ground state, we will obtain a good approximate solution for the TSP. Intuitively, this *approximate AQC* (AAQC) approach should have an advantage over the usual quantum annealing in obtaining an approximate solution, if only because in the

---

[12] There remains yet another slight problem: since there is only one electron per $(M+1)$ columns in each qubit-representing row, with the probability $\sim M/(M+1)$ it will not be present. It is, though, easily resolved by decreasing the potential on this column and thus attracting the electrons there.

[13] To be precise, the proof of Mizel et al. (2007) is valid not for the Hamiltonian of the "canonical AQC" form (6.49), but rather for

$$H = H_0 + \lambda(t)H_1 + \lambda^2(t)H_2,$$

where $[H_1, H_2] \neq 0$.

[14] To identify the allowed maximal value of $d\lambda/dt$ near the dangerous values of λ, where the gap is small, with an average $d\lambda/dt = 1/\tau$, the inverse of the computation time, would be the same as to estimate the journey time by the speed with which your car can safely go around a sharp turn on a slippery road. Of course, with the AQC we do not know where the turn is.

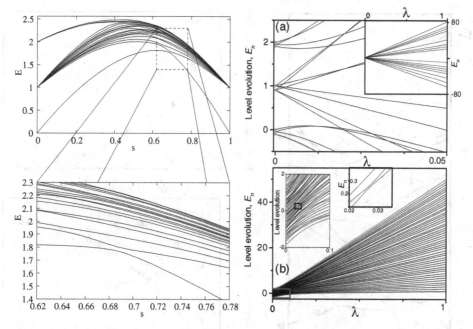

Fig. 6.9. Examples of adiabatic-level evolution. (Left) Lowest 18 out of 16 384 levels in a 3-SAT problem (reprinted with permission from Znidaric, 2005, © 2005 American Physical Society; the adiabatic parameter s was introduced via $H(s) = sH_{3-\text{SAT}} + (1-s)H_{\text{perturbation}}$). (Right) Reprinted with permission from Zagoskin et al., 2007, © 2007 American Physical Society. (a) CNOT gate simulation for the operation $|00\rangle \to |00\rangle$, described by the Hamiltonian (6.60). Some levels cross due to the symmetry of the problem. (b) Evolution of energy levels of a model Hamiltonian (from the Gaussian unitary ensemble (GUE) of random matrix theory; see Stöckmann (1999)).

beginning the system is already in the ground state, and for as long as it does not deviate too far away from it.

To estimate the efficiency of AAQC let us consider the mechanisms by which the excited states are populated: noise (including thermal fluctuations) and Landau–Zener tunnelling. The noise and temperature effects can be, in principle, reduced to zero. We will, therefore, first consider Landau–Zener tunnelling, which cannot be eliminated at any finite speed of adiabatic evolution. Near every anticrossing of energy levels we can neglect the existence of the other anticrossings and work in the two-level approximation of § 1.4.6. Then the probability of switching states is given by Eq. (1.167),

$$P_{LZ} = e^{-\frac{\pi \Delta^2}{2\hbar v}},$$

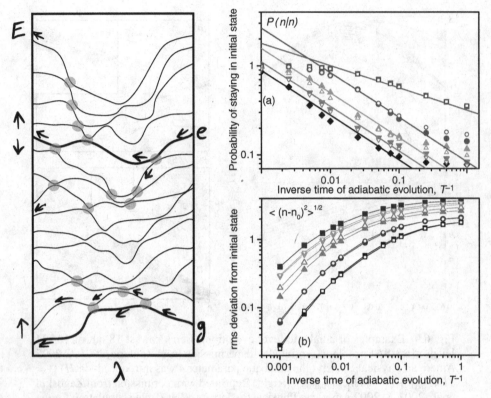

Fig. 6.10. Approximate adiabatic quantum computing (AAQC). (Left) During quasiadiabatic evolution the system can deviate from the initial ground (g) or excited (e) state via a series of Landau–Zener transitions (grey circles), in a process similar to a random walk. (Right) Probability of staying in the same state (a) and the r.m.s. deviation from the initial state (b) as a function of the inverse evolution time for Hamiltonians from the GUE of random matrix theory (reprinted with permission Zagoskin et al., 2007, © 2007 American Physical Society; cf. Eq. (6.63)). Different symbols correspond to different initial energy eigenstates.

where Δ is the minimal value of the interlevel spacing, $v = d\epsilon/dt$ (assuming $v = $ constant) and $\epsilon(e)$ is the energy difference between the diabatic states. We can estimate the result of the Landau–Zener effect using a simple heuristic picture. Consider the energy levels of the Hamiltonian (6.49) as paths along which a point, representing our system, can move, as shown in Fig. 6.10. Suppose the system is initially in the eigenstate $|n(\lambda)\rangle$ of the Hamiltonian. When reaching the first anticrossing (with the state $|n \pm 1\rangle$), with probability $(1 - P_{LZ})$ it will keep to its original path (i.e. stay in the nth state), and with probability P_{LZ} will switch to the other one. At the next anticrossing of the currently occupied level the same thing happens (with a different value of P_{LZ}), etc. The evolution of the system's state is, therefore, a random walk, with variable time intervals and asymmetric right–left (or rather up–down) step probabilities, but where the level number can only change by one (or not at all) at each step. The result will be a slow, diffusion-like displacement of the system from its initial state.

A very rough estimate of this displacement can be made as follows. Denote by $N \gg 1$ the average number of anticrossings per energy level, which happen as the control parameter changes from one to zero, by $P(n, k = \lambda N)$ the probability of finding the system in the nth eigenstate of $H(\lambda)$, and by p the average Landau–Zener transition probability. We will also assume that $n \gg 1$. Then

$$P(n, k+1) = (1 - p)P(n, k) + \frac{p}{2}[P(n-1, k) + P(n+1, k)]; \qquad (6.61)$$

i.e., if after an anticrossing we find the system in state $|n\rangle$, it either stayed in this state, or underwent a Landau–Zener transition from the neighbouring higher or lower energy level (assuming equal probabilities of either event). This is almost the equation for the standard random walk (e.g., Gardiner, 2003, 3.8.2), and explicit expression for the dispersion can be found using standard methods:[15]

$$\left\langle (n - n_0)^2 \right\rangle_{\lambda=1} = pk|_{\lambda=1} = pN. \qquad (6.62)$$

Because of the assumption $n \gg 1$, Eq. (6.62) does not apply to the ground state, but it is at least plausible that the displacement scales with the Landau–Zener transition rate and the number of anticrossings. A literal application of (6.62) to the case of constant evolution speed, $|\dot{\lambda}| = t/T_A$, would give for the average accuracy of AAQC

$$\Delta n = \sqrt{\langle n^2 \rangle} \propto e^{-\alpha T_A}, \qquad (6.63)$$

where T_A is the running time of the algorithm, and α is some non-universal rate. This admittedly oversimplified picture is still in qualitative agreement with numerical

---

[15] To get this result, we introduce the characteristic function $G(s, k) = \langle \exp(ins) \rangle_k = \sum_{n=-\infty}^{\infty} \exp(ins) P(n, k)$ (so that $P(n, k) = \int_{-\pi}^{\pi} (ds/2\pi) \exp(-ins) G(s, k)$). Then, iterating Eq. (6.61), we find

$$G(s, k+1) = [p \cos s + (1 - p)]^{k+1} e^{in_0 s},$$

where $|n_0\rangle$ is the state of the system at the start of the adiabatic evolution. Then the dispersion is

$$\left\langle (n - n_0)^2 \right\rangle_k = \sum_n (n - n_0)^2 P(n, k) = \int_{-\pi}^{\pi} \frac{ds}{2\pi} [p \cos s + (1 - p)]^{k+1} \sum_n (n - n_0)^2 e^{-i(n-n_0)s}.$$

Using the identity

$$\sum_n (n - n_0)^2 e^{-i(n-n_0)s} = -\frac{d^2}{ds^2} \sum_n e^{-i(n-n_0)s} = -\frac{d^2}{ds^2} \left(2\pi \sum_m \delta(s - 2\pi m) \right)$$

and integrating by parts, we arrive at Eq. (6.62).

simulations by Zagoskin et al. (2007) (Fig. 6.10). Eq. (6.63) indicates a tantalizing possibility that the accuracy of AAQC may increase exponentially with the running time (or, conversely, that the time required for an AAQC run with the accuracy Δn only grows as $\ln(1/\Delta n)$). Whether (6.63) or a refinement of it applies to AAQC, and if so to what classes of algorithms and under what conditions, is still an open question.[16]

If the problem Hamiltonian H_0 contains any symmetries, the levels can cross; in particular, if the ground state can cross the first excited state the adiabatic theorem becomes useless (or rather it applies separately to the groups of states of different symmetry). Our AAQC reasoning also stops working. Here, unexpectedly, the situation can be improved by the presence of noise. First, in the presence of a bosonic heat bath with temperature

$$\min \Delta \ll k_B T \ll \langle \Delta \rangle, \tag{6.64}$$

where Δ is the gap between the ground and excited states, the thermally activated transitions between these states tend to equipopulate them, while the system, sufficiently slowly, passes through the (anti)crossing region. Therefore, the system stays in the ground state with probability $\sim 1/2$ even if the minimal gap is zero. If the gap is not zero, and the bosonic bath is super-ohmic ($s > 1$, § 5.3.2), one can even increase the speed of adiabatic evolution compared to that imposed by the adiabatic theorem: $T_A \sim (\min \Delta)^{-4/(4-s)}$ rather than $T_A \sim (\min \Delta)^{-2}$. After the anticrossing, the coupling to the bath allows the system to relax to the ground state, if it underwent a Landau–Zener transition; this also increases the probability of success of the AQC run (Amin et al., 2008). Second, the noise terms can break the symmetries of H_0 and eliminate level crossings; there is actually an optimal level of noise, which could be artificially introduced into the system (Wilson et al., 2010).

If, to be after what we believe to be adiabatic evolution, the system ends up in its ground state, or close enough to it, it is still too early to celebrate, since there

---

[16] For example, an Ising spin glass in a transverse field is modelled by the Edwards–Anderson Hamiltonian

$$H = -\sum_{\langle ij \rangle} J_{ij} \sigma_z^i \sigma_z^j - h \sum_i \sigma_x^i,$$

where $\langle ij \rangle$ are nearest neighbours on a D-dimensional cubic lattice, and J_{ij} are random. Eq. (16) with $D > 2$ is often used as a representation of an NP-complete computational problem. Theoretical consideration of *quantum annealing* in this system (i.e., the slow reduction to zero of the transverse field h; more generally, switching off the kinetic energy term in the Hamiltonian) suggests that to arrive within Δn of the ground state the required annealing time is exponential in $1/\Delta n$, the same as in the case of classical annealing: $T_A \sim \exp[(\Delta n)^{-1/\zeta}]$, where ζ is some positive parameter. Quantum annealing is still numerically faster, since in the classical case $\zeta_{cl} \leq 2$, while in the quantum case $\zeta_{cl} < \zeta_{quant} \leq 6$. This slowing down of quantum annealing is ascribed to the large number of anticrossings of the ground state, becoming denser as $h \to 0$ and may be only specific for the Ising spin glass (Santoro et al., 2002).

is always a possibility that this happened through normal annealing (maybe, at some points assisted by quantum tunnelling). This presents yet another problem – how to recognize whether a device which is supposed to be an (approximate) adiabatic quantum computer is actually an (approximate) adiabatic quantum computer. Setting up an appropriate experimental protocol is a nontrivial problem.[17]

6.3 Quantum engines, fridges and demons

6.3.1 Quantum thermodynamic cycles

It is hard to imagine a branch of theoretical physics farther removed from quantum mechanics than classical thermodynamics, with its coolers, heaters, quasistatic processes, ideal cycles and heat engines. Nevertheless, quantum heat engines (QHE), i.e., engines with a quantum coherent working body, were introduced quite early on (Scovil and Schulz-DuBois, 1959; Geusic et al., 1967) in the context of lasing. Since then they have been extensively used in *Gedankenexperiments* helping to clarify subtle points of statistical mechanics like the role of fluctuations and the operation of Maxwell's demon in the quantum regime.[18] The arrival of macroscopic devices based on individually controlled qubits, which can maintain quantum coherence over relatively long time intervals, made the experimental realization of many such *Gedankenexperiments* feasible. This completes the circle, formed by the links between classical thermodynamics, statistical mechanics, information theory and quantum computing.

To begin with, consider a textbook situation, where there are two thermal reservoirs at temperatures T_c, T_h, and a system that can be periodically put in thermal equilibrium with either of them (e.g., Fermi, 1956). In its quantum analogue, we employ as reservoirs and the working body three harmonic oscillators, with frequencies $\omega_c, \omega_h, \omega$, respectively, assuming that the frequency ω of the "working body" oscillator can be changed at will. To be specific, suppose that these are superconducting LC circuits, with the "working body" containing a Josephson junction; the nonlinearity due to this junction is neglected (Zagoskin et al., 2010). Biasing the Josephson junction changes its effective inductance and, therefore, allows us to control the frequency ω, as we have seen in § 2.5.1. We assume that

$$\omega_h > \omega_c \tag{6.65}$$

(but not necessarily $T_h > T_c$). The mechanical work produced by, or applied to, the system is determined by the quantum state-dependent energy required to change

---

[17] This is one of the reasons the results of recent attempts to realize AQC or AAQC in 16-qubit and 28-qubit superconducting processors were inconclusive (van Dam, 2007; Guizzo, 2010).

[18] See, e.g., Geva and Kosloff, 1994; Scully et al., 2003; Kieu, 2004; Humphrey and Linke, 2005; Quan et al., 2007; Quan, 2009; Maruyama et al., 2009 and Gemmer et al., 2009, Ch. 20.

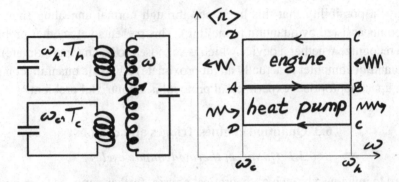

Fig. 6.11. LC-circuit-based realization of a quantum heat engine/heat pump and an ideal quantum Otto cycle.

the bias current in the Josephson junction. Electrical energy can be, in principle, converted into mechanical work and back without loss, so we need not concern ourselves about it.

Let us run the tunable oscillator through the following cycle (Fig. 6.11, where the average number of quanta is plotted versus its resonant frequency). (AB) Tune it to resonance with the cold reservoir, $\omega = \omega_c$, and allow it to equilibrate. (BC) Infinitesimally slowly increase the frequency until $\omega = \omega_h$. According to the quantum adiabatic theorem, the occupation numbers of its energy states will not change during this process (of course, we are neglecting the relaxation in the system). Therefore, it will be out of equilibrium with the hot reservoir. Now (CD) we will allow it to equilibrate, and then (D) slowly lower its frequency to ω_c. Here the system is again out of equilibrium, and will equilibrate with the cold reservoir (DA), which completes the cycle.[19] The work performed by the system against the external forces (i.e., the current source, which biases the Josephson junction and, thus, changes the resonance frequency) is given by the area enclosed by the contour ABCD:

$$R = -\int_A^B \langle n \rangle \mathrm{d}\omega + \int_C^D \langle n \rangle \mathrm{d}\omega = \pm A_{\mathrm{ABCD}}. \tag{6.66}$$

If the contour is traversed anticlockwise, R is positive, and our device is a heat engine; otherwise, R is negative, and the device works as a heat pump (refrigerator), transferring energy from the cold to the hot reservoir. If we compare the workings of our device (Fig. 6.11) with textbook thermodynamic cycles of idealized heat engines, we see that the frequency ω is analogous to the inverse volume, and the product $\langle n \rangle \omega^2$ to the pressure in the system. The cycle ABCD then consists of two

---

[19] We implicitly assumed that the resonances in the heater, cooler and the working body oscillators are infinitely sharp; otherwise, the energy exchange will continue until the detuning exceeds the linewidth. But this correction can be made arbitrarily small.

adiabatic (AB, CD) and two isochoric (BC, DA) sections and corresponds to Otto cycle, the same as in an internal combustion engine.[20] This cycle is not reversible due to the two isochoric stages, but it is the simplest quantum cycle to implement. It is also straightforward to find its efficiency. Denoting by $|n, c(h)\rangle$ the nth energy eigenstate of an oscillator with frequency $\omega_c(\omega_h)$, and by $Z_{c(h)}$ the corresponding partition function, we can write for the density matrix of the working body at points A, B, C, D

$$\rho_A = Z_c^{-1} \sum_n e^{-\frac{\hbar\omega_c}{k_B T_c}} |n, c\rangle\langle n, c|, \quad \rho_B = Z_c^{-1} \sum_n e^{-\frac{\hbar\omega_c}{k_B T_c}} |n, h\rangle\langle n, h|,$$

$$\rho_C = Z_h^{-1} \sum_n e^{-\frac{\hbar\omega_h}{k_B T_h}} |n, h\rangle\langle n, h|, \quad \rho_A = Z_c^{-1} \sum_n e^{-\frac{\hbar\omega_h}{k_B T_h}} |n, c\rangle\langle n, h|, \quad (6.67)$$

which immediately yields the expressions for the energy of the system at the corresponding points and for the work done and the energy received by the system:

$$R_1 = E_A - E_B = -\frac{\hbar(\omega_h - \omega_c)}{2} \coth\frac{\hbar\omega_c}{2k_B T_c},$$

$$R_2 = E_C - E_D = \frac{\hbar(\omega_h - \omega_c)}{2} \coth\frac{\hbar\omega_h}{2k_B T_h},$$

$$R = R_1 + R_2 = \frac{\hbar(\omega_h - \omega_c)}{2} F\left(\frac{\hbar\omega_c}{2k_B T_c}, \frac{\hbar\omega_h}{2k_B T_h}\right); \quad (6.68)$$

$$Q_1 = E_C - E_B = \frac{\hbar\omega_h}{2} F\left(\frac{\hbar\omega_c}{2k_B T_c}, \frac{\hbar\omega_h}{2k_B T_h}\right);$$

$$Q_2 = E_A - E_D = -\frac{\hbar\omega_c}{2} F\left(\frac{\hbar\omega_c}{2k_B T_c}, \frac{\hbar\omega_h}{2k_B T_h}\right),$$

where

$$F(x, y) = \frac{\sinh(x - y)}{\sinh x \sinh y}. \quad (6.69)$$

As expected, $R_1 R_2 < 0$ and $Q_1 Q_2 < 0$. The direction of the cycle is determined by the parameter

$$\lambda = \frac{\hbar\omega_c}{k_B T_c} - \frac{\hbar\omega_h}{k_B T_h}. \quad (6.70)$$

For $\lambda > 0$ the net work $R = R_1 + R_2 > 0$, and the device works as a heat engine with efficiency

$$\eta = \frac{R}{Q_1} = 1 - \frac{\omega_c}{\omega_h}. \quad (6.71)$$

---

[20] For a detailed review of quantum analogues to different classical thermodynamic cycles, including the quantum Carnot cycle, see Quan et al. (2007) and Quan (2009).

This seems to allow a violation of the Carnot inequality,

$$\eta \leq \eta_C = 1 - \frac{T_c}{T_h}, \tag{6.72}$$

by choosing $(\omega_c/\omega_h) < (T_c/T_h)$, i.e., $\lambda < 0$. But in this case our heat engine simply becomes a refrigerator with efficiency

$$\zeta = \frac{-Q_2}{R} = \frac{\omega_c}{\omega_h - \omega_c}. \tag{6.73}$$

The second law of thermodynamics cannot be broken that easily, or at all. We can directly check that its other formulation, the Clausius inequality,

$$C \equiv \Delta S - \Delta Q/T \geq 0, \tag{6.74}$$

is also satisfied during the non-adiabatic stages of the cycle. Here ΔS is the change of the entropy of the working body, and ΔQ is the heat transferred to it. Indeed, using the standard expression for the entropy of a quantum oscillator in equilibrium (e.g., Landau and Lifshitz, 1980, Chapter V), the entropy of the working body during the adiabatic compression stage is

$$S_{AB} = -\ln\left(2\sinh\frac{\hbar\omega_c}{2k_BT_c}\right) + \frac{\hbar\omega_c}{2k_BT_c}\coth\frac{\hbar\omega_c}{2k_BT_c}, \tag{6.75}$$

and for the compression stage, CD, we replace ω_c/T_c with ω_h/T_h. From here, using (6.69) and the inequality

$$\ln(\sinh x/\sinh y) \leq (x - y)\coth y, \quad (x, y > 0), \tag{6.76}$$

we find that, indeed

$$C_{BC} = S_{CD} - S_{AB} - \frac{Q_2}{k_BT_h} \geq 0; \quad C_{DA} = S_{AB} - S_{CD} - \frac{Q_1}{k_BT_c} \geq 0. \tag{6.77}$$

Building such an Otto engine is well within current experimental capabilities. Producing electricity with its help would not be at all cost-effective, but it may prove useful as a way of cooling a quantum system (e.g., an adiabatic quantum computer, or a nanomechanical beam) during its operation. The cooling limit of such a fridge is given by the condition $\lambda = 0$, i.e.,[21]

$$T_c = \frac{\omega_c}{\omega_h}T_h. \tag{6.78}$$

---

[21] Eq. (6.78) actually holds for any kind of cooling based on resonant transitions (Grajcar et al., 2008a).

Its cooling efficiency ζ is temperature-independent, but as the temperature difference drops, the heat Q_2 extracted per cycle from the cold reservoir will go to zero according to (6.69). There will also be the usual loss of efficiency due to the finite speed of operation – in our case, during the adiabatic stages AB and CD. The loss (i.e., the increase of entropy) is due to the parametric squeezing of the oscillator states (Zagoskin et al., 2010) – the effect considered in § 4.4.6.

What changes if the roles of the heater, cooler and the working body are played by qubits, instead of harmonic oscillators? Actually, nothing much. It is straightforward to see, for example, that Eqs (6.70, 6.71, 6.73) remain the same, with ω_c and ω_h replaced by the corresponding interlevel distances, $\hbar\Omega_c$ and $\hbar\Omega_h$ (Quan et al., 2007). We can play with different combinations of qubits and linear/nonlinear oscillators, to find the easiest one to realize experimentally, but the essential results will stay the same.

6.3.2 Sisyphus effect: cooling and amplification in qubit devices

The operation of a quantum Otto cycle between "hot" and "cold" resonators can be approximated by a simple sinusoidal drive applied to the working body. This will, of course, reduce the efficiency of the cycle, because of the irreversible processes during the variable-speed "adiabatic" stages as well as the shorter effective duration of the isochoric stages (which would have altogether shrunk to zero at the moments when $\omega = \omega_{c,h}$, if not for the finite linewidth of the resonators and the working substance). The advantage of such an approach is in its simplicity, which allowed Grajcar et al. (2008b) to realize an experimental version (Fig. 6.12). The setup is by now familiar: a persistent current flux qubit fabricated of aluminium was placed inside a niobium pickup coil, which formed part of a high-quality LC tank circuit, which ensured inductive coupling between them. The qubit was driven by a high-frequency signal, with frequency $\hbar\Omega$ close to the qubit interlevel spacing, Δ. The large mismatch between the tank frequency, ≈ 20 MHz, and $\Delta/h \geq 3.5$ GHz, ensured that direct energy transfer between the two was suppressed (as for the Rabi spectroscopy experiment, §5.5.5). Depending on the choice of the working point of the qubit, with respect to the resonance between the qubit and the high-frequency drive, the effect of the interplay between these two signals will be either damping or amplification of the low-frequency current in the tank circuit. The temperature of the qubit $T_q < 100$ mK, and, by manipulating the qubit, the effective temperature of the tank (initially ≈ 4.2 K) could be decreased by $\approx 8\%$ or increased by $\approx 36\%$. Since the working point of the qubit stays the same, the gradual change of the current amplitude will move the maximum of the tank oscillations away from resonance with the high-frequency drive, thus rapidly decreasing the efficiency of the process (by reducing the time the qubit spends near resonance). It is also clear that the

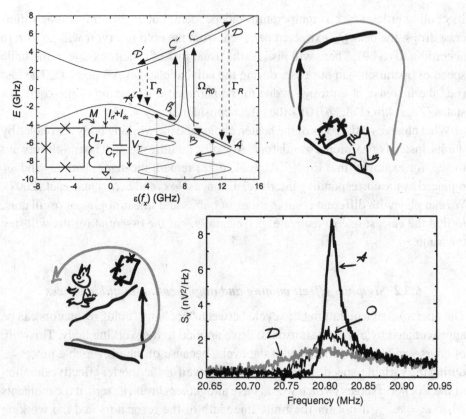

Fig. 6.12. Sisyphus cooling and signal amplification in a flux qubit coupled to a tank circuit (adapted by permission from *Nature Physics* Grajcar et al., 2008b, © 2008 Macmillan Publishers Ltd). The flux qubit, driven by a harmonic signal, is analogous to the working body, and the LC tank circuit to the source of mechanical work *and* to the cold/hot reservoir, depending on the choice of the working point of the qubit and the amplitude of the low-frequency signal in the tank. The equivalent of the other reservoir is the qubit's environment. (Top) The cooling/heating cycle. The energy levels of the flux qubit ($|0\rangle$, $|1\rangle$) are driven by the low-frequency harmonic signal in the tank (sinusoids) to the point near resonance with the high-frequency signal (indicated by the resonance curve). Near resonance the qubit, initially in state $|0\rangle$, is excited by the high-frequency signal to state $|1\rangle$ and keeps being driven by the tank signal until it relaxes back to the ground state. The cycle ABCD corresponds to the cooling Otto cycle of Fig. 6.11. The tank circuit cools down, because during both the "adiabatic expansion" stage, AB, and "compression", CD, it *increases* the energy of the qubit at the expense of the energy of the tank circuit. The transferred energy is equal to the difference between the energy of the photon absorbed from the high-frequency source in resonance and the energy lost to the environment when the qubit relaxes to its ground state. The cycle A′B′C′D′ is a counterpart of the heating Otto cycle. The tank circuit obtains per cycle the energy difference between the qubit's energy loss to the environment and the gain from the resonant field. (Bottom) Spectral density of voltage fluctuations in the LC tank circuit off-resonance (O) and in the damping (D) and amplifying (A) regimes.

optimal efficiency will be reached when the relaxation rate along CD or C'D' is close to the frequency of the tank circuit drive – otherwise, either the qubit will undergo many cycles without transferring energy to/from the tank (if $\Gamma \ll \omega_T$), or the effective area of the cycle loop and the energy transfer per cycle will be too small (if $\Gamma \gg \omega_T$).

Note that, unlike a "real" Otto cycle, discussed above, here the energy transfer between the working body (qubit) and the "cold/hot reservoir" (LC tank circuit) is reversible. (Another, minor, difference is that the relaxation in this case is not a resonant process.) It is then more accurate to talk about attenuation and amplification rather than cooling and heating.

A quantitative analysis can be conveniently conducted starting from the Hamiltonian

$$H = -\left[\frac{1}{2}\epsilon + \hbar\Omega_0 \cos\Omega t\right]\sigma_z - \frac{\Delta}{2}\sigma_x + \hbar\omega_T a^\dagger a + g\sigma_z(a^\dagger + a), \qquad (6.79)$$

where the terms in the brackets represent the dc bias and the high-frequency drive of the qubit, and with the current in the tank circuit quantized. Switching to the qubit's eigenbasis, applying the RWA and the Schrieffer–Wolff transformation, we finally obtain (Grajcar et al., 2008b, Methods)

$$H = -\left[\frac{E}{2} - g\sin\zeta(a^\dagger + a) + \frac{g^2}{E}\cos^2\zeta(a^\dagger + a)^2\right]\tilde{\sigma}_z$$
$$+ \hbar\Omega_0 \cos\Omega t \cos\zeta\tilde{\sigma}_x + \hbar\omega_T a^\dagger a, \qquad (6.80)$$

where $E = \sqrt{\epsilon^2 + \Delta^2}$ is the interlevel spacing, and $\tan\zeta = \epsilon/\Delta$. Substituting this Hamiltonian into the master equation with standard Lindblad terms for the qubit and tank circuit (§1.3.2) produces a quantitative match to the experimental data (e.g., to the voltage amplitude in the tank circuit, Fig. 6.12,c bottom).

The whole scheme is reminiscent of so-called *Sisyphus cooling* (Wineland et al., 1992), where atoms are forced to lose their energy to the laser light field. In our case, in the cooling regime Sisyphus (the tank circuit) keeps pushing uphill (along AB and CD) the stone (qubit), which keeps dropping down (DA).[22] For the lack of an appropriate mythological figure we can venture calling the opposite effect, A'B' to C'D', "Indiana Jones heating". This latter process is actually a predecessor of lasing. Given our ability to manipulate the qubit state, it is only to be expected that population inversion and lasing should be achievable in qubit systems. Indeed, lasing has been observed in a charge qubit coupled to a resonator (Astafiev et al.,

---

[22] The idea of using qubit-based refrigerators for local cooling of mesoscopic devices has attracted much theoretical attention (Hauss et al., 2008a,b; You et al., 2008; Jaehne et al., 2008; Wang et al., 2009; Macovei, 2010; Linden et al., 2010), but its full potential as not yet been realized experimentally.

2007). Instead of a large number of oscillators interacting with the cavity field simultaneously, here there is a single "atom" being repeatedly re-excited in order to pump up the electromagnetic field. This is a situation characteristic of a *micromaser* (e.g., Orszag, 1999, 11.1); a true qubit laser would require a quantum metamaterial as the working substance – something which has not yet been achieved.[23]

6.3.3 *Quantum Maxwell's demon*

The quantum heat engines we have considered so far contain just a single quantum object – a qubit or a tunable oscillator. We should expect that adding more quantum objects can produce something more interesting than just increasing the total output per cycle, and this expectation would be quite right. We will show this by adding just one more qubit – "demon qubit" - into the quantum Otto cycle of § 6.3.1. It is called so because it will play the role of *Maxwell's demon* (see Leff and Rex, 1990, 2002; Maruyama et al., 2009) – the famous hypothetical being seemingly capable of breaking the second law of thermodynamics – in the Maxwell's demon-assisted quantum Otto heat engine (Quan et al., 2006, 2007).

First, let us see how the second law might be broken in the quantum case. Let the "working body" qubit be in *permanent* contact with a thermal reservoir at a temperature T_W (Fig. 6.13a). After waiting long enough for it to reach thermal equilibrium, we find the qubit in the state

$$\rho_W^0 = n_{0W}|0_W\rangle\langle 0_W| + n_{1W}|1_W\rangle\langle 1_W|, \tag{6.81}$$

where n_{jW} is the equilibrium occupation number of the working qubit's state $|j\rangle$. Now if we measure the state of the qubit and find it in its excited state (with probability n_{1W}), we can flip it and perform work equal to the interlevel spacing Δ, e.g., by exciting an extra photon in a resonator (we can always make the time to reach thermal equilibrium, that is the thermalization time τ_T, long enough, to make the influence of the reservoir negligible during the measurement and flipping processes). If, on the contrary, it is in the ground state, we do nothing and wait for another τ_T, after which we repeat the cycle. On average, we will be extracting work $n_{1W}\Delta$ per cycle from the single reservoir at temperature T_W – something expressly contradicting the second law of thermodynamics in Lord Kelvin's formulation: "A transformation whose only final result is to transform into work heat extracted from a source which is at the same temperature throughout is impossible" (see Fermi, 1956, III,7). This law-breaking can be performed by the demon qubit in two steps. First, a CNOT operation is performed on the demon qubit with the working

---

[23] Theoretical analysis of single- and few-qubit lasing is given in Hauss et al. (2008a,b), Ashhab et al. (2009), André et al. (2009a,b).

qubit in control. This is a unitary operation, so there will be no irreversibility or losses. Second, if the demon qubit *was* flipped, this means that the working qubit is in the excited state and can perform work, and flipping the demon qubit back can then be used to trigger the extraction of work from the working qubit. Since flipping the demon qubit up and down leaves it in a state with the same energy, the energy extracted from the heat bath will be, indeed, completely transformed into work.

All we need is two qubits with switchable coupling, which would allow us to realize CNOT operations, e.g.,

$$H = -\frac{\Delta_W}{2}\sigma_z^W - \frac{\Delta_D}{2}\sigma_z^D + \frac{g_1}{4}(1+\sigma_z^W)\sigma_x^D + \frac{g_2}{4}(1+\sigma_z^D)\sigma_x^W. \qquad (6.82)$$

The CNOT with the working qubit control and demon target can be performed by switching on the coupling g_1 for the period τ_{1CNOT} such that $g_1\tau_{1CNOT} = \pi/2$; similarly, a demon-controlled CNOT gate requires switching on of the coupling g_2. We have seen enough of different qubit coupling schemes to see that realizing (6.82) is feasible, and we are just a step away from the *perpetuum mobile* of the second kind.

This is too good to be true, of course. A quantum Maxwell's demon can break the second law no better than the classical one, and for the same reason. As was pointed out by Landauer (1961), the demon does not violate Kelvin's postulate, because its memory *changes* in the process of work extraction (in our case, the quantum state of the demon qubit will depend on the outcome of each CNOT-CNOT cycle), and the transformation of heat into work is not the *only* outcome. The process of erasing the demon's memory – i.e., bringing it into some "standard" state, usually a thermal state at some temperature – is obviously irreversible, and if it is taken into account, the second law stands. This is known as *Landauer's erasure principle*.[24] Incidentally, *gaining* information can be reversible (Bennett, 1982), and so this process cannot be used to defeat Maxwell's demon (see Maruyama et al., 2009, and references therein).

To see how this works, let us look at the cycle of Fig. 6.13a in more detail (Quan et al., 2006, 2007). The "working body" qubit is still in permanent contact with its bath; but we need as well a thermal reservoir for the demon at a temperature T_D (Fig. 6.13b), which will be necessary to erase its memory. Assuming for simplicity that the thermalization time is the same for both qubits, we decouple them from each other and wait for τ_T. Now both qubits are in thermal equilibrium, and the

---

[24] This is the same Landauer of the conductance formula.

density matrix of the quantum heat engine is

$$\rho(A) = (n_{0W}|0_W\rangle\langle 0_W| + n_{1W}|1_W\rangle\langle 1_W|)(n_{0D}|0_D\rangle\langle 0_D| + n_{1D}|1_D\rangle\langle 1_D|)$$

$$\equiv \sum_{i,j=0,1} P_{ij}|iw\, jD\rangle\langle iw\, jD|. \tag{6.83}$$

Next we apply a CNOT operation to the demon qubit, which will only flip its state if the working qubit is in the excited state. This produces the density matrix

$$\rho(B) = P_{11}|1_W 0_D\rangle\langle 1_W 0_D| + P_{10}|1_W 1_D\rangle\langle 1_W 1_D|$$

$$+ P_{01}|0_W 1_D\rangle\langle 0_W 1_D| + P_{00}|0_W 0_D\rangle\langle 0_W 0_D|. \tag{6.84}$$

So far the state of the working qubit has not changed: its reduced density matrix is

$$\rho_W(B) \equiv \mathrm{tr}_D \rho(B) = \rho_W(A) \equiv \mathrm{tr}_D \rho(A), \tag{6.85}$$

while the state of the demon, $\rho_D = \mathrm{tr}_W \rho$, has changed: the demon acquired information about the state of the working qubit. The demon's entropy, $S_D = \mathrm{tr}[\rho_D \ln \rho_D]$, changed accordingly. The internal energy of our heat engine also changed due to the demon-performed work:

$$W_D = \mathrm{tr}[H\rho(B)] - \mathrm{tr}[H\rho(A)] = \Delta_D(n_{1D} - P_{10} - P_{01}). \tag{6.86}$$

Remarkably, this intelligence gathering was reversible, in full agreement with Bennett (1982): the entropy of the heat engine stayed constant,

$$S(B) = -k_B \mathrm{tr}[\rho(B)\ln\rho(B)] = -k_B \sum_{ij} P_{ij} \ln P_{ij} \tag{6.87}$$

$$= -k_B(n_{0W}\ln n_{0W} + n_{1W}\ln n_{1W}) - k_B(n_{0D}\ln n_{0D} + n_{1D}\ln n_{1D}) = S(A).$$

The decrease in the demon's entropy was exactly balanced by the decrease in the mutual entropy of the demon and the working system, $S_M \equiv S_W + S_D - S$.

Now we use the demon-gathered information about the state of the system to make it perform work, this time applying a CNOT operation to the working qubit, with the demon qubit as control. This operation does not change the state of the demon. The work performed by the working qubit is

$$W_W = \mathrm{tr}[H\rho(B)] - \mathrm{tr}[H\rho(C)] = \Delta_W(n_{1W} - P_{11} - P_{01}). \tag{6.88}$$

The density matrix of the working qubit after the operation is

$$\rho_W(C) = n_{0D}|1\rangle\langle 1| + n_{1D}|0\rangle\langle 0|, \tag{6.89}$$

and its entropy is obviously the same as the initial entropy of the demon:

$$S_W(C) = S_D(A). \tag{6.90}$$

The total entropy of the heat engine and the entropy of the demon have not changed at this stage (the former, because the CNOT operation is unitary, the latter, because it did not change the state of the demon):

$$S(C) = S(B) = S(A); \quad S_D(C) = S_D(B). \tag{6.91}$$

Finally, the working and demon qubits are decoupled and allowed to reach thermal equilibrium. *This* is an irreversible process. The heat engine absorbs the heat

$$Q = \text{tr}[H\rho(A)] - \text{tr}[H\rho(C)] = \Delta_W(P_{10} - P_{01}), \tag{6.92}$$

while the demon releases the heat

$$Q_D = \text{tr}[H\rho_D(C)] - \text{tr}[H\rho_D(A)] = \Delta_D(P_{10} + P_{01} - n_{1D}). \tag{6.93}$$

The work performed during the cycle is

$$W = W_W - W_D = Q - Q_D = \Delta_W(P_{10} - P_{01}) - \Delta_D(P_{10} - P_{11}). \tag{6.94}$$

Expressing P_{ij} using equilibrium occupation numbers, we find the condition that the work is positive (i.e., our quantum contraption actually works as a heat engine):

$$\frac{T_W}{T_D} \geq \frac{\Delta_D}{\Delta_W}. \tag{6.95}$$

Its efficiency is

$$\eta = \frac{W}{Q} = 1 - \frac{\Delta_D}{\Delta_W}\frac{P_{11} - P_{10}}{P_{10} - P_{00}} \leq 1 - \frac{\Delta_D}{\Delta_W}. \tag{6.96}$$

The latter value is achieved if the demon's bath is at absolute zero and the demon is every time brought to its ground state; and even then the efficiency of the demon-assisted quantum Otto cycle just reaches the efficiency of the simple quantum Otto cycle, Eq. (6.71). If instead of the CNOT gate we used a different demon-controlled operation in order to extract work from the working qubit, the efficiency would have been even less (Quan et al., 2007).

Concluding remarks

The above examples represent just a few of the possibilities presented by solid-state-based qubits, be it superconducting, 2DEG- or based on any of numerous

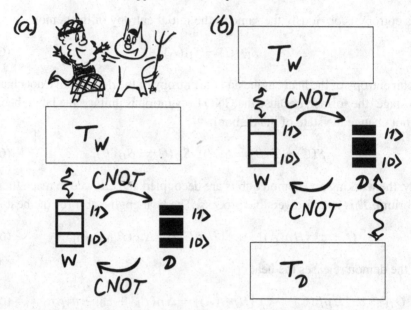

Fig. 6.13. Demon-assisted quantum Otto cycle. (a) Using a Maxwell's demon qubit (D) we can read out the state of the working qubit (W) after it equilibrated with the bath, using a CNOT gate with D as target. Applying another CNOT gate, with W as target, we flip the working qubit, but only if it is in the excited state. Repeating this process, we seem to extract work from a single thermal bath without compensation, since CNOT operations are isentropic. (b) In order to make the process cyclic, it is necessary to restore the demon qubit to its initial state after each CNOT-CNOT operation, e.g., by bringing it in contact with another thermal reservoir. Taking this into account shows that even a quantum Maxwell's demon cannot break the second law of thermodynamics.

other approaches currently being proposed and tested. For the near future, I would expect results primarily in one of the three areas we have discussed – quantum metamaterials, adiabatic quantum computing and quantum cooling – but only time will tell what particular applications beyond pure research are best suited to qubit-based devices. What is certain, is that some such applications will be found; and that the development of multiqubit devices for whatever purpose will allow us to test the limits of quantum mechanics and understand the nature of the quantum–classical transition better than ever.

Appendix A: Quantum gates

Quantum gates are unitary operations performed on quantum states of a given number of qubits. The most widely used ones are one- and two-qubit gates, since a combination of one-qubit rotations (i.e., arbitrary unitary transformations of a single qubit state) and certain two-qubit gates (e.g., the CNOT gate) is universal, i.e., it allows us to realize any arbitrary quantum algorithm (e.g., Le Bellac, 2006, 5.3). Here we give a list of the most common two-qubit gates written in the basis $\{|00\rangle, |01\rangle, |10\rangle, |11\rangle\}$.

The CNOT ("control-NOT") gate flips the second (target) qubit's state if the first (control) one is in state $|1\rangle$ and does nothing otherwise:

$$\text{CNOT} = \begin{pmatrix} 1 & 0 & 0 & 0 \\ 0 & 1 & 0 & 0 \\ 0 & 0 & 0 & 1 \\ 0 & 0 - 1 & 0 \end{pmatrix}. \tag{A.1.1}$$

A generalization of the CNOT is the CU ("control-U") gate, which instead of flipping the second qubit performs on its state some one-qubit unitary operation U:

$$\text{CU} = \begin{pmatrix} 1 & 0 & 0 & 0 \\ 0 & 1 & 0 & 0 \\ 0 & 0 & U_{11} & U_{12} \\ 0 & 0 & U_{21} & U_{22} \end{pmatrix}. \tag{A.1.2}$$

Examples are control-Z $(U = \sigma_z)$ and CROT $(U = -i\sigma_y)$ gates.

The SWAP gate swaps the qubits' states:

$$\text{SWAP} = \begin{pmatrix} 1 & 0 & 0 & 0 \\ 0 & 0 & 1 & 0 \\ 0 & 1 & 0 & 0 \\ 0 & 0 & 0 & 1 \end{pmatrix}, \tag{A.1.3}$$

and is therefore especially useful if one wants to implement quantum algorithms with qubits, which have nearest-neighbour couplings only (e.g., in the GSQC

Fig. A.1.1.

approach). We can introduce $\sqrt{\text{SWAP}}$ so that $\text{SWAP} = \left(\sqrt{\text{SWAP}}\right)^2$:

$$\sqrt{\text{SWAP}} = \frac{1}{1+i}\begin{pmatrix} 1+i & 0 & 0 & 0 \\ 0 & 1 & i & 0 \\ 0 & i & 1 & 0 \\ 0 & 0 & 0 & 1+i \end{pmatrix}. \qquad (A.1.4)$$

Similarly,

$$\text{iSWAP} = \begin{pmatrix} i & 0 & 0 & 0 \\ 0 & 0 & i & 0 \\ 0 & i & 0 & 0 \\ 0 & 0 & 0 & i \end{pmatrix}; \qquad (A.1.5)$$

$$\sqrt{\text{iSWAP}} = \frac{1}{\sqrt{2}}\begin{pmatrix} 1+i & 0 & 0 & 0 \\ 0 & 1 & i & 0 \\ 0 & i & 1 & 0 \\ 0 & 0 & 0 & 1+i \end{pmatrix}. \qquad (A.1.6)$$

References

Abdalla, M. S. and Colegrave, P. K. (1993). *Phys. Rev. A*, **48**, 1526.

Abdumalikov, A. A., Astafiev, O., Nakamura, Y., Pashkin, Yu. A. and Tsai, J.-S. (2008). *Phys. Rev. B*, **78**, 180502(R).

Abdumalikov, Jr., A. A., Astafiev, O., Zagoskin, A. M., Pashkin, Yu. A., Nakamura, Y. and Tsai, J. S. (2010). *Phys. Rev. Lett.*, **104**, 193601.

Abraham, R. J., Fisher, J. and Loftus, P. (1988). *Introduction to NMR Spectroscopy*, 2nd edn. John Wiley and Sons.

Abramowitz, M. and Stegun, I. A. (eds). (1964). *Handbook of Mathematical Functions*. National Bureau of Standards.

Agarwal, G. S. and Kumar, S. A. (1991). *Phys. Rev. Lett.*, **67**, 3665.

Agraït, N., Levy Yeyati, A. and van Ruitenbeek, J. M. (2003). Quantum properties of atomic-sized conductors. *Physics Reports*, **377**, 81.

Aguado, R. and Kouwenhoven, L. P. (2000). *Phys. Rev. Lett.*, **84**, 1986.

Aharonov, D., van Dam, W., Kempe, J., Landau, Z., Lloyd, S. and Regev, O. (2004). *Adiabatic Quantum Computation is Equivalent to Standard Quantum Computation*. quant-ph/0405098.

Aharonov, Y. and Bohm, D. (1959). *Phys. Rev.*, **115**, 485.

Allman, M. S., Altomare, F., Whittaker, J. D., Cicak, K., Li, D., Sirois, A., Strong, J., Teufel, J. D. and Simmonds, R. W. (2010). *Phys. Rev. Lett.*, **104**, 177004.

Amin, M. H. S. (2009). *Phys. Rev. Lett.*, **102**, 220401.

Amin, M. H. S., Love, P. J. and Truncik, C. J. S. (2008). *Phys. Rev. Lett.*, **100**, 060503.

André, S., Brosco, V., Marthaler, M., Shnirman, A. and Schön, G. (2009a). *Phys. Scr.*, **T137**, 014016.

André, S., Brosco, V., Shnirman, A. and Schön, G. (2009b). *Phys. Rev. A*, **79**, 053848.

Andreev, A. F. (1964). *Sov. Phys. JETP*, **19**, 1228.

Aristotle (2008). *Physics*. Oxford University Press.

Ashhab, S., Johansson, J. R., Zagoskin, A. M. and Nori, F. (2009). *New J. Phys.*, **11**, 023030.

Astafiev, O., Pashkin, Y. A., Yamamoto, T., Nakamura, Y. and Tsai, J. S. (2004). *Phys. Rev. B*, **69**, 180507.

Astafiev, O., Inomata, K., Niskanen, A. O., Yamamoto, T., Pashkin, Yu. A., Nakamura, Y. and Tsai, J. S. (2007). *Nature*, **449**, 588.

Astafiev, O., Zagoskin, A. M., Abdumalikov, A. A., Pashkin, Yu. A., Yamamoto, T., Inomata, K., Nakamura, Y. and Tsai, J. S. (2010a). *Science*, **327**, 840.

Astafiev, O. V., Abdumalikov, Jr., A. A., Zagoskin, A. M., Pashkin, Yu. A., Nakamura, Y. and Tsai, J. S. (2010b). *Phys. Rev. Lett.*, **104**, 183603.

Averbukh, I., Sherman, B. and Kurizki, G. (1994). *Phys. Rev. A*, **50**, 5301.

Averin, D. V. (2003). Linear quantum measurements. In *Quantum Noise in Mesoscopic Systems*, ed. Yu. Nazarov. Kluwer Academic Publishers pp. 205–228.

Averin, D. V. and Bruder, C. (2003). *Phys. Rev. Lett.*, **91**, 057003.

Averin, D. V. and Likharev, K. K. (1991). *Mesoscopic Phenomena in Solids*, eds B. L. Altshuler, P. A. Lee and R. A. Webb. North-Holland p. 167.

Averin, D. V. and Sukhorukov, E. V. (2005). *Phys. Rev. Lett.*, **95**, 126803.

Balescu, R. (1975). *Equilibrium and Nonequilibrium Statistical Mechanics*. John Wiley & Sons.

Bardeen, J. and Johnson, J. L. (1972). *Phys. Rev. B*, **5**, 72.

Barone, A. and Paterno, G. (1982). *Physics and Applications of the Josephson Effect*. Wiley-VCH.

Barzykin, V. and Zagoskin, A. M. (1999). *Superlattices and Microstructures*, **25**, 797.

Beenakker, C. and Schönenberger, C. (2003). *Physics Today*, May, 37.

Bender, C. M. and Orszag, S. A. (1999). *Advanced Mathematical Methods for Scientists and Engineers I: Asymptotic Methods and Perturbation Theory*. Springer.

Bennett, C. H. (1982). *Int. J. Theor. Phys.*, **21**, 905.

Berkley, A. J., Xu, H., Ramos, R. C., Gubrud, M. A., Strauch, F. W., Johnson, P. R., Anderson, J. R., Dragt, A. J., Lobb, C. J. and Wellstood, F. C. (2003). *Science*, **300**, 1548.

Bertaina, S., Gambarelli, S., Mitra, T., Tsukerblat, B., Müller, A. and Barbara, B. (2008). *Nature*, **453**, 203.

Bladh, K., Gunnarson, D., Johansson, G., Käck, A., Wendin, G., Aassime, A., Taskalov, M. and Delsing, P. (2002). *Phys. Scr.*, **T102**, 167.

Blais, A., Maassen van den Brink, A. and Zagoskin, A. M. (2003). *Phys. Rev. Lett.*, **90**, 127901.

Blais, A., Huang, R.-S., Wallraff, A., Girvin, S. M. and Schoelkopf, R. J. (2004). *Phys. Rev. B*, **69**, 062320.

Blais, A., Gambetta, J., Wallraff, A., Schuster, D. I., Girvin, S. M., Devoret, M. H. and Schoelkopf, R. J. (2007). *Phys. Rev. A*, **75**, 032329.

Blum, J. J. (1969). *Introduction to Analog Computation*. Harcourt.

Blum, K. (2010). *Density Matrix Theory and Applications*. 2nd edn. Springer US.

Bochkov, G. N. and Kuzovlev, Yu. E. (1983). *Uspekhi Fizicheskih Nauk*, **141**, 151.

Bohr, N. (1935). *Phys. Rev.*, **48**, 696–702.

Bolotovskii, B. M. (1985). *Oliver Heaviside*. Nauka (in Russian).

Bourassa, J., Gambetta, J. M., Abdumalikov, A. A., Astafiev, O., Nakamura, Y. and Blais, A. (2009). *Phys. Rev. A*, **80**, 032109.

Braginsky, V. B. and Khalili, F. Ya. (1992). *Quantum Measurement*. Cambridge: Cambridge University Press.

Buckingham, M. J. (1983). *Noise in Electronic Devices and Systems*. Ellis Horwood Ltd. (John Wiley and Sons).

Burkard, G. (2005). Theory of solid state quantum information processing. In *Handbook of Theoretical and Computational Nanotechnology*, vol. 2, eds M. Rieth and W. Schommers. American Scientific Publishers.

Büttiker, M. (1986). *Phys. Rev. Lett.*, **57**, 1761.

Büttiker, M. (1990). *Phys. Rev. Lett.*, **65**, 2901.

Byrd, M. S. and Lidar, D. A. (2002). *Quantum Information Processing*, **1**, 19.

Caldeira, A. O. and Leggett, A. J. (1983). *Ann. Phys.*, **149**, 374.

Cao, Z. and Elgart, A. (2010). Adiabatic quantum computation: Enthusiast and Sceptic's perspectives. arXiv:1004.5409v1.

Carruthers, P. and Nieto, M. M. (1968). Phase and angle variables in quantum mechanics. *Rev. Mod. Phys.*, **40**, 411.

Casey, H. C. and Panish, M. B. (1978). *Heterojunction*. Academic Press p. 191.

Castellanos-Beltran, M. A., Irwin, K. D., Hilton, G. C., Vale, L. R. and Lehnert, K. W. (2008). *Nature Physics*, **4**, 928.

Caves, C. M. (1982). *Phys. Rev. D*, **26**, 1817.

Chaize, J. (2001). *Quantum Leap: Tools for Managing Companies in the New Economy*. Palgrave Macmillan.

Chapman, G., Cleese, J., Idle, E., Gilliam, T., Jones, T. and Palin, M. (1975). *Monty Python and the Holy Grail*.

Clerk, A. A., Girvin, S. M. and Stone, A. D. (2003). *Physical Review B*, **67**, 165324.

Clerk, A. A., Devoret, M. H., Girvin, S. M., Marquardt, F. and Schoelkopf, R. J. (2010). Introduction to quantum noise, measurement, and amplification. *Rev. Mod. Phys.*, **82**, 1155, (including online notes).

Cooper, K. B., Steffen, M., McDermott, R., Simmonds, R. W., Oh, S., Hite, D. A., Pappas, D. P. and Martinis, J. M. (2004). *Phys. Rev. Lett.*, **93**, 180401.

Cottet, A., Vion, D., Aassime, A., Joyez, P., Esteve, D. and Devoret, M. H. (2002). *Physica C*, **367**, 197.

Courant, R. and Hilbert, D. (1989). *Methods of Mathematical Physics*, vol. 2. Wiley-VCH.

Das, A. and Chakrabarti, B. K. (2008). Colloquium: Quantum annealing and analog quantum computation. *Rev. Mod. Phys.*, **80**, 1061.

Deutsch, D. (1985). *Proceedings of the Royal Society of London; Series A, Mathematical and Physical Sciences*, **400**(1818), 97.

Devoret, Michel H. (1997). Quantum fluctuations in electrical circuits. In *Les Houches Session LXIII, 1995*, eds S. Reynaud, E. Giacobino and J. Zinn-Justin. Elsevier Science B. V.

Devoret, M. H. and Martinis, J. M. (2004). Implementing Qubits with superconducting integrated circuits. *Quantum Information Processing*, **3**, 163.

Devoret, M. H., Wallraff, A. and Martinis, J. M. (2004). *Superconducting Qubits: A Short Review*, unpublished: cond-mat/0411174.

Dewdney, A. K. (1984). *Scientific American*, **250**, 19.

DiCarlo, L., Lynch, H. J., Johnson, A. C., Childress, L. I., Crockett, K., Marcus, C. M., Hanson, M. P. and Gossard, A. C. (2004). *Phys. Rev. Lett.*, **92**, 226801.

Dirac, P. A. M. (2001). *Lectures on Quantum Mechanics*. Dover Publications.

DiVincenzo, D. P. (2000). The Physical Implementation of Quantum Computation. *arXiv:quant-ph/0002077*. unpublished: quant-ph/0002077.

DiVincenzo, D. P. and Loss, D. (1998). *Superlatt. & Microstruct.*, **23**, 419.

Dodonov, V. V. (2002). 'Nonclassical' states in quantum optics: a 'squeezed' review of the first 75 years. *J. Opt. B – Quantum and Semiclassical Optics*, **4**, R1.

Dutta, P. and Horn, P. M. (1981). Low-frequency fluctuations in solids: $1/f$ noise. *Rev. Mod. Phys.*, **53**, 497.

Duty, T., Johansson, G., Bladh, K., Gunnarsson, D., Wilson, C. and Delsing, P. (2005). *Phys. Rev. Lett.*, **95**, 206807.

Dykman, M. I. (1978). *Sov. Phys. Solid State*, **20**, 1306.

Ekert, A. and Josza, R. (1996). Shor's factoring algorithm. *Rev. Mod. Phys.*, **68**, 733.

Elzerman, J. M., Hanson, R., Greidanus, J. S., Willems van Beveren, L. H., De Franceschi, S., Vandersypen, L. M. K., Tarucha, S. and Kouwenhoven, L. P. (2003). *Phys. Rev. B*, **67**, 161308.

Elzerman, J. M., Hanson, R., van Beveren, L. H. W., Witkamp, B., Vandersypen, L. M. K. and Kouwenhoven, L. P. (2004). *Nature* **430**, 431.

Engheta, N. and Ziolkowski, R. W. (eds). (2006). *Electromagnetic Metamaterials: Physics and Engineering Explorations*. Wiley-IEEE Press.

Farhi, E., Goldstone, J., Gutmann, S. and Sipser, M. (2000). *Quantum Computation by Adiabatic Evolution*. quant-ph/0001106.

Farhi, E., Goldstone, J., Gutmann, S., Laplan, J., Lundgren, A. and Preda, D. (2001). *Science*, **292**, 472.

Fermi, Enrico (1956). *Thermodynamics*. Dover Publications.

Feynman, R. (1985). Quantum mechanical computers. *Optics News*, Feb., 11.

Feynman, R. (1996). *Feynman Lectures on Computation*. Addison-Wesley.

Feynman, R. P. and Hibbs, A. R. (1965). *Quantum Mechanics and Path Integrals*. McGraw-Hill Companies.

Feynman, R. P. and Vernon, F. L. (1963). *Ann. Phys.*, **24**(1), 118.

Fink, J. M., Goeppl, M., Baur, M., Bianchetti, R., Leek, P. J., Blais, A. and Wallraff, A. (2008). *Nature*, **454**, 315.

Fisher, D. S. and Lee, P. A. (1981). *Phys. Rev. B*, **23**, 6851.

Friedman, J. R., Patel, V., Chen, W., Tolpygo, S. K. and Lukens, J. E. (2000). *Nature*, **406**, 43.

Furusaki, A. (1999). *Superlattices and Microstructures*, **25**, 809.

Gambetta, J., Blais, A., Schuster, D. I., Wallraff, A., Frunzio, L., Majer, J., Devoret, M. H., Girvin, S. M. and Schoelkopf, R. J. (2006). *Phys. Rev. A*, **74**, 042318.

Gamelin, T. W. (2001). *Complex Analysis*. Springer.

Garanin, D. A. and Schilling, R. (2002). *Phys. Rev. B*, **66**, 174438.

Gardiner, C. W. (2003). *Handbook of Stochastic Methods*, 3rd edn. Springer.

Gardiner, C. W. and Zoller, P. (2004). *Quantum Noise*, 3rd edn. Springer.

Gatteschi, D., Sessoli, R. and Villain, J. (2006). Molecular Nanomagnets. *Mesoscopic Physics and Nanotechnology*. Oxford University Press.

Gavish, U., Yurke, B. and Imry, Y. (2000). *Phys. Rev. B*, **62**, (R) 10637.

Gemmer, J., Michel, M. and Mahler, G. (2009). *Quantum Thermodynamics: Emergence of Thermodynamic Behavior Within Composite Quantum Systems*, 2nd edn. Lecture Notes in Physics. Springer.

Geusic, J. E., Schulz-DuBois, E. O. and Scovil, H. E. D. (1967). *Phys. Rev.*, **156**, 343.

Geva, E. and Kosloff, R. (1994). *Phys. Rev. E*, **49**, 3903.

Glauber, R. J. (1963). *Phys. Rev. E*, **10**, 277.

Glazman, L. I., Lesovik, G. B., Khmelnitskii, D. E. and Shekhter, R. I. (1988). *JETP Lett.*, **48**, 238.

Goldstein, H. (1980). *Classical Mechanics*. Addison-Wesley.

Gradshteyn, L. S. and Ryzhik, I. M. (2000). *Table of Integrals, Series and Products*. Academic Press.

Graham, R. (1987). *J. Mod. Opt.*, **34**, 873.

Grajcar, M., Izmalkov, A., Il'ichev, E., Wagner, Th., Oukhanski, N., Hübner, U., May, T., Zhilyaev, I., Hoenig, H. E., Greenberg, Ya. S., Shnyrkov, V. I., Born, D., Krech, W., Meyer, H.-G., Maassen van den Brink, A. and Amin, M. H. S. (2004). *Phys. Rev. B*, **69**, 060501(R).

Grajcar, M., Izmalkov, A. and Il'ichev, E. (2005). *Phys. Rev. B*, **71**, 144501.

Grajcar, M., Izmalkov, A., van der Ploeg, S. H. W., Linzen, S., Plecenik, T., Wagner, Th., Hubner, U., Il'ichev, E., Meyer, H.-G., Smirnov, A. Yu., Love, P. J., Maassen van den Brink, A., Amin, M. H. S., Uchaikin, S. and Zagoskin, A. M. (2006). *Phys. Rev. Lett.*, **96**, 047006.

Grajcar, M., Asshab, S., Johansson, J. R. and Nori, F. (2008a). *Phys. Rev. B*, **78**, 035406.

Grajcar, M., van der Ploeg, S. H. W., Izmalkov, A., Il'ichev, E., Meyer, H.-G., Fedorov, A., Shnirman, A. and Schön, G. (2008b). *Nature Physics*, **4**, 612.

Greenberg, Ya. S., Izmalkov, A., Grajcar, M., Il'ichev, E., Krech, W., Meyer, H.-G., Amin, M. H. S. and Maassen van den Brink, A. (2002). *Phys. Rev. B*, **66**, 214525.

Grover, L. K. (1997). *Phys. Rev. Lett.*, **78**, 325.

Grover, L. K. (2001). From Schrödinger's equation to quantum search algorithm. *American Journal of Physics*, **69**, 769.

Guizzo, E. (2010). *IEEE Spectrum*, January.

Gutmann, H., Wilhelm, F. K., Kaminsky, W. M. and Lloyd, S. (2005). *Phys. Rev. A*, **71**, 020302.

Hahn, E. L. (1950). *Phys. Rev.*, **80**, 580.

Hanson, R., van Beveren, L. H. W., Vink, I. T., Elzerman, J. M., Naber, W. J. M., Koppens, F. H. L., Kouwenhoven, L. P. and Vandersypen, L. M. K. (2005). *Phys. Rev. Lett.*, **94**, 196802.

Hanson, R., Kouwenhoven, L. P., Petta, J. R., Tarucha, S. and Vandersypen, L. M. K. (2007). Spins in few-electron quantum dots. *Rev. Mod. Phys.*, **79**, 1217.

Harris, R., Berkley, A. J., Johnson, M. W., Bunyk, P., Govorkov, S., Thom, M. C., Uchaikin, S., Wilson, A. B., Chung, J., Holtham, E., Biamonte, J. D., Smirnov, A. Yu., Amin, M. H. and Maassen van den Brink, A. (2007). *Phys. Rev. Lett.*, **98**, 177001.

Hauss, J., Fedorov, A., André, S., Brosco, V., Hutter, C., Kothari, R., Yeshwant, S., Shnirman, A. and Schön, G. (2008a). *New J. Phys.*, **10**, 095018.

Hauss, J., Fedorov, A., Hutter, C., Shnirman, A. and Schön, G. (2008b). *Phys. Rev. Lett.*, **100**, 037003.

Hayashi, T., Fujisawa, T., Cheong, H. D., Jeong, Y. H. and Hirayama, Y. (2003). *Phys. Rev. Lett.*, **91**, 226804.

Heida, J. P., van Wees, B. J., Klapwijk, T. M. and Borghs, G. (1998). *Phys. Rev. B*, **57**, R5618.

Hiyamizu, S. (1990). *Semiconductors and Semimetals*, **30**, 53.

Hofheinz, M., Weig, E. M., Ansmann, M., Bialczak, R. C., Lucero, E., Neeley, M., O'Connell, A. D., Wang, H., Martinis, J. M. and Cleland, A. N. (2008). *Nature*, **454**, 310.

Hofheinz, M., Wang, H. M., Ansmann, M., Bialczak, R. C., Lucero, E., Neeley, M., O'Connell, A. D., Sank, D., Wenner, J., Martinis, J. M. and Cleland, A. N. (2009). *Nature*, **459**, 546.

Hood, C. J., Lynn, T. W., Doherty, A. C., Parkins, A. S. and Kimble, H. J. (2000). *Science*, **287**, 1447.

Hooge, F. N. (1969). *Phys. Lett. A*, **29**, 139.

Horodecki, R., Horodecki, P., Horodecki, M. and Horodecki, K. (2009). Quantum entanglement. *Rev. Mod. Phys.*, **81**, 865.

Houck, A. A., Schuster, D. I., Gambetta, J. M., Schreier, J. A., Johnson, B. R., Chow, J. M., Frunzio, L., Majer, J., Devoret, M. H., Girvin, S. M. and Schoelkopf, R. J. (2007). *Nature*, **449**, 328.

Houck, A. A., Schreier, J. A., Johnson, B. R., Chow, J. M., Koch, J., Gambetta, J. M., Schuster, D. I., Frunzio, L., Devoret, M. H., Girvin, S. M. and Schoelkopf, R. J. (2008). *Phys. Rev. Lett.*, **101**, 080502.

Houck, A. A., Koch, J., Devoret, M. H., Girvin, S. M. and Schoelkopf, R. J. (2009). *Quantum Information Processing*, **8**, 105.

Humphrey, T. E. and Linke, H. (2005). *Physica E*, **29**, 390.

Il'ichev, E., Oukhanski, N., Izmalkov, A., Wagner, Th., Grajcar, M., Meyer, H.-G., Smirnov, A. Yu., Maassen van den Brink, A., Amin, M. H. S. and Zagoskin, A. M. (2003). *Phys. Rev. Lett.*, **91**, 097906.

Il'ichev, E., Oukhanski, N., Wagner, Th., Meyer, H.-G., Smirnov, A. Yu., Grajcar, M., Izmalkov, A., Born, D., Krech, W. and Zagoskin, A. (2004). *Low Temperature Physics*, **30**, 620.

Imry, Y. (2002). *Introduction to Mesoscopic Physics*. Oxford University Press.

Ishii, G. (1970). *Progr. Theor. Phys.*, **44**, 1525.

Ithier, G., Collin, E., Joyez, P., Meeson, P. J., Vion, D., Esteve, D., Chiarello, F., Shnirman, A., Makhlin, Y., Schriefl, J. and Schön, G. (2005). *Phys. Rev. B*, **72**, 134519.

Izmalkov, A., Grajcar, M., Il'Ichev, E., Oukhanski, N., Wagner, T., Meyer, H. G., Krech, W., Amin, M. H. S., Maassen van den Brink, A. and Zagoskin, A. M. (2004a). *Europhys. Lett.*, **65**, 844.

Izmalkov, A., Grajcar, M., Il'ichev, E., Wagner, Th., Meyer, H.-G., Smirnov, A. Yu., Amin, M. H. S., Maassen van den Brink, A. and Zagoskin, A. M. (2004b). *Phys. Rev. Lett.*, **93**, 037003.

Izmalkov, A., Grajcar, M., van der Ploeg, S. H. W., Hübner, U., Il'ichev, E., Meyer, H.-G. and Zagoskin, A. M. (2006). *Europhys. Lett.*, **76**, 533.

Jackson, A. S. (1960). *Analog Computation*. New York: McGraw-Hill.

Jaehne, K., Hammerer, K. and Wallquist, M. (2008). *New J. Phys.*, **10**, 095019.

Janszky, J. and Adam, P. (1992). *Phys. Rev. A*, **46**, 6091.

Jardine, L. (2003). *The Curious Life of Robert Hooke: The Man who Measured London*. HarperCollins.

Johansson, G., Tornberg, L., Shumeiko, V. S. and Wendin, G. (2006a). *J. Phys.: Condens. Matter*, **18**, S901.

Johansson, G., Tornberg, L., Shumeiko, V. S. and Wendin, G. (2006b). *J. Phys.: Condens. Matter*, **18**, S901.

Johansson, J., Saito, S., Meno, T., Nakano, H., Ueda, M., Semba, K. and Takayanagi, H. (2006c). *Phys. Rev. Lett.*, **96**, 127006.

Johnson, D. S. and McGeoch, L. A. (1997). In *Local Search in Combinatorial Optimization*, pp. 215–310, eds E. H. L. Aarts and J. K. Lenstra. J. Wiley and Sons, London.

Johnson, J. B. (1925). *Phys. Rev.*, **26**, 71.

Johnson, J. B. (1927). *Nature*, **119**, 50.

Johnson, J. B. (1928). *Phys. Rev.*, **32**, 97.

Joyez, P., Lafarge, P., Filipe, A., Esteve, D. and Devoret, M. H. (1994). *Phys. Rev. Lett.*, **72**, 2458.

Kafanov, S., Brenning, H., Duty, T. and Delsing, P. (2008). *Phys. Rev. B*, **78**, 125411.

Ketterson, J. B. and Song, S. N. (1999). *Superconductivity*. Cambridge University Press.

Khlus, V. A. (1987). *Sov. Phys. JETP*, **66**, 1243.

Kholevo, A. S. (1982). *Probabilistic and Statistical Aspects of Quantum Theory*. North-Holland Publishing Company.

Kieu, T. D. (2004). *Phys. Rev. Lett.*, **93**, 140403.

Kiss, T., Janszky, J. and Adam, P. (1994). *Phys. Rev. A*, **49**, 4935.

Kittel, C. (1987). *Quantum Theory of Solids*, 2nd edn. Wiley.

Kittel, C. (2004). *Introduction to Solid State Physics*, 8th edn. John Wiley and Sons.

Kleinert, H. (2006). *Path Integrals in Quantum Mechanics, Statistics, Polymer Physics, and Financial Markets*, 4th edn. World Scientific.

Klich, I. (2003). In *Quantum Noise in Mesoscopic Physics*, p. 397, ed. Yu. V. Nazarov. NATO Science Series. Kluwer Academic Publishers.

Klimontovich, Yu. L. (1986). *Statistical Physics*. New York: Harwood.

Kline, J. S., Wang, H., Oh, S., Martinis, J. M. and Pappas, D. P. (2009). *Supercond. Sci. Tech.*, **22**, 015004.

Koch, J., Yu, T. M., Gambetta, J., Houck, A. A., Schuster, D. I., Majer, J., Blais, A., Devoret, M. H., Girvin, S. M. and Schoelkopf, R. J. (2007). *Phys. Rev. A*, **76**, 042319.

Kogan, Sh. M. (1985). *Uspekhi Fizicheskikh Nauk*, **145**, 285.

Kozyrev, A. B. and van der Weide, D. W. (2008). *J. Phys. D: Appl. Phys.*, **41**, 173001.

Kulik, I. O. (1970). *Sov. Phys. JETP*, **30**, 944.

Kulik, I. O. and Omelyanchuk, A. N. (1977). *Fiz. Nizk. Temp.*, **3**, 945.

Kumar, A., Laux, S. E. and Stern, F. (1990). *Phys. Rev. B*, **42**, 5166.

Kuzmin, L. S. (1993). *IEEE Trans. Appl. Superconductivity*, **3**, 1983.

Kuzmin, L. S. and Haviland, D. B. (1991). *Phys. Rev. Lett.*, **67**, 2890.

Lambert, C. J. (1991). *J. Phys.: Cond. Matter*, **3**, 6579.

Landau, L. D. (1932). *Phys. Z. Sowjetunion*, **2**, 46.

Landau, L. D. and Lifshitz, E. M. (1976). *Mechanics*, 3rd edn. Butterworth-Heinemann.

Landau, L. D. and Lifshitz, E. M. (1980). *Statistical Physics Part I*, 3rd edn. Butterworth-Heinemann.

Landau, L. D. and Lifshitz, E. M. (2003). *Quantum Mechanics (Non-relativistic Theory)*, 3rd edn. Butterworth-Heinemann.

Landau, L. D., Pitaevskii, L. P. and Lifshitz, E. M. (1984). *Electrodynamics of Continuous Media*, 2nd edn. Butterworth-Heinemann.

Landauer, R. (1957). *IBM J. Res. Dev.*, **1**, 223.

Landauer, R. (1961). *IBM J. Res. Dev.*, **5**, 183.

Lax, M. (1963). *Phys. Rev.*, **129**, 2342.

Lax, M. (1968). *Phys. Rev.*, **172**, 350.

Le Bellac, M. (2006). *A Short Introduction to Quantum Information and Quantum Computation*. Cambridge University Press.

Leff, H. S. and Rex, A. F. (eds). (1990). *Maxwell's Demon: Entropy, Information, Computing*. Adam Hilger.

Leff, H. S. and Rex, A. F. (eds). (2002). *Maxwell's Demon 2: Entropy, Information, Computing*. Taylor & Francis.

Leggett, A. J. (1980). Macroscopic quantum systems and the quantum theory of measurement. *Suppl. Prog. Theor. Phys., No. 69*, 80.

Leggett, A. J. (2002a). *J. Phys.: Condens. Matter*, **14**, R415–51.

Leggett, A. J. (2002b). *Science*, **296**, 861.

Leggett, A. J., Chakravarty, A. T., Dorsey, M. P. A., Fisher, A. G. and Zwerger, W. (1987). Dynamics of the dissipative two-state system. *Rev. Mod. Phys.*, **59**, 1.

Lesovik, G. B. (1989). *JETP Lett.*, **49**, 594.

Levitov, L. S. and Lesovik, G. B. (1992). *JETP Lett.*, **55**, 555.

Levitov, L. S. and Lesovik, G. B. (1993). *JETP Lett.*, **58**, 230.

Levitov, L. S., Lee, H. and Lesovik, G. B. (1996). *J. Math. Phys.*, **37**, 4845–66.

Likharev, K. K. (1986). *Dynamics of Josephson Junctions and Circuits*. CRC Press.

Likharev, K. K. and Zorin, A. B. (1985). *Journal of Low Temperature Physics*, **59**, 347.

Lindblad, G. (1976). *Commun. Math. Phys.*, **48**, 119.

Linden, N., Popescu, S. and Skrzypczyk, P. (2010). *Phys. Rev. Lett.*, **105**, 130401.

Lupascu, A., Verwijs, C. J. M., Schouten, R. N., Harmans, C. J. P. M. and Mooij, J. E. (2004). *Phys. Rev. Lett.*, **93**, 177006.

Lupascu, A., Driessen, E. F. C., Roschier, L., Harmans, C. J. P. M. and Mooij, J. E. (2006). *Phys. Rev. Lett.*, **96**, 127003.

Maassen van den Brink, A. (2005). *Phys. Rev. B*, **71**, 064503.

Maassen van den Brink, A. and Zagoskin, A. M. (2002). *Quantum Information Processing*, **1**, 55.

Maassen van den Brink, A., Berkley, A. J. and Yalowsky, M. (2005). *New J. Phys.*, **7**, 230.

Macovei, M. A. (2010). *Phys. Rev. A*, **81**, 043411.

Majer, J. B., Butcher, J. R. and Mooij, J. E. (2002). *Appl. Phys. Lett.*, **80**, 3638.

Majer, J. B., Paauw, F. G., ter Haar, A. C. J., Harmans, C. J. P. M. and Mooij, J. E. (2005). *Phys. Rev. Lett.*, **94**, 090501.

Majer, J., Chow, J. M., Gambetta, J. M., Koch, J., Johnson, B. R., Schreier, J. A., Frunzio, L., Schuster, D. I., Houck, A. A., Wallraff, A., Blais, A., Devoret, M. H., Girvin, S. M. and Schoelkopf, R. J. (2007). *Nature*, **449**, 443.

Makhlin, Yu. and Shnirman, A. (2004). *Phys. Rev. Lett.*, **92**, 178301.

Makhlin, Yu., Schön, G. and Shnirman, A. (1999). *Nature*, **398**, 305.

Makhlin, Yu., Schon, G. and Shnirman, A. (2001). Quantum-state engineering with Josephson-junction devices. *Reviews of Modern Physics*, **73**, 357.

Mallet, F., Ong, F. R., Palacios-Laloy, A., Nguyen, F., Bertet, P., Vion, D. and Esteve, D. (2009). *Nature Physics*, **5**, 791.

Mandelbrot, B. B. (1982). *The Fractal Geometry of Nature*. W. H. Freeman & Co.

Mansoori, G. Ali. (2005). *Principles of Nanotechnology*. World Scientific.

Martinis, J. M., Nam, S., Aumentado, J. and Urbina, C. (2002). *Phys. Rev. Lett.*, **89**, 117901.

Martinis, J. M., Nam, S., Aumentado, J., Lang, K. M. and Urbina, C. (2003). *Phys. Rev. B*, **67**, 094510.

Martinis, J. M., Cooper, K. B., McDermott, R., Steffen, M., Ansmann, M., Osborn, K. D., Cicak, K., Oh, S., Pappas, D. P., Simmonds, R. W. and Yu, C. C. (2005). *Phys. Rev. Lett.*, **95**, 210503.

Martonák, R., Santoro, G. E. and Tosatti, E. (2004). *Phys. Rev. E*, **70**, 057701.

Maruyama, K., Nori, F. and Vedral, V. (2009). The physics of Maxwell's demon and information. *Reviews of Modern Physics*, **81**, 1.

Matveev, K. A., Gisselfält, M., Glazman, L. I., Jonson, M. and Shekhter, R. I. (1993). *Phys. Rev. Lett.*, **70**, 2940.

Mayer, J. E. and Mayer, M. Goeppert. (1977). *Statistical Mechanics*, 2nd edn. John Wiley and Sons.

McDermott, R., Simmonds, R. W., Steffen, M., Cooper, K. B., Cicak, K., Osborn, K. D., Oh, Seongshik, Pappas, D. P. and Martinis, J. M. (2005). *Science*, **307**, 1299.

Menand, L. (2010). *The Marketplace of Ideas: Reform and Resistance in the American University (Issues of Our Time)*. W. W. Norton & Co.

Messiah, A. (2003). *Quantum Mechanics*. Dover Publications.

Mizel, A. (2010). Fixed-gap adiabatic quantum computation. arXiv:1002.0846v2.

Mizel, A., Mitchell, M. W. and Cohen, M. L. (2001). *Phys. Rev. A*, **63**, 040302(R).

Mizel, A., Lidar, D. A. and Mitchell, M. W. (2007). *Phys. Rev. Lett.*, **99**, 070502.

Mooij, J. E., Orlando, T. P., Levitov, L., Tian, L., van der Wal, C. H. and Lloyd, S. (1999). *Science*, **285**, 1036.

Nakamura, Y., Pashkin, Yu. A. and Tsai, J. S. (1999). *Nature*, **398**, 786.

Namiki, M., Pascazio, S. and Nakazato, H. (1997). *Decoherence and Quantum Measurements*. World Scientific.

Neeley, M., Ansmann, M., Bialczak, R. C., Hofheinz, M., Katz, N., Lucero, E., O'Connell, A., Wang, H., Cleland, A. N. and Martinis, J. M. (2008). *Nature Physics*, **4**, 523.

Nussenzveig, H. M. (1972). *Causality and dispersion relations*. Academic Press.

Nyquist, H. (1928). *Phys. Rev.*, **32**, 110–13.

Oh, S., Cicak, K., Kline, J. S., Sillanpaa, M. A., Osborn, K. D., Whittaker, J. D., Simmonds, R. W. and Pappas, D. P. (2006). *Phys. Rev. B*, **74**, 100502.

Olariu, S. and Popescu, I. I. (1985). The quantum effects of electromagnetic fluxes. *Rev. Mod. Phys.*, **57**, 339.

Orlando, T. P., Mooij, J. E., Tian, L., van der Wal, Caspar H., Levitov, L. S. and Mazo, J. J., Lloyd, S. (1999). *Phys. Rev. B*, **60**, 15398.

Orszag, M. (1999). *Quantum Optics*. Springer.

Paauw, F. G., Fedorov, A., Harmans, C. J. P. M. and Mooij, J. E. (2009). *Phys. Rev. Lett.*, **102**, 090501.

Penrose, R. (2004). *The Road to Reality: A Complete Guide to the Laws of the Universe*. London: Jonathan Cape.

Percival, I. (2008). *Quantum State Diffusion*. Cambridge University Press.

Perelomov, A. (1986). *Generalized Coherent States and Their Applications*. Springer.

Petta, J. R., Johnson, A. C., Marcus, C. M., Hanson, M. P. and Gossard, A. C. (2004). *Phys. Rev. Lett.*, **93**, 186802.

Petta, J. R., Johnson, A. C., Taylor, J. M., Laird, E. A., Yacoby, A., Lukin, M. D., Marcus, C. M., Hanson, M. P. and Gossard, A. C. (2005). *Science*, **309**, 2180.

Pfannkuche, D., Blick, R. H., Haug, R. J., von Klitzing, K. and Eberl, K. (1998). *Superlatt. Microstruct.*, **23**, 1255.

Pitaevskii, L. P. and Lifshitz, E. M. (1980). *Statistical Physics, Part 2: Theory of the Condensed State*, 3rd edn. Butterworth-Heinemann.

Pitaevskii, L. P. and Lifshitz, E. M. (1981). *Physical Kinetics*. Butterworth-Heinemann.

Price, P. J. (1998). *Am. J. Phys.*, **66**, 1119.

Prokof'ev, N. V. and Stamp, P. C. E. (2000). Theory of the spin bath. *Rep. Prog. Phys.*, **63**, 669.

Purcell, E. M. (1946). *Phys. Rev.*, **69**, 681.

Quan, H. T. (2009). *Phys. Rev. E*, **79**, 041129.

Quan, H. T., Wang, Y. D., Liu, Y., Sun, C. P. and Nori, F. (2006). *Phys. Rev. Lett.*, **97**, 180402.

Quan, H. T., Liu, Y., Sun, C. P. and Nori, F. (2007). *Phys. Rev. E*, **76**, 031105.

Raimond, J., Brune, M. and Haroche, S. (2001). Manipulating quantum entanglement with atoms and photons in a cavity. *Rev. Mod. Phys.*, **73**, 565.

Rakhmanov, A. L., Zagoskin, A. M., Savel'ev, S. and Nori, F. (2008). *Phys. Rev. B*, **77**, 144507.

Reimann, S. M. and Manninen, M. (2002). Electronic structure of quantum dots. *Rev. Mod. Phys.*, **74**, 1283.

Richtmyer, R. D. (1978). *Principles of Advanced Mathematical Physics (Vol. 1)*. Springer-Verlag.

Rifkin, R. and Deaver, B. S. (1976). *Phys. Rev. B*, **13**, 3894.

Röpke, G. (1987). *Statistische Mechanik für das Nichtgleichgewicht*. Wiley VCH Verlag.

Ryder, L. (1996). *Quantum Field Theory*. Cambridge University Press.

Rytov, S. M., Kravtsov, Yu. A. and Tatarskii, V. I. (1987). *Principles of Statistical Radiophysics: Elements of Random Process Theory*, vol. 1. Springer.

Rytov, S. M., Kravtsov, Yu. A. and Tatarskii, V. I. (1988). *Principles of Statistical Radiophysics: Correlation Theory of Random Processes*, vol. 2. Springer.

Rytov, S. M., Kravtsov, Yu. A. and Tatarskii, V. I. (1989). *Principles of Statistical Radiophysics: Elements of Random Fields*, vol. 3. Springer.

Saleh, B. E. A. and Teich, M. C. (2007). *Fundamentals of Photonics*, 2nd edn. Wiley-Interscience.

Santoro, G. E., Martonak, R., Tosatti, E. and Car, R. (2002). *Science*, **295**, 2427.

Sarandy, M. S. and Lidar, D. A. (2005a). *Phys. Rev. Lett.*, **95**, 250503.

Sarandy, M. S. and Lidar, D. A. (2005b). *Physical Review A*, **71**, 012331.

Scappucci, G., Gaspare, L., Di, Giovine, E., Notargiacomo, A., Leoni, R. and Evangelisti, F. (2006). *Phys. Rev. B*, **74**, 035321.

Schleich, W. P. (2001). *Quantum Optics in Phase Space*. Wiley-VCH.

Schmidt, V. V. (2002). *The Physics of Superconductors*. Springer.

Schoelkopf, R. J., Clerk, A. A., Girvin, S. M., Lehnert, K. W. and Devoret, M. H. (2003). In *Quantum Noise in Mesoscopic Physics*, p. 175, ed. Y. V. Nazarov. Kluwer Academic Publishers.

Schoeller, H. and Schön, G. (1994). *Phys. Rev. B*, **50**, 18436.

Schönhammer, K. (2007). *Phys. Rev. B*, **75**, 205329.

Schottky, W. (1918). *Ann. d. Phys. (Leipzig)*, **57**, 541.

Schottky, W. (1926). *Phys. Rev.*, **28**, 74.

Schreier, J. A., Houck, A. A., Koch, J., Schuster, D. I., Johnson, B. R., Chow, J. M., Gambetta, J. M., Majer, J., Frunzio, L., Devoret, M. H., Girvin, S. M. and Schoelkopf, R. J. (2008). *Phys. Rev. B*, **77**, 180502.

Schrieffer, J. R. and Wolff, P. A. (1966). *Phys. Rev.*, **149**, 491.

Schuster, D. I., Wallraff, A., Blais, A., Frunzio, L., Huang, R.-S., Majer, J., Girvin, S. M. and Schoelkopf, R. J. (2005). *Phys. Rev. Lett.*, **94**, 123602.

Schuster, D. I., Houck, A. A., Schreier, J. A., Wallraff, A., Gambetta, J. M., Blais, A., Frunzio, L., Majer, J., Johnson, B., Devoret, M. H., Girvin, S. M. and Schoelkopf, R. J. (2007). *Nature*, **445**, 515.

Schwinger, J. (1961). *J. Math. Phys.*, **2**, 407.

Scovil, H. E. D. and Schulz-DuBois, E. O. (1959). *Phys. Rev. Lett.*, **2**, 262.

Scully, M. O. and Zubairy, M. S. (1997). *Quantum Optics*. Cambridge University Press.

Scully, M. O., Zubairy, M. S., Agarwal, G. A. and Walther, H. (2003). *Science*, **299**, 862.

Shapiro, B. (1983). *Phys. Rev. Lett.*, **50**, 747.

Shaw, M. D., Schneiderman, J. F., Palmer, B., Delsing, P. and Echternach, P. M. (2007). *IEEE Trans. Appl. Supercond.*, **17**, 109.

Shinkai, G., Hayashi, T., Ota, T. and Fujisawa, T. (2009). *Phys. Rev. Lett.*, **103**, 056802.

Shnirman, A. and Schön, G. (2003). Dephasing and renormalization in quantum two-level systems. In *Quantum Noise in Mesoscopic Systems*, p. 357, ed. Yu. V. Nazarov. Kluwer Academic Publishers.

Shnirman, A., Schön, G. and Hermon, Z. (1997). *Phys. Rev. Lett.*, **79**, 2371.

Shor, P. W. (1997). *SIAM J. Comput.*, **26**, 1484.

Sillanpää, M. A., Lehtinen, T., Paila, A., Makhlin, Yu., Roschier, L. and Hakonen, P. J. (2005). *Phys. Rev. Lett.*, **95**, 206806.

Simmonds, R. W., Lang, K. M., Hite, D. A., Nam, S., Pappas, D. P. and Martinis, J. M. (2004). *Phys. Rev. Lett.*, **93**, 077003.

Simmonds, R. W., Allman, M. S., Altomare, F., Cicak, K., Osborn, K. D., Park, J. A., Sillanpaa, M., Sirois, A., Strong, J. A. and Whittaker, J. D. (2009). *Quantum Information Processing*, **8**, 117.

Smirnov, A. Yu. (2003). *Phys. Rev. B*, **68**(13), 134514.

Smirnov, D. F. and Troshin, A. S. (1987). New phenomena in quantum optics: photon antibunching, sub-Poisson photon statistics, and squeezed states. *Sov. Phys. – Uspekhi*, **30**, 851.

Smith, C. G. (1996). Low-dimensional quantum devices. *Rep. Prog. Phys.*, **59**, 235.

Smith, D. R., Padilla, W. J., Vier, D. C., Nemat-Nasser, S. C. and Schulz, S. (2000). *Phys. Rev. Lett.*, **84**, 4184.

Stamp, P. C. E. (2008). *Nature*, **453**, 168.

Steffen, M., van Dam, W., Hogg, T., Breyta, G. and Chuang, I. (2003). *Phys. Rev. Lett.*, **90**, 067903.

Steffen, M., Ansmann, M., Bialczak, R. C., Katz, N., Lucero, E., McDermott, R., Neeley, M., Weig, E. M., Cleland, A. N. and Martinis, J. M. (2006). *Science*, **313**, 1423.

Stöckmann, H.-J. (1999). *Quantum Chaos: An Introduction.* Cambridge University Press.

Stone, M. (ed.) (1992). *Quantum Hall Effect.* World Scientific.

Takayanagi, H., Hansen, J. B. and Nitta, J. (1995a). *Phys. Rev. Lett.*, **74**, 166.

Takayanagi, H., Akazaki, T. and Nitta, J. (1995b). *Phys. Rev. Lett.*, **75**, 3533.

Tan, Sze M. (2002). *Quantum Optics Toolbox: www.qo.phy.auckland.ac.nz/qotoolbox.html.*

Tatarskii, V. I. (1987). Example of the description of dissipative processes in terms of reversible dynamic equations and some comments on the fluctuation-dissipation theorem. *Sov. Phys. – Uspekhi*, **30**, 134.

Teich, M. C. and Saleh, B. E. A. (1989). *Quantum Optics*, **1**, 153.

Tinkham, M. (2004). *Introduction to Superconductivity*, 2nd edn. Dover.

Titulaer, U. M. (1975). *Phys. Rev. A*, **11**, 2204.

Umezawa, H., Matsumoto, H. and Tachiki, M. (1982). *Thermo Field Dynamics and Condensed States.* North-Holland Publishing Company.

van Dam, W. (2007). *Nature Physics*, **3**, 220.

van der Ploeg, S. H. W., Izmalkov, A., Grajcar, M., Huebner, U., Linzen, S., Uchaikin, S., Wagner, Th., Smirnov, A. Yu., Maassen van den Brink, A., Amin, M. H. S., Zagoskin, A. M., Il'ichev, E. and Meyer, H.-G. (2006). In: *Applied Superconductivity Conference 2006.* http://arxiv.org/abs/cond-mat/0702580, *IEEE Trans. Appl. Supercond.* **17**, 113 (2007).

van der Ploeg, S. H. W., Izmalkov, A., Maassen van den Brink, A., Hübner, U., Grajcar, M., Il'ichev, E., Meyer, H.-G. and Zagoskin, A. M. (2007). *Phys. Rev. Lett.*, **98**, 057004.

van der Wal, C. H., ter Haar, A. C. J., Wilhelm, F. K., Schouten, R. N., Harmans, C. J. P. M., Orlando, T. P., Lloyd, S. and Mooij, J. E. (2000). *Science*, **290**(5492), 773.

van der Wiel, W. G., De Franceschi, S., Elzerman, J. M., Fujisawa, T., Tarucha, S. and Kouwenhoven, L. P. (2002). Electron transport through double quantum dots. *Rev. Mod. Phys.*, **75**, 1.

van der Wiel, W. G., Nazarov, Yu. V., De Franceschi, S., Fujisawa, T., Elzerman, J. M., Huizeling, E. W. G. M., Tarucha, S. and Kouwenhoven, L. P. (2003). *Phys. Rev. B*, **67**, 033307.

Van Der Ziel, A. (1986). *Noise in Solid State Devices and Circuits.* Wiley-Interscience.

van Ruitenbeek, J. M. (2003). Shot noise and channel composition of atomic-sized contacts. In *Quantum Noise in Mesoscopic Physics*, ed. Y. V. Nazarov. Kluwer Academic Publishers.

van Wees, B. J., van Houten, H., Beenakker, C. W. J., Williamson, J. G., Kouwenhoven, L. P., van der Marel, D. and Foxon, C. T. (1988). *Phys. Rev. Lett.*, **60**, 848.

Veselago, V. G. (1964). *Usp. Fiz. Nauk*, **92**, 517.

Veselago, V. G. (1968). *Sov. Phys. Uspekhi*, **10**, 509.

Viola, L. and Lloyd, S. (1998). *Phys. Rev. A*, **58**, 2733.

Viola, L., Knill, E. and Lloyd, S. (1999a). *Phys. Rev. Lett.*, **82**, 2417.

Viola, L., Lloyd, S. and Knill, E. (1999b). *Phys. Rev. Lett.*, **83**, 4888.

Vion, D., Aassime, A., Cottet, A., Joyez, P., Pothier, H., Urbina, C., Esteve, D. and Devoret, M. H. (2002). Manipulating the quantum state of an electrical circuit. *Science*, **296**, 886.

Wallraff, A., Schuster, D. I., Blais, A., Frunzio, L., Huang, R.-S., Majer, J., Kumar, S., Girvin, S. M. and Schoelkopf, R. J. (2004). *Nature*, **431**, 162.

Wallraff, A., Schuster, D. I., Blais, A., Frunzio, L., Majer, J., Devoret, M. H., Girvin, S. M. and Schoelkopf, R. J. (2005). *Phys. Rev. Lett.*, **95**, 060501.

Walther, H., Varcoe, B. T. H., Englert, B.-G. and Becker, T. (2006). Cavity quantum electrodynamics. *Rep. Prog. Phys.*, **69**, 1325.

Wang, Y.-D., Li, Y., Xue, F., Bruder, C. and Semba, K. (2009). *Phys. Rev. B*, **80**, 2009.

Weiss, U. (1999). *Quantum Dissipative Systems*, 2nd edn. World Scientific.

Weissman, M. B. (1988). $1/f$ noise and other slow, nonexponential kinetics in condensed matter. *Rev. Mod. Phys.*, **60**, 537.

Wells, D. A. (1967). *Schaum's Outline of Theory and Problems of Lagrangian Dynamics*. NY: McGraw-Hill.

Wendin, G. and Shumeiko, V. S. (2005). Superconducting quantum circuits, qubits and computing. In: *Handbook of Theoretical and Computational Nanotechnology*, vol. 3, eds M. Rieth and W. Schommers. American Scientific Publishers.

Wharam, D. A., Pepper, M., Ahmed, H., Frost, J. E. F., Hasko, D. G., Peacock, D. C., Ritchie, D. A. and Jones, G. A. C. (1988). *J. Phys. C*, **21**, L887.

White, R. M. (1983). *Quantum Theory of Magnetism*, 2nd edn. Springer.

Wigner, E. P. (1932). *Phys. Rev.*, **40**, 749.

Wilson, R. D., Zagoskin, A. M. and Savel'ev, S. (2010) *Phys. Rev. A*, **82**, 052328.

Wineland, D. J., Dalibard, J. and Cohen-Tannoudji, C. (1992). *J. Opt. Soc. Am. B*, **9**, 32.

Wittig, C. (2005). *Journal of Physical Chemistry B*, **109**, 8428.

Yamamoto, T., Pashkin, Yu. A., Astafiev, O., Nakamura, Y. and Tsai, J. S. (2003). *Nature*, **425**, 941.

Yang, C. N. (1962). Concept of off-diagonal long-range order and the quantum phases of liquid He and of superconductors. *Rev. Mod. Phys.*, **34**, 694.

You, J. Q., Liu, Y.-X. and Nori, F. (2008). *Phys. Rev. Lett.*, **100**, 047001.

Zagoskin, A. M. (1990). *JETP Lett.*, **52**, 435.

Zagoskin, A. M. (1998). *Quantum Theory of Many-Body Systems: Techniques and Applications*. Springer.

Zagoskin, A. M. and Saveliev, S. S. (2010). *Ambidextrous Quantum Metamaterials*, (to be published).

Zagoskin, A. M. and Shekhter, R. I. (1994). *Phys. Rev. B*, **50**, 4909.

Zagoskin, A. M., Ashhab, S., Johansson, J. R. and Nori, F. (2006). *Phys. Rev. Lett.*, **97**, 077001.

Zagoskin, A. M., Savel'ev, S. and Nori, F. (2007). *Phys. Rev. Lett.*, **98**, 120503.

Zagoskin, A. M., Il'ichev, E., McCutcheon, M. W., Young, J. F. and Nori, F. (2008). *Phys. Rev. Lett.*, **101**, 253602.

Zagoskin, A. M., Rakhmanov, A. L., Saveliev, S. and Nori, F. (2009). *Physica Status Solidi B*, **246**, 955.

Zagoskin, A. M., Savel'ev, S., Nori, F. and Kusmartsev, F. V. (2010). *Squeezing as the Source of Inefficiency in Quantum Otto Cycle*. (to be published).

Zakosarenko, V., Bondarenko, N., van der Ploeg, S. H. W., Izmalkov, A., Linzen, S., Kunert, J., Grajcar, M., Il'ichev, E. and Meyer, H.-G. (2007). *Appl. Phys. Lett.*, **90**, 022501.

Zener, C. (1932). *Proc. Roy. Soc. A*, **137**, 696.

Zhou, L., Gong, Z. R., Sun, C. P., Liu, Y. and Nori, F. (2008). *Phys. Rev. Lett.*, **101**, 100501.

Ziman, J. M. (1979). *Principles of the Theory of Solids*, 2nd edn. Cambridge University Press.

Znidaric, M. (2005). *Phys. Rev. A*, **71**, 062305.

Zubarev, D., Morozov, V. and Röpke, G. (1996). *Statistical Mechanics of Nonequilibrium Processes 1. Basic Concepts, Kinetic Theory*. Wiley-VCH.

Zurek, W. H. (2003). Decoherence, einselection, and the quantum origins of the classical. *Rev. Mod. Phys.*, **75**, 715.

Index

Printed in the United States
by Baker & Taylor Publisher Services